Concepts and Values in Biodiversity

'Biodiversity' may refer to the diversity of genes, species or ecosystems in general. These varying concepts of biodiversity occasionally lead to conflicts among researchers and policy makers, as each of them requires a customized type of protection strategy. This book addresses the questions surrounding the merits of conserving an existing situation, evolutionary development or the intentional substitution of one genome, species or ecosystem for another. Any practical steps towards the protection of biodiversity demand a definition of that which is to be protected and, in turn, the motivations for protecting biodiversity. Is biodiversity a necessary model which is also useful, or does it carry intrinsic value? Debates like this are particularly complex when interested parties address them from different conceptual and moral perspectives. Composed of three parts, each complemented by a short introductory paragraph, this collection presents a variety of approaches to this challenge.

The chapters cover the perspectives of environmental scientists with expertise in evolutionary and environmental biology, systematic zoology and botany, as well as those of researchers with expertise in philosophy, ethics, politics, law and economics. This combination facilitates a truly interdisciplinary debate by highlighting hitherto unacknowledged implications that inform current academic and political debates on biodiversity and its protection. The book should be of interest to students and researchers of environment studies, biodiversity, environmental philosophy, ethics and management.

Dirk Lanzerath is Executive Officer of the German Reference Centre for Ethics in the Life Sciences (DRZE) and Lecturer for Philosophy and Ethics at the University of Bonn, Germany.

Minou Friele is Head of the Junior Research Group Medicine and Ethics at the Institute for Science and Ethics (IWE) at the University of Bonn, Germany.

Routledge Studies in Biodiversity Politics and Management

Concepts and Values in Biodiversity
Edited by Dirk Lanzerath and Minou Friele

Concepts and Values in Biodiversity

Edited by Dirk Lanzerath and Minou Friele

LONDON AND NEW YORK

First published 2014
by Routledge
2 Park Square, Milton Park, Abingdon, Oxon, OX14 4RN

and by Routledge
711 Third Avenue, New York, NY 10017

Routledge is an imprint of the Taylor & Francis Group, an informa business

British Library Cataloguing in Publication Data
A catalogue record for this book is available from the British Library

Library of Congress Cataloging-in-Publication Data
Concepts and values in biodiversity / edited by Dirk Lanzerath and Minou Friele.
pages cm. – (Routledge studies in biodiversity politics and management)
Includes bibliographical references and index.
1. Biodiversity. 2. Biodiversity conservation. I. Lanzerath, Dirk. II. Friele, Minou Bernadette.
QH541.15.B56C633 2014
333.95′16–dc23 2013027921

ISBN13: 978-0-415-66057-0 (hbk)
ISBN13: 978-0-203-07396-4 (ebk)

Typeset in Baskerville by
HWA Text and Data Management, London

Printed and bound in Great Britain by
TJ International Ltd, Padstow, Cornwall

Contents

Figures

Tables

Contributors

Professor Wilhelm Barthlott is Professor of Botany at Nees Institute for Biodiversity of Plants, University of Bonn.

Professor Dieter Birnbacher is Professor of Practical Philosophy at Heinrich-Heine University of Düsseldorf.

Dr. Ines Bruchmann is Lecturer in Botany, Landscape Ecology and Aspects of Nature Conservation at the University of Flensburg.

Professor Karen Bubna-Litic is Associate Professor at the School of Law, University of South Australia.

Professor Walter R. Erdelen is Assistant Director-General for Natural Sciences at UNESCO.

Dr. Minou Friele is Head of Junior Research Team on Medicine and Ethics at the Institute of Science and Ethics, University of Bonn, and Lecturer in Philosophy at the University of Bonn.

Dr. Tobias A. M. Gulder is Research Group Leader at the Kekulé-Institute of Organic Chemistry and Biochemistry at the University of Bonn.

Professor Mathias Gutmann is Professor of Philosophy at the Karlsruhe Institute of Technology (KIT), University of Karlsruhe.

Dr. Ulrich Heink is a Scientific Assistant at the Department of Conservation Biology, Helmholtz Centre for Environmental Research, Leipzig.

Professor Manfred O. Hinz is Professor of Law at the Faculty of Law, University of Namibia.

Professor Kurt Jax is Deputy Department Head of the Department of Conservation Biology, Helmholtz Centre for Environmental Research, Leipzig.

Dr. Thomas Kirchhoff is a Scientific Assistant at the Protestant Institute for Interdisciplinary Research (FEST), Heidelberg.

Dr. Aline Kühl-Stenzel is Associate Scientific and Technical Officer at the Regional Activity Centre for Specially Protected Areas, Mediterranean Action Plan, United Nations Environment Programme.

Professor Dirk Lanzerath is Executive Manager of the Reference Centre for Ethics in the Life Sciences (DRZE), Bonn.

Professor Thomas Potthast is Professor of Philosophy at the International Centre for Ethics in the Sciences and Humanities (IZEW), Eberhard Karls University Tübingen.

Dr. M. Daud Rafiqpoor is Scientific Collaborator at the Academy of Sciences and Literature, Mainz, and the Nees Institute for Biodiversity of Plants, University of Bonn.

Mr. René Richarz is a PhD student at the Kekulé-Institute of Organic Chemistry and Biochemistry at the University of Bonn.

Professor Oliver C. Ruppel is Professor of Law at the Faculty of Law, University of Stellenbosch, South Africa.

Ms Gesine Schepers is a PhD student and Lecturer for Special Duties at the Faculty of History, Philosophy and Theology / Department of Philosophy, University of Bielefeld.

Dr. Tade M. Spranger is Head of Junior Research Team on Medicine and Law at the Institute of Science and Ethics, University of Bonn, and Lecturer in Law at the University of Bonn.

Ms Klara H. Stumpf is a PhD student and Scientific Research Associate at the Institute of Sustainability Governance (INSUGO), Leuphana University of Lüneburg.

Professor Dieter Sturma is Professor of Philosophy, Director of the Institute of Science and Ethics and the Reference Centre for Ethics in the Life Sciences (DRZE), University of Bonn.

Professor Fred Van Dyke is Executive Director at Au Sable Institute of Environmental Studies, Mancelona, USA.

Professor Johann-Wolfgang Wägele is Head of the Department for Systematic Zoology, University of Bonn, and Director of the Zoological Research Museum König, Bonn.

Preface

Minou Friele

In recent years the protection of biodiversity has become subject to numerous national and international academic debates and bio-political regulations. The frequency with which the term biodiversity is used in science, politics and the media could be construed as evidence for the existence of clear scientific, empirical and practical conceptions of its meaning as well as a consensus of its value. Taking a closer look, however, one sees that much remains unclear in respect of a practical understanding of biodiversity and of the criteria and principles that may be employed towards assessing its value and usefulness. 'Biodiversity' may refer to the diversity of genes, of species and of ecosystems in general. These varying concepts of biodiversity occasionally lead to conflicts among researchers and policy makers, as each of them requires a customized type of protection strategy; the strategy to protect a genotype is quite different from that to protect a habitat or an ecosystem, for instance. The term 'protection' itself is, in turn, also contested. In this context, questions arise concerning the merits of the conservation of an existing situation vs. an evolutionary development vs. the intentional substitution of one genome (or species, or ecosystem) for another. Does protection require conserving a particular situation, or does it rather necessitate the promotion of certain developments along the lines of evolution; does it perhaps even require the deliberate substitution of one genome, species or ecosystem by another?

Questions like the abovementioned indicate the close interactions between conceptual questions and questions of value assignment. Any practical steps towards the protection of biodiversity demand a definition of that which is to be protected and, in turn, the motivations for protecting biodiversity. Why should we aim at protecting biodiversity after all? Because particularity or diversity of species, genomes, or ecosystems as such is instrumentally valuable, i.e. valuable in terms of providing us and future generations with new and often yet unknown interesting materials and models relevant for the further development of our practical knowledge? Or do we have reasons to consider biological diversity as intrinsically valuable irrespective of such instrumental – and perhaps even irrespective of any aesthetic – values? Debates on these questions are particularly complex and interested parties address them from different conceptual and moral perspectives. Often the protection of both an ecosystem and a certain species, or the preservation of a certain habitat without hindering its evolutionary processes, turn out to be incompatible goals. As a result, debates about biodiversity present a variety of dilemmas to their researchers.

Further challenges arise if protection goals conflict with practical interests, such as the traditional and sometimes existential interests of the indigenous populations of a habitat, with the interests of other people living in completely different regions of the world or with the (presumed) interests of future generations. Policies and legislations need to balance these and further interests in an acceptable and fair manner. And since biodiversity issues often range beyond national borders, these aims have to be pursued on an international rather than a mere national level. Without the support by supra- or internationally operating agents and agencies such as the UN, the solution of intertwined and complex legal challenges will not be possible. Questions of ownership, of patent laws, the terms of access and benefit sharing are questions that need to be discussed on an international level if we as the human race want to come up with solutions that have a real chance of implementation and respect.

This anthology addresses the named questions and challenges in a series of articles by researchers working in various disciplines and areas related to biodiversity evaluation, description, policies and a number of particular practical questions. Their articles are complemented by a series of further articles dealing with exemplary problems and with particular challenges agents working in the area find themselves confronted with. Plant endemism, migratory species, the relevance of biodiversity for various practical and theoretical research purposes… these are but a few examples for the more specific problems this anthology deals with.

In line with what has been said so far, this anthology comprises three sections. The first and most comprehensive section addresses the relation of biological concepts and values in the debate about biodiversity, the examples of biodiversity in the context of evolution, substitutability and biodiversity protection, and the economic valuation of ecosystem services. The second section considers questions of legislation and justice, such as international conventions concerning biodiversity, the challenges arising in the context of access and benefit sharing, and the 'biopiracy' debate. The third section presents a number of typical practical problems of biodiversity protection, including incompatible interests of different local groups affected by biodiversity protection policies, and difficulties connected with plant endemism as an indicator of biodiversity.

With this anthology, we want to present an exemplary collection of the different approaches and challenges in today's practice and theory of biodiversity protection. Its articles cover the perspectives of biologists with a particular expertise in evolutionary and environmental biology and systematic zoology, as well as those of philosophers with an expertise in philosophy of science, in ethics and political philosophy; of social scientists, legal experts, and economists with an expertise in ecological economics. These articles continue discussions initiated by an international and interdisciplinary workshop held in March 2011 by the Institute of Science and Ethics (IWE) and the German Reference Centre for Ethics in the Life Sciences (DRZE), which was funded by the German Federal Ministry of Education and Research (BMBF). This workshop was the first to bring together distinguished experts and junior scientists (doctoral students and postdoctoral researchers) from the United States of America, South Africa and all over Europe. As a combination of articles written by renowned experts and junior scientists, the volume aims at encouraging communication between different scientific areas as well as between diverse groups of researchers. In particular, the book aims at facilitating future interdisciplinary debates by highlighting hitherto unacknowledged implications (both conceptual and qualitative) that inform current academic and political debates on biodiversity and its protection. We very much hope that this anthology will turn out to be relevant for advanced undergraduate and postgraduate students in the fields of environmental studies, applied philosophy and other subject areas as well as for academics, political agents, and an educated general public interested in environmental issues.

Bonn, 17 June 2013

Acknowledgements

The editors would like to thank all authors for their contributions to this anthology. We hope there will be many more opportunities for further cooperation in the future! Further gratitude is owed to Professor Dr. Dieter Sturma for supervising a workshop on Biodiversity: Concept and Values, which provided the basis for this publication, and to the German Federal Ministry of Education and Research (BMBF) for its financial support of this workshop. Special thanks also go to Ole Höffken for his helpful assistance and comments, and to Dorothee Güth for her assistance in obtaining permissions for reprinting the figures in this book. We would like to extend our appreciation and sincere thanks to Helen Bell and Louisa Earls for their assistance throughout the editing process.

Biodiversity as an ethical concept

An introduction

Dirk Lanzerath

Introduction

Since the Convention on Biological Diversity (CBD) came into force, the term 'biodiversity' has advanced to become a central notion and keyword in international environmental politics. The protection of biological diversity in various regions is the subject of a large number of national strategies and regulations as well as international agreements. 'Biodiversity' has become a significant element of those political decision-making processes that in one way or another affect natural spaces and resources. To name just a few examples, these processes range from the construction of roads, airports and residential developments to negotiations on fisheries quotas, to the exploitation of raw materials, to the touristic or commercial use of forest areas.

The frequency with which the term biodiversity is used could be construed as evidence for the existence of clear scientific, empirical and also practical conceptions of its meaning as well as clear-cut agreement as to its intrinsic value. Taking a closer look, however, one sees that much remains unclear in respect of a practical understanding of biodiversity and of the criteria and principles that may be employed towards assessing its value and usefulness. The light in which it is viewed may depend on the *discipline* within whose purviews it is being investigated, on a particular *ethical approach* or on a given political strategy. The way biodiversity is valued may differ not only from one discipline to another – be it biology, economics, jurisprudence, philosophy or any other – but also on quite a different level regarding the way society, politics and the media give weight to its significance (Cf. Koricheva and Siipi 2004: 43–6). For instance, it is quite easy to convey the importance of protecting large and aesthetically pleasing species such as panda bears and blue whales to the public, although less conspicuous species may be of much greater importance for maintaining the stability of an ecosystem (Cf. for details Ehrenfeld 1997, esp. 137f.).

Biodiversity as an ethical concept

The questions as to whether biodiversity can represent a *value* and whether there is an obligation to conserve it can provoke extremely divergent intuitive answers. On the one hand, we clearly find the diversity of nature important, be it on account of its economic usefulness or our aesthetic appreciation. On the other hand, neither of these categories can be readily applied to a deep-sea fish population that no-one has ever seen, and they seem to be at odds with the idea of demanding protection in principle for pathogenic organisms – even though these may also be useful in some way. Not only is the establishment of an ethical framework that encompasses biodiversity still subject to much debate, but also the *empirical* questions concerning the *extent* and the *rate* at which it is being *diminished*. Finally, the very term 'biodiversity' is subject to a fundamental degree of uncertainty. It is used in different ways not only in the empirical, natural scientific context, but also in the field of ethics (Cf. Takacs 1996).

Biodiversity and man's treatment of nature

The debates currently being conducted concerning the usefulness and value of biodiversity are closely associated with the more fundamental question: exactly what *is* this 'nature' that we are attempting to protect (Cf. for details Honnefelder 1998)? Do we mean by this the 'natural' nature, so to speak – that which has developed and evolved independently of mankind? The evolutionary processes that have brought about complex ecosystems such as coral reefs and tropical rainforests? Or do we mean that which man has wrought as his living space since his appearance on the earth: gardens, parks and cultivated forests? Perhaps the term is meant to describe the 'ecosystem services' and 'natural capital' upon which we and future generations depend, including highly specialized, cultivated plants and animals that have been bred over thousands of years? To what extent is man's intervention into nature itself a 'natural' component of the evolutionary development of species and habitats? How, in this case, are secondary habitats to be rated in comparison with primary habitats? And how can the truly 'natural' be substituted by 'close-to-natural' without the latter being merely a falsification (Cf. Elliot 2003)? The practical implications that arise from the *differing concepts and views of nature* regarding the protection of biodiversity are many and various.

Man's relationship to nature always has been and remains ambivalent from two points of view. For mankind, nature is at once a threat and means of survival; it is the Familiar, and it is at the same time the Foreign. The threat manifests itself in the form of natural disasters (volcanic eruptions, earthquakes, floods), in diseases caused by the lethal activities of other organisms such as pathogenic bacteria, and not least through man's own imperfection, contingencies rendering his existence vulnerable and finite. As his knowledge about nature has increased, so has man's power over it and therefore his ability to shape it and change it; time and again, he has overcome 'natural' limits to his capabilities.

The ancient understanding of nature (physis) subsumed not only 'becoming' and 'growth', but also an element of structure: the order of the cosmos with its plurality of elements. Already in Plato's writings, we find emphasis being laid on the principle of natural multiplicity (esp. Plato 1997, Timaeus 300c–d), on its ontological significance and on its importance for the stability of the natural order, because diversity and stability go hand in hand (Cf. Pietarinen 2004).

Aristotle and Thomas Aquinas also emphasise the significance and the value of a diversity of living beings, whereby Aquinas' considerations need to be seen in the light of his theology of creation and the assumption of a divine order. In their reflections on organisms, they concentrate their attention on the internal process and dynamic of every living being. This is seen as an autopoietic process towards a natural goal that is defined at the organism's very base (entelechy). This is why there is no need anymore for the assumption of a metaphysical, external world of platonic ideas. A purposeful (teleological), structured nature in the Aristotelian sense is to be understood as being destined – predetermined – and as such, its peculiar being is to be protected. However, at the same time, nature (physis) is something fatefully 'given' and a demand to 'modify', which implies that nature is neither in itself normative nor to be subjected to arbitrary change.

During the Renaissance, a far greater aesthetic moment was ushered in man's attitude towards nature. Nature was far from being seen as an 'end in itself' yet, but it began to play a vital role in man's inquiry into his own 'nature'. A key text in this context can be seen in Petrarch's description of his climb to the summit of Mont Ventoux, which he did voluntarily, out of curiosity and pure desire. Petrarch develops a new experience of nature – albeit he concludes with an energetic eulogy of the soul to the disadvantage of nature (Petrarch 2005: 172–80). An altered attitude towards nature is also reflected in contemporary paintings: aesthetic and contemplative attitudes combine with each other. With the advance of modern natural sciences, the idea of gaining control over nature is becoming a stronger focus of man's thoughts. Francis Bacon emphasises the social benefit of having the capacity to rule over nature (Bacon 1902) and Descartes describes mankind as the 'lord and proprietor of nature' (Descartes 1998, Discourse: 6,2 (AT 62)).

Increasingly, modern man sees nature simply as an object of research and manipulation (Cf. for the historical development of the concept of 'nature' Hager et al. 1984, Weber 1989, Honnefelder 1992, Schwemmer 1987, Rapp 1981, Schäfer and Ströker 1996, Dreyer and Fleischhauer 1998). At the same time, his progressive achievement of control over nature exorcises its element of demonic threat. Concomitant with the liberation from this menace, he is able to give more space to his aesthetic experience of nature. Notwithstanding this, nature's dangerous aspects continue to confound even the most modern, technologically versed peoples. For man's *capacity* to form nature has *limits*, and this means that the nature's manipulation itself may conceal a threat. In contrast to the threat that emanates from 'natural nature', the hazards immanent in the 'artificial nature', resulting from technological intervention, are always in part the responsibility of the causal agent, man himself. The list of these anthropogenic 'natural catastrophes' is growing: from rock and snow avalanches caused by mountain deforestation, flooding on account of ecologically unsound water management measures, overfishing of the high seas with dragnets, up to unintentional and perhaps irreversibly damaging consequences of speculative and risky genetic engineering (Cf. Lanzerath 2001: 148–53).

Nowadays, people tend to have little awareness of their lives being dependent on nature. By inhaling air, eating food and metabolising it, they are in a constant state of exchange with nature. Therefore, not only man's very survival, but also every cultural activity is contingent on an intact nature. To this extent, it is not in any way wrong but rather indispensable that man forms his natural surroundings and uses natural resources. Naturally, the same applies to all other living beings. In contrast to them, however, man has the *option to decide*. It devolves upon him to determine *how* and *to what extent* he exploits other species. Insofar as nature, including living beings, is seen by man only as a set of foreign things to be subjected, rather than the fabric of those metabolic interdependencies maintaining his very humanity, that same man loses touch with 'naturalness'. In turn, this itself becomes a threat; man becomes a threat to himself and to the natural 'system earth'. As Holmes Rolston III puts it: 'each species made extinct is forever slain, and each extinction incrementally erodes the regenerative powers on our planet' (Rolston 1988: 158). Therefore, the questions as to *which nature* should be protected and to the *scope of protection* of biological diversity are closely interwoven.

The protection of biodiversity and nature

The United Nations Conference on Environment and Development (UNCED) environmental protection conferences have popularised the term 'biodiversity' (Cf. for a systematic overview of the conceptual history Piechocki 2007: 14, Eser 2007: 44–8). The documents indicate that the concept is not to be restricted to the empirical and scientific understanding of biological diversity, but that it rather already contains a value-related dimension and is therefore suited for use in the context of environmental politics. By no means are all biologists in agreement with this development and intention (Cf. Takacs 1996). However, the American Forum on Biodiversity has acknowledged the necessity to draw the attention of politicians to the consequences of the enormous rate of species extinction. The invention of the new catchphrase 'biodiversity' can be seen as the result of a deliberate politicisation of scientific insights (Cf. Piechocki 2007: 13–15). The progressing global destruction of the natural environment is to be combated through more generous research funding of the systematic branches of biology, resulting in an upturn in their participation in decision-making processes in environmental politics. Concurrently, this will provide evidence for the economic and aesthetic significance of the ecosystems and their species for our societies (Cf. also Piechocki 2007: 14f.). As a result, biological taxonomy that almost disappeared at universities in the wake of the trend towards the molecularisation of biology is experiencing at least a slight renaissance. It is a little chance to renew and to expand our knowledge and expertise about species and how they evolved.

To sum up: 'Biodiversity' appears as an 'epistemic-moral hybrid' (Potthast 2007: 68–76). It should not be overlooked that two different notions are being designated here, even though their combination is intentional. As an illustration of this, it is necessary to distinguish between the question as to whether a high degree of diversity contributes to the stability of an ecosystem on the one hand, and whether such stability is actually desirable on the other. Having said that, there is of course no reason why scientists should not be able to voice their opinions on the latter point. However, if they do so, they must make very clear the 'selection' on which they base their judgement (Cf. Eser 2007: 52). The discussion on biodiversity highlights the fact that these two elements must indeed be distinguished from each other, but at the same time, they are mutually dependent on each other, especially when it comes to practical political action. Where the conservation of nature and protection of biodiversity are regarded as constituting elements of political practice, it is not possible to derive categorical requirements for action on the basis of empirical findings alone. This requires consideration of the ethical implications. The question as to *which* nature is to be protected is not primarily one that involves scientific substantiation, but rather ethical discourse and its implied moral convictions.

Moral claims for the protection of biodiversity in different ethics approaches

If the normative questions arising from the way we deal with biodiversity are discussed from an ethical point of view, then much depends on the way in which the relationship between man as a moral agent and nature as the embodiment of the structural conditions constituting biodiversity can be described.

Notwithstanding the societal relevance of the concept of biodiversity in environmental politics, its *normative* relevance needs to be analysed within the context of a certain ethical position. Biodiversity can then be seen as a second-order value (Birnbacher 2006: 182) that must be related to higher-order values such as health, beauty and so forth.

Biodiversity protection is frequently articulated in exclusively pragmatic arguments, where exaggerated moral reflection has not been considered as helpful, because it is perceived as purely subjective. Doubts are expressed as to whether ethics as a discipline can in any way contribute a worthwhile appraisal in respect of nature conservation and environmental protection. Instead, moral intuition or attitudes are placed on the same level as axioms of religious faith transcending rational argument or as ideologies representing irrational views (Cf. Galert 1998: 18–22). Such modes of argument refute the possible existence of rational standards – comparable with scientific standards – that can provide a basis for moral reflection and argument prior to the adoption of final justifications within the framework of religious or ideological convictions. In contrast to this, ethical argumentation is to be employed to show that protective measures for the preservation of biodiversity are contingent on reflection that takes adequate account of man not only as an economically active subject, but also as a *morally active subject*.

Approaches adopted in the ethics of nature and the environment include biodiversity either implicitly or explicitly (Cf. Lanzerath et al. 2008). They set both positive and negative limits concerning the ways we treat nature and its diversity, thereby making use of normative guiding concepts such as 'mankind', 'life', 'suffering', 'natural goals', 'interests', 'needs', etc.

Many overviews of contributions to the ethics of nature and the environment categorize the various positions – depending on the scope of the beings to be appraised from a moral point of view – into *anthropocentric* (concerning man alone), *pathocentric* (concerning beings that try to avoid pain), *biocentric* (concerning all living beings), *physiocentric* (concerning all natural beings) or *holistic*[1] (concerning the totality of nature) approaches (Cf. e.g. Birnbacher 1986, Krebs 1997a, v.d. Pfordten 1996). Particularly the positions of anthropocentricity and biocentricity appear to be completely in opposition, accusing each other of 'exploitive morals' or 'specism', on the one hand, or 'irrationalism' or 'head-in-the-sky morals' on the other (Cf. Birnbacher 1990: 66). However, in the clear light of day, they frequently lead to similar results or decisions in practice. With a view to obtaining a generally acceptable consensus, it is therefore wise to avoid a polarising 'centric' mode of thought as well as negating the differences between the approaches. Instead, the respective argumentative strains should be subjected to pragmatic analysis (Cf. Ott 1996: 99) in order to enrich the debate on biodiversity.

Within the framework of a typology of paradigmatic approaches in the current ethical appraisal of nature, two initial basic types can be distinguished. As human activity and man's striving to master reality are always involved, and all decisions depend on his own experience of reality (Cf. Gethmann 1996: 34, Galert 1998: 34), the conceptions are always *anthroporelational*. But we must distinguish between *exclusively* anthroporelational conceptions, on the one hand, and *trans*-anthroporelational conceptions on the other (Cf. for this distinction, e.g. v.d. Pfordten 1996, Regan 1995: 159, Krebs 1997b: 342, likewise Ricken 1987: 1–3, Schlitt 1992: 19f., Siep 1998: 17f.). The former do accept the existence of ethical limits to man's dominion over nature and non-human living beings, but they deny that these are defined in terms of obligations to protect, or as claims on behalf of or even being lodged by the living beings themselves, thereby reserving possible moral claims *exclusively to mankind* itself (i).

The latter, in contrast, are characterised specifically by their explicit recognition of a direct onus of responsibility and duty to provide protection for non-human organisms. Proceeding from this admission, different approaches draw different demarcation lines in defining the subset of organisms that qualify for such asymmetrical moral recognition. The transition from exclusively anthroporelational reasoning to trans-anthroporelational reasoning is to be found in those approaches dealing with the ethics of equity generally associated with the tradition of deontological approaches in ethics. They raise biodiversity protection to a component of equity-oriented social ethics, and include non-human organisms – in gradations – for their own sake (ii).

Such an extension of the scope of moral recognition is also called for in interest-based ethical systems through the instrument of a *graded degree of protection for the interests of different living beings* (iii).

Whereas a pathorelational extension in (ii) and (iii) can already be seen to play an important role, it takes on the status of a central argument in approaches that make animal rights a central issue within the scope of a '*rights view*'. Here, a normative justification for biodiversity protection is derived from an *inherent value residing within every thing that experiences a 'subject-of-life'*. Some jurisprudential-philosophical contributions even demand a '*fundamental right of existence for nature*' altogether (iv).

As the avoidance of suffering and pain constitutes an elementary part of the utilitarian tradition, modern preference-utilitarian approaches afford protection of nature to the degree in which it serves the *needs of all living beings of preferential status*, i.e. not only mankind (v).

The appetitive nature of species is at the focus of more elaborate ethical approaches. Some such see in *man's striving towards basic resources or values* the *culturally invariant common denominator* of mankind. These basic resources or values are in many ways associated with the issue of upholding biodiversity (vi).

The fact that all beings pursue their natural *goals* provides the basis on which *teleologically* defined approaches take account not only of human striving, but also the fulfilment of biological functionality by organisms, as *purpose-analogous appetites* from the point of view of moral philosophy. Thus, biodiversity protection can also be derived from the needs of non-human living beings (vii).

Whereas biological diversity remains only an indirect issue in the approaches up to and including (vii), these being very largely derived from the needs of the individual, the holistic and value-ethical approaches place the greatest emphasis on the multiplicity of the entire systemic whole. *Holistic and value-ethical* approaches see nature's value-as-such to be grounded especially in the *order of life* and the *naturally engendered multiplicity* and derive from this the comprehensive, normative inclusion of biodiversity (viii).

As the perspectives of Judaic-Christian tradition play an important role in Western culture, this is included in the list, despite the fact that it is not susceptible to philosophical treatment and represents only one religious tradition amongst a whole range in which the attitudes to nature are many and very various (Cf. Hamilton 1993). As traditional Christian theology sees all *beings as having a common origin in God*, this perspective can also, in principle, be regarded as being trans-anthroporelational (ix).

Qualitative practical standards and preferential rules for protecting biodiversity: relevant aspects for a societal-ethical discussion

By way of providing a justification for biodiversity protection, these ethical and meta-ethical reflections have a constitutive significance. Discussion within society is dependent on this reflection. The dissent resulting unavoidably touches the practicalities of environmental politics as well. However, this need not necessarily lead to the blockage of all decision-making processes and modes of action. The question as to whether one recognizes non-instrumental or intrinsic values in nature, construed holistic or ecocentric, is less a question of the practical search for moral standards than of the various underlying assumptions and basic convictions

(Cf. also Birnbacher 2006: 94f.). These do not need to be exchanged anew every time technical questions are discussed. Ecocentric axiology can coexist peacefully with meta-ethical subjectivism, because the two theories express notions lying on two different levels. Value statements convey information as to what is of value, and in what connection. Meta-ethical statements concern the status applying to such value statements (Cf. Birnbacher 2006: 96). If one regards the usefulness of biodiversity, i.e. the number and heterogeneity of species and ecosystems, not merely as being of a utilitarian nature, but also trans-utilitarian (i.e. pertaining to aesthetic or transcendental benefit), manifold arguments for comprehensive protection of biodiversity become accessible across all the different ethical approaches.

The ethical considerations concerning the value of biodiversity lead to the requirement that the morally correct way of preserving essential life-processes in the biosphere in their richness and diversity depends on the acceptability of preferential rules or practical norms, together with their associated compatibilities, that apply to all the various approaches and are bound to grounds not easily rejected. As well as general *quantitative preferential rules* or practical norms, such as ones giving preference to actions with a low risk of undesirable consequences or to a quantitatively low degree of detriment, ones seeking to reduce negative side-effects to the smallest possible amount or ones determining that the probability of a positive or negative consequence is to be multiplied by the extent of the consequence (Cf. Ricken 2003: 248), the evaluation of biodiversity in moral-philosophical discussion also requires *qualitative preferential rules* or *practical norms* in order to be able to serve a useful purpose in decision-making processes in society. These can then provide structure to a societal-ethical discussion without the need to achieve consensus regarding all values, ideals, norms or normative attitudes.

To this extent, the following list is intended as a survey of differing ethical approaches that are indeed subject to various conditions of validity, but at the same time can be justified to those of different persuasions, so that, in principle, broad agreement may be expected. This creates the conditions necessary for achieving cooperation between groups without the need to remove disagreement in all areas (Cf. Eser 2007: 51). There will, of course, be differences in the ranking of the resources employed and the values called on. This applies to the list as such, but in particular also to competing resources and areas of benefit (such as mobility, energy use, agricultural appropriations, etc.). Lists of this kind can generally only be seen as 'open-ended'.

Securing survival

Biodiversity protection as a measure towards sustainably securing natural living conditions for mankind and other organisms

Even though modern man has achieved ever greater emancipation from nature and natural forces, as a natural being *and* 'cultural animal' he remains to a great extent interwoven with and dependent on nature. Respiration and the intake of food represent a continuous metabolic exchange between man and non-human nature. Thus, an intact nature is a necessary condition for man's basic survival, as well as that of other organisms. The highly fragile interdependencies between different species, the webs of dependency between organisms and abiotic elements as well as the entire state of equilibrium of an ecosystem are all immensely susceptible to destruction through human intervention. The double-tracked preservation of stability and diversity is a precondition for sustainability of ecosystems as well as human well-being. Insofar as biological diversity proves to be a necessary criterion for an intact nature, its protection becomes part of nature conservation founded on that premise. Even if an ecosystem or biotope contains 'redundant' or 'passenger' species that may not be relevant to the stability of such a system, any species loss may herald the onset of instability and in turn lead to the loss of other species.

Biodiversity protection as a measure towards securing a multiplicity of foodstuffs for all beings

Natural biodiversity can be designated as the origin and 'raw material' of the variety of foodstuffs on which mankind, but also other living beings, are dependent. Many substances in plants and animals relevant for nutrition remain unknown to this day. In view of the fact that 90 per cent of cultivated plants that supply global food requirements can be traced back to only 20 wild species, the plant kingdom with its huge genetic variability represents a rich reservoir for securing the world's basic food needs. For instance, a genetic disposition towards immunity against parasites and pests may prove to be a valuable resource for plant cultivation. Thus, wild species represent a productive potential through which cultured and in-bred animals or plants can be 'refreshed' as required as a measure towards providing sustainable food supplies throughout the world.

Securing health and economic well-being

Biodiversity protection as a means of maintaining a supply of substances for medicines and other products

Scientific studies frequently reveal the existence of substances in plants, animals or bacteria that have pharmacological or therapeutic value. To name just two examples, substances obtained from marine gastropod molluscs are being used in the development of new cancer therapies, and studies on the slow coagulation of blood in sea cows (sirenians) help scientists to understand the nature of haemophilia. The full potential of these sources of natural drugs is still a matter of sheer speculation. Wild species also supply raw materials for which no synthetic replacement is known, e.g. oils, paints, fibres and so forth. Especially the – in many cases still unknown – value of the substances is of great significance for the traditional economic structures in developing and emerging countries. Thus, the conservation of 'natural capital' and ecosystem services leads indirectly to biodiversity protection.

The potential of biodiversity protection for local economies and social structures

Traditional agricultural societies and local biological diversity can represent a great potential for ecologically sustainable management or 'soft' ecological tourism, promoting cultural diversity with a significant social amenity value above and beyond economic concerns. Such amenity value represents an important, but in view of the difficult conditions of economic globalization also endangered, benefit.

Securing natural elements for the socio-cultural surroundings as a space for human achievement

Biodiversity protection as a contribution towards sustaining the space we live in

Man's natural surroundings, of which he feels a part, are with few exceptions surroundings formed by human intervention. Parks and land used for agriculture and forests are cultivated, but so are most of the 'nature' protection areas in the Western hemisphere; some are the result of extensive human exploitation (e.g. in Germany: Lüneburg Heath, the calcareous grasslands of the Eifel, etc.) and therefore rather 'near-to-nature' than 'natural' – insofar as one wishes to reserve the term 'natural' for areas completely untouched by man. As such, these areas must be maintained. Left to 'pure nature' they would, in time, no longer retain their particular character. This 'nature' represents part of the *living space* that we have created for our society and culture. This is where man leads his life and does what he does. With its *simultaneity of diversity and unity*, this living space represents not only the basic framework for human survival, but must also be seen as the source from which man draws what he needs for his personal, social, cultural and economic activities. It is the natural basis for human flourishing. If this be granted, then not only is the peculiar phenomenon of living nature qua nature to be regarded as being worthy of protection, but also that of the socio-cultural synthesis in its capacity as a unit of meaning having a specific appreciable value to mankind. This can be seen as an argument that places natural biodiversity and the diversity of near-natural cultural landscapes on a par with each other. Biodiversity becomes the *symbol* for the interweaving of nature and culture.

Biodiversity protection as a means of sustaining homeland and cultural identity

Regional plants and animals represent part of that cultural and natural specificity of human living spaces that allows for identification. There may indeed be no such thing as a cultural tradition that specifically promotes biodiversity, but it is clear that certain species and habitats are associated with such a thing as a 'homeland', and are therefore part of this form of cultural identity. This is appreciated by the people concerned. It even applies in cases where the commercial exploitation of species is seen as being part of a cultural identity; and this attitude is carried to the point of decimating or even extinguishing species, as we see in the example of whaling in Japan or Norway. This form of cultural identity, too, would be lost, if the species figuring as its object disappears. Thus, even where nature is exploited in such a way, the obligation to use the resource sustainably makes sense from both the commercial as well as cultural points of view. Generally, species can be imbued with a considerable aesthetic or symbolic significance by a culture. This 'added value' is irreversibly lost where ecosystems are destroyed and species exterminated.

Biodiversity protection as a contribution towards sustaining the inexhaustibility of nature in view of increasingly scarce natural resources

The conditions determining the engendering and development of biological diversity, even today, represent an open horizon and an inexhaustible, irreplaceable source of experience for our epistemic, technological and economic interests on account of its dynamic character and autopoietic potential. The fewer natural resources there are, including biological species, the more valuable the remaining natural elements will appear.

Securing trans-utilitarian appreciation of nature

Biodiversity protection as a means of sustaining our sensory and aesthetic appreciation of nature

Our aesthetic and emotional perception of biological diversity can be conceived as a trans-utilitarian benefit of biodiversity, in contrast to immediate, utilitarian usefulness of an economic nature or for survival. Above and beyond the valuable achievements of civilization and the cultivation of nature, the experience of more 'raw' and diversified nature provides a great deal of contentment, and this has deep anthropological roots. This can be seen not least in the growing interest being displayed in safaris, recreational diving, whale watching, etc. The experience of untouched nature and the inspiration to be drawn from it have been the subject of artistic, musical and literary activities for hundreds of years, giving rise to the practice of aesthetic contemplation. In this way, the capacity to experience nature in its multiplicity and totality advances to an aesthetic and emotional value. The aesthetics of nature cannot be equated with the aesthetics of artefacts. Nature means process, growth and development without artificial influences; this form of 'natural art' represents the essence of a biological system that does not come from without. This is the sense in which we appreciate the existence of individuals with differing characters; they symbolize modes of expression that are entirely lost, when the species concerned will go extinct or ecosystems will be destroyed.

The potential of biodiversity protection for attaining religious-contemplative experiences with regard to nature and its regime

In the light of human activity, nature appears neither as brute material nor as a blueprint providing normative instructions. The first case would deny experience of nature's self-evidence. The second case would deny the development of nature in its ingenuousness and complexity. What remains is the possibility of grasping nature as a dimension of significance and open-endedness. Traditionally, this is given expression in the understanding of nature as being something *created* and with the secular term *cosmos*. In other words: order exists not only in the sense of a series of causal processes, but also in a sense that can even be seen in analogy to a social order, i.e. in an equilibrium in the existence and well-being of various forms of life. As such, diversified nature becomes a foundation for (religious) contemplative experience. Quite apart from the various forms of expression and rituals associated with confessional religious belief, the search for a consciousness closely bound to nature, the search for a fundamentally objective consciousness as a natural being and as such as a being *that is a part of the natural community with other species* represents a value in itself. In this respect, nature and its diversified structure can be a source supplying both symbols and meaning.

Securing an appreciation of nature in its immediacy

Biodiversity protection is in accordance with the desire to live in a community of living beings

Humans share their living space with other natural species. Natural beings have in common their single evolutionary origin. Non-human organisms do not make decisions concerning their behaviour and their actions, and certainly not their raison d'être; rather, they pursue inherent 'goals'. As a first approximation, these can be described quite non-metaphysically as biological functions that can be seen in analogy to human biological functions. Thus, the protection of natural diversity stimulates the establishment of a human sense of being a member of a community of living beings, each with its own dynamics and drives – be it as individuals or as species.

Biodiversity protection enables the experience of nature in the wild

The increasing degree of control man gains over nature corresponds with nature losing its demonic, terrifying might. This rawness of nature is disappearing more and more from the scope of man's awareness. The aesthetic-contemplative dimension of human perception and experience is highly dependent on such wildness. Therefore the desire to provide for refuges of *true wildness* with their rich diversity of species is grounded not just in scientific interest or for utilitarian purposes. It also has to do with preserving the possibility of experiencing it. Wild nature becomes a symbol of human-ness in the sense of a being that is part of a common history of nature. It is *that other* that is at once foreign and familiar. We appreciate the Arctic and tropical rainforests, even though few of us will ever take the opportunity of actually travelling to these regions. That which has evolved outside the sphere of man's influence stands for natural heterogeneity. This is precisely because in relation to the dominating cultivated landscapes of the earth, it represents a scarce commodity, and serves as a criterion for the protection of biodiversity as an instrument for preserving *natural-historical authenticity*.

Securing scope for research and knowledge pools

Biodiversity protection as a contribution towards maintaining biological models for technological developments and as ecological indicators: bionics and bioindication

Some species represent a model for technological systems; time and again engineering innovations and improvements are made on this basis (bionics). To name a few examples: humpback whales show us how underwater communication works, deep-sea fish have developed bioenergetic illuminants, stalks of grass provide insights into flexural strength that allow construction of television towers. Also, various species of a given habitat may serve as biological sensors that yield qualitative information about the state of certain ecosystems (bioindicators). They can also be used as indicators to detect the presence of certain minerals or toxic substances in various media. The behaviour of a number of species is used in the monitoring of global warming as an indicator of resultant climatic changes.

Biodiversity protection as a contribution towards sustaining biological research objects

According to Aristotle, man has a natural tendency to acquire knowledge, irrespective of questions as to its usefulness and relevance for the survival of the human race. The primary goal of research into biodiversity, which itself depends on the existence of biodiversity, is to obtain information without reference to any concrete usage. It has the purely scientific value inherent in learning about life, biological systems, their origins, developments and interrelationships. Research into biodiversity supplies important background information required for research in molecular biology leading to a deeper understanding of the evolution of living beings including man. Therefore, the extermination of species leads to an irreversible loss that circumscribes the thirst for knowledge on the part of modern man and future generations.

Conclusion

If one regards the value of biodiversity, i.e. the number and heterogeneity of species and habitats, not merely as being of a utilitarian nature, but also trans-utilitarian (i.e. pertaining to aesthetic or transcendental benefit), many cogent reasons present themselves speaking out for comprehensive protection of biodiversity across all the different ethical approaches. This can then be articulated in the form of preferential rules and practical standards and increase the weight given to biodiversity in respect of other (competing) goods and areas of benefit.

In future, as now, the 'ethics of biodiversity protection' will not represent an autonomous approach, but must be embedded in a social-ethical approach and/or one of natural ethics so as to shape our understanding of fairness and proper living as members of a worldwide community of living beings which share a common history of nature and a global habitat.

Note

1 The terms are derived from the Greek: anthropos = human/man; pathos = suffering, pain; bios = life; physis = nature, holos = whole, entirety.

References

Bacon, F. (1902) *Novum Organum scientarum*, ed. J. Devey, New York: P.F. Collier.

Birnbacher, D. (1986) *Ökologie und Ethik*, Stuttgart: Reclam.

Birnbacher, D. (1990) 'Rechte des Menschen oder Rechte der Natur? Die Stellung der Freiheit in der ökologischen Ethik', in H. Hozhey and J.-P. Leyvranz (eds) *Persönliche Freiheit*, Bern: Paul Haupt, 61–80.

Birnbacher, D. (2006) *Natürlichkeit*, Berlin: de Gruyter.

Descartes, R. (1998) *Discourse on Method*, trans. D.A. Cress, Indianapolis, IN: Hackett Publishing.

Dreyer, M. and Fleischhauer, K. (eds) (1998) *Natur und Person im ethischen Disput*, Freiburg: Alber.

Ehrenfeld, D. (1997) 'Das Naturschutzdilemma', in D. Birnbacher (ed.) *Ökophilosophie*, Stuttgart: Reclam, 135–77.

Elliot, R. (2003) 'Faking Nature', in A. Light and H. Rolston III (eds) *Environmental Ethics. An Anthology*, Oxford: Blackwell, 381–9.

Eser, U. (2007) '"Biodiversität" und der Wandel im Wissenschaftsverständnis', in T. Potthast (ed.) *Biodiversität – Schlüsselbegriff des Naturschutzes im 21. Jahrhundert*, Bonn: Bundesamt für Naturschutz, 41–55.

Galert, T. (1998) *Die Bewahrung von Biodiversität als Problem der Naturethik. Literaturreview und Bibliographie*, ed. Europäische Akademie zur Erforschung von Folgen wissenschaftlich-technischer Entwicklungen, Bad Neuenahr-Ahrweiler.

Gethmann, C.F. (1996) 'Zur Ethik des umsichtigen Naturumgangs', in P. Janich and C. Rüchardt (eds) *Natürlich, technisch, chemisch: Verhältnisse zur Natur am Beispiel der Chemie*, Berlin: de Gruyter, 27–46.

Hager, F.P., Gregory, T., Maierù, A., Stabile, G. and Kaulbach, F. (1984) 'Natur', in J. Ritter and K. Gründer (eds) *Historisches Wörterbuch der Philosopie*, vol. 6, Darmstadt: Wissenschaftliche Buchgesellschaft, 421–78.

Hamilton, L.S. (1993) *Ethics, Religion and Biodiversity. Relations Between Conservation and Cultural Values*, Cambridge: White Horse Press.

Honnefelder, L. (ed.) (1992) *Natur als Gegenstand der Wissenschaften*, Freiburg: Alber.

Honnefelder, L. (1998) 'Welche Natur sollen wir schützen?', in Bundesministerium für Umwelt, Naturschutz und Reaktorsicherheit (ed.) *Ziele des Naturschutzes und einer nachhaltigen Naturnutzung in Deutschland*, Bonn, 29–41.

Koricheva, J. and Siipi, H. (2004) 'The Phenomenon of Biodiversity', in M. Oksanen and J. Pietarinen (eds) *Philosophy and Biodiversity*, Cambridge: Cambridge University Press, 27–53.

Krebs, A. (ed.) (1997a) *Naturethik. Grundtexte der gegenwärtigen tier- und ökoethischen Diskussion*, Frankfurt am Main: Suhrkamp.

Krebs, A. (1997b) 'Naturethik im Überblick', in A. Krebs (ed.) *Naturethik. Grundtexte der gegenwärtigen tier- und ökoethischen Diskussion*, Frankfurt am Main: Suhrkamp, 337–9.

Lanzerath, D. (2001) 'Gentechnik. Ethische Kriterien bei der Beurteilung ihrer Anwendungsfelder', in W. Fricke (ed.) *Jahrbuch Arbeit und Technik 2001/2002*, Bonn: Dietz, 138–56.

Lanzerath, D., Mutke, J., Barthlott, W., Baumgärtner, S., Becker, C. and Spranger, T.M. (2008) *Biodiversität*, Freiburg: Alber.

Ott, K. (1996) 'Wie ist eine diskursethische Begründung von ökologischen Rechts- und Moralnormen möglich?', in K. Ott (ed.) *Vom Begründen zum Handeln. Aufsätze zur angewandten Ethik*, Tübingen: Atempto, 86–128.

Petrarch (2005) *Letters on Familiar Matters*, vol. 1: books I–VIII, trans. A.S. Bernardo, New York: Italica Press.

v.d. Pfordten, D. (1996) *Ökologische Ethik. Zur Rechtfertigung menschlichen Verhaltens gegenüber der Natur*, Reinbek: Rowohlt.

Piechocki, R. (2007) '"Biodiversität" – Zur Entstehung und Tragweite eines neuen Schlüsselbegriffs', in T. Potthast (ed.) *Biodiversität – Schlüsselbegriff des Naturschutzes im 21. Jahrhundert*, Bonn: Bundesamt für Naturschutz, 11–24.

Pietarinen, J. (2004) 'Plato on Diversity and Stability in Nature', in M. Oksanen and J. Pietarinen (eds) *Philosophy and Biodiversity*, Cambridge: Cambridge University Press, 85–100.

Plato (1997) *Complete Works*, ed. John M. Cooper, Indianapolis, IN: Hackett Publishing.

Potthast, T. (ed.) (2007) *Biodiversität – Schlüsselbegriff des Naturschutzes im 21. Jahrhundert*, Bonn: Bundesamt für Naturschutz.

Rapp, F. (ed.) (1981) *Naturverständnisse und Naturbeherrschung*, München: Fink.

Regan, T. (1995) 'Animal Welfare Rights – I. Ethical Perspectives on the Treatment and Status of Animals', in W.T. Reich (ed.) *Encyclopedia of Bioethics*, rev. edn, New York: Macmillan, 159–71.

Ricken, F. (1987) 'Anthropozentrismus oder Biozentrismus? Begründungsprobleme der ökologischen Ethik', *Theologie und Philosophie* 62: 1–21.

Ricken, F. (2003) *Allgemeine Ethik*, 4th edn, Stuttgart: Kohlhammer.

Rolston, H. III (1988) *Environmental Ethics. Duties to and Values in the Natural World*, Philadelphia, PA: Temple University Press.

Schäfer, L. and Ströker, E. (eds) (1996) *Naturauffassungen in Philosophie, Wissenschaft, Technik*, vol. IV, Freiburg: Alber.

Schlitt, M. (1992) *Umweltethik. philosophisch-ethische Reflexionen – theologische Grundlagen – Kriterien*, Paderborn: Schöningh.

Schwemmer, O. (ed.) (1987) *Über Natur*, Frankfurt am Main: Klostermann.

Siep, L. (1998) 'Bioethik', in A. Pieper and U. Thurnherr (eds) *Angewandte Ethik. Eine Einführung*, München: Beck, 16–36.

Takacs, D. (1996) *The Idea of Biodiversity. Philosophy of Paradise*, Baltimore, MD: Johns Hopkins University Press.

Weber, H.-D. (ed.) (1989) *Vom Wandel des neuzeitlichen Naturbegriffs*, Konstanz: Konstanz University Press.

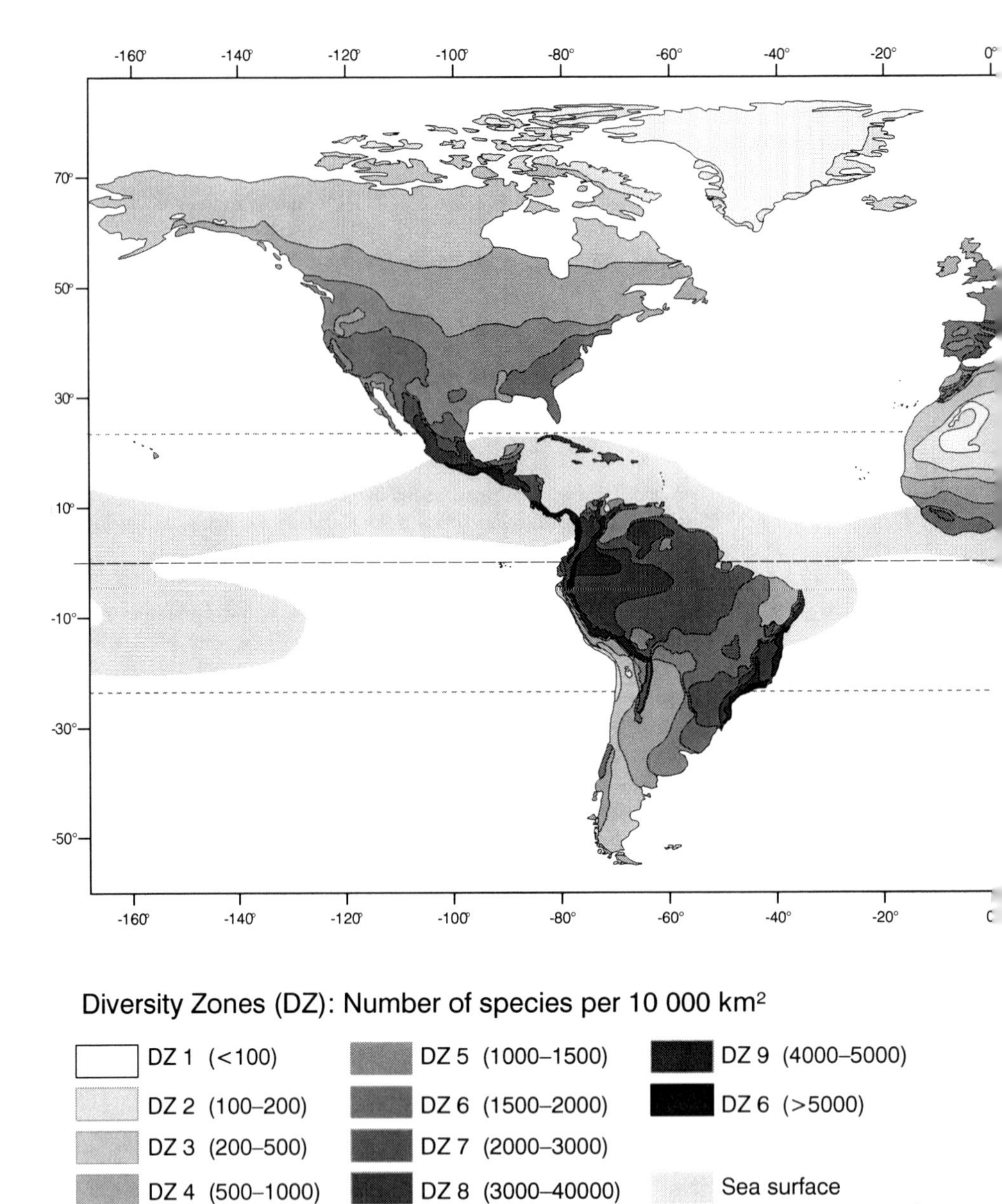

Figure 0.1 Uneven distribution of global biodiversity: species numbers of plants per 10,000 km^2
Source: W. Barthlott, M.D. Rafiqpoor, J. Mutke 2013 Nees Institute for Biodiversity of Plants, University of Bonn ©Barthlott

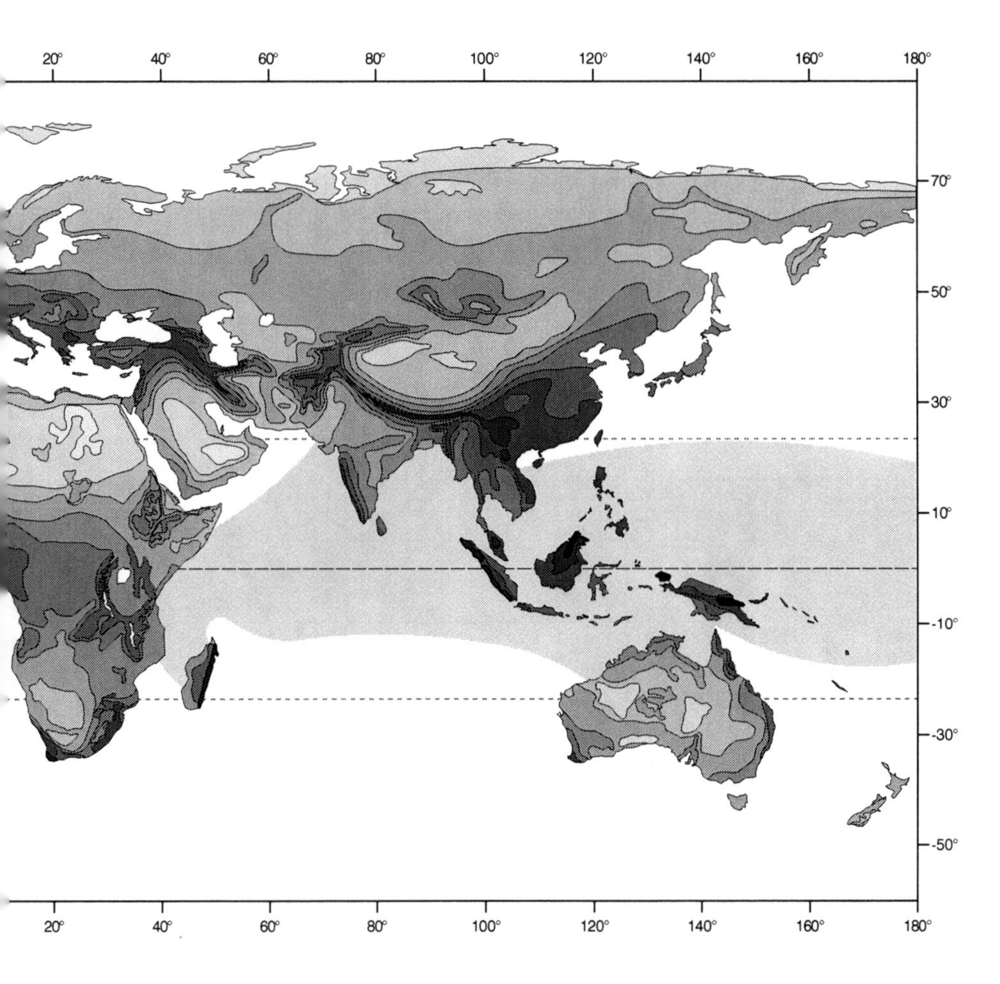
20°
40°
60°
80°
100°
120°
140°
160°
180°
70°
50°
30°
10°
-10°
-30°
-50°

Part I

Concepts and values

One of the core objectives of this book is to facilitate future interdisciplinary debates by raising awareness of the conceptual and evaluative premises that – often subliminally – inform academic and political debates on biodiversity protection. Analysis and recommendations are often based on different, and sometimes mutually exclusive, concepts and values. It might be reasonably assumed that many of these debates would benefit if tacit assumptions as well as conflicting positions were made explicit. Hence, the first section of this anthology highlights and discusses some of the prototypical concepts and values underlying current scientific and political debates in biodiversity-related matters. Conceptual questions and questions of value assignment hereby turn out to be quite closely intertwined with each other. Accordingly, even experts who agree that biodiversity is to be protected may, and often do, disagree about what exactly is required in order to achieve which objectives and goals.

Johann-Wolfgang Wägele, for instance, admonishes that major changes in biodiversity are likely to be irreversible and that the present extent and pace of changes in biodiversity are not to be mistaken to be on a par with evolutionary processes, but that current omissions will cause future generations to suffer from a loss of biodiversity. Philosophers like Dieter Birnbacher, however, point out that we might first of all have to clarify the meaning of expressions such as 'diverse' in the context of a natural system. Is it the mere number of items that counts, or should diversity be defined by some qualitative measure, e.g. by the amount of difference between the items considered, provided these differences are in some way significant? In the latter case, i.e. provided that the overall value of nature remains the same or is even increased, any replacement of one entity by another may in fact turn out to be legitimate, perhaps even required. It may hence turn out that at least some of the arguments put forward by preservationists are, at best, incomplete. Birnbacher takes on this problem and discusses the provocative claim that any entity can be replaced by another entity exhibiting the same valuable properties, without diminishing the overall value of nature.

Mathias Gutmann seems to go even further as he discusses whether or not the typical terms referred to in the biodiversity protection debates are to be seen as metaphorical rather than conceptual terms in the strict sense. From the perspective of philosophy of science, seemingly clear-cut concepts such as the concept of species, Gutmann shows, in fact all too often turn out to be rather ambiguous. Similarly, Ulrich Heink and Kurt Jax point out that our conceptual system is fundamentally metaphorical in nature. They discuss this problem by taking a closer look at one of the most common concepts in biodiversity protection: the concept of invasive species. Both the terms 'invasion' and 'alien' are metaphors. Their article shows that ecological science and environmental policy alike frame the concept of invasive alien species. Scientific debates on such species are heavily influenced by ideas relevant to policy, i.e. on normative assumptions.

Similar criticism is formulated in respect to the concept of biodiversity as such. As Thomas Kirchhoff critically outlines, the term was coined in the context of conservation biology to draw attention to species loss. Since species are primarily protected as characteristic elements of landscapes, the concept of a unique local biodiversity turns out to be a cultural concept. The assessment of this level of biodiversity – which comprises within-community diversity and between-communities diversity at different scales – is subject to even more difficulties than that of species or genetic diversity, Kirchhoff argues.

Dieter Sturma highlights a further philosophical challenge as he draws attention to the fact that no consensus exists to date in respect of the normatively relevant meaning of the term 'nature'. This makes for a difficult starting point when it comes to extending the scope of ethical acknowledgement beyond that of human individuals. Ethical reasons for imposing limitations to human actions can be approached from at least two fundamentally different directions: the anthropocentric approach on the one hand and the non-anthropocentric or physiocentric approach on the other. Sturma discusses in how far normative statements concerning biodiversity will always be made from a certain point of view characterized by narrow epistemic boundaries as well as specific interests. Political and pragmatic debates might be dominated by considerations of short-term benefits. But in ethical reasoning the problem of justification is further complicated by epistemic and semantic problems in identifying biodiversity and by exceedingly asymmetric forms of recognition. Recognition takes on a particular asymmetric form if those addressed in normative deliberations are living beings but not persons. This is to suggest how a selfish and unjustified practice can be replaced by revising our attitude toward nature, including its biological biodiversity.

Thomas Potthast explores in his chapter the diverse notions of values related to biodiversity from an epistemological and an ethical perspective. As a practical term and policy-related concept, 'biodiversity' is a kind of epistemic-moral-hybrid; it comprises descriptive as well as normative elements and – to complicate matters – the latter are connected with a diversity of value ascriptions. Sometimes these values are seen to be reducible to individual preferences or to mere economic values. Potthast argues against such a reduction. Due to the opaqueness of the notion of values itself, he states, they do not provide the necessary means for weighing up different values in cases of conflict. If specific (political) decisions are to be taken, also moral rules and principles are required – and these can be contested too. However, the multiple values of biodiversity still often provide a common moral ground – despite theoretical differences in detail – and allow for joint environmental decisions and governance. The author concludes that investigating the diversity of values of biodiversity may turn out to provide an important link between normative discussions and practical (political) activities.

Many scientists and many philosophers are aiming at and arguing for the protection of a multiplicity of life forms. Bringing together papers by scientists and by philosophers, this chapter aims to provide a scientific as well as philosophical elucidation and clarification as to how and why we can and should give good reasons to biodiversity protection efforts.

1 The necessity for biodiversity research

We are responsible for the quality of life of coming generations

Johann-Wolfgang Wägele

Introduction

We see a strong conflict between – on one side – the realization of human potentials to shape the world according to our wishes, to seek pleasure, to increase personal and national power and property, and on the other side the laws of ecology that prohibit endless growth, the ongoing loss of ecosystem functions that are the foundation for human quality of life, and the loss of diversity in nature, which is not only an economically important resource but also a source of joy, inspiration and tranquillity. It has been claimed that 'elements of nature or natural systems have a value only when man attributes it to them, they do not have a value *in se*' (Negri 2005: 19, cf. 3–25). If this is so, we have to discuss the anthropocentric valuation of nature.

The civilised community is in its majority aware of some of the man-made global changes, such as the connection between an increase of land use and global warming. It is furthermore generally acknowledged that we might run out of some resources like mineral oil and some rare chemical elements.

In contrast to the effects of global warming, the melting of glaciers or the sealing of soil with asphalt and cement, we usually neither see nor feel the loss of species, with the consequence that the political support for actions that can mitigate biodiversity losses is half-hearted and inadequate. This is extremely dangerous, because the loss of species is *irreversible*.

Our planet has undergone many cycles of warming and cooling periods. In the past, glaciers have been growing and shrinking; trees have been prospering in the high Alps and even in Antarctica, where today we find large ice sheets, to name but two examples of biosphere changes. However, these processes of climate and geological change did not cause any permanent loss of water, oxygen or basic minerals.

Major changes in biodiversity, however, are *irreversible*. Here, a previous state is never recapitulated; life forms, once lost, cannot regenerate. Of course, species also become extinct under natural conditions and other ones evolve anew, and every period of earth history is different from the one before it. Man evolved in ecological systems and is highly adapted to a changing biosphere. However, evolution of new life forms and adaptation are very slow processes. Evolution cannot match the current speed of biodiversity destruction. It is our responsibility to avoid further losses in the interest of the quality of life of our descendants.

Since many facts about evolutionary processes and about past and present biodiversity changes are not common knowledge, it is relevant to briefly summarize what we know in this context and to stress that our knowledge is still too meagre for a sound management of biodiversity.

What is biodiversity?

The term 'biodiversity' was coined by E.O. Wilson within a forum held in Washington, D.C. (September 1986) under the auspices of the U.S. National Academy of Sciences and the Smithsonian Institution (see Wilson 1997: 1–3). The forum's proceedings were published in 1988 under the title BioDiversity (Wilson and Peters 1988). Since then, the term has become very popular and 'biodiversity research' often replaces the term 'ecology', which is in many cases a misuse driven by the desire to get access to financial resources.

Generally speaking, the term biodiversity refers to the variety of life at all levels of biological organisation. This includes the diversity of genes, of body constructions, of ecological preferences, the diversity of animal and plant communities, of habitat types and, last but not least, of landscapes, which constitute the 'homes' of different life forms. All these facets of biodiversity are correlated with species diversity.

Different places on this planet harbour different quantities of diversity in varying patterns. Therefore, scientists aim at measuring biodiversity to offer sound data for decision making (for environmental audits, landscape management, nature protection, species conservation). There are many different ways to estimate aspects of biodiversity (cf. Colwell 2009: 257–63). There are numerical indices to indicate species numbers per area, considering not only species presence but also abundances, while others quantify species differences between habitats. There are measurements for homogeneity of abundances of different species and measurements for functional diversity, for the correlation between diversity and the size of an area (e.g. Colwell and Coddington 1995: 101–18; Colwell 2009: 257–63; Faith 1995: 45–58; Gray 2001: 41–56; Hellmann and Fowler 1999: 824–34; May 1995: 13–20).

However, these indices tell us nothing about the properties, function or significance of a species within particular systems. Neither do they inform about a species' life cycles, its ecological needs or whether its occurrence at a global scale is a unique one. The identification is the key to access all species-specific metadata we can have, including information about the unique genetic and ecological properties of each species. Hence, it will not be far-fetched to claim that it is scientifically essential *to know the different species*, not necessarily by name, but identifying each of them (either by the Linnean name, according to the traditional taxonomic nomenclature, or e.g. with the help of a unique genetic marker).

What is taxonomy?

Taxonomy is a discipline of biological systematics, the science of classification of organisms based on hypotheses about their relationships (Schuh and Brower 2009: 1–311). In biology, taxonomy (from ancient Greek τάξις *taxis, arrangement,* and νομία *nomia, method)* is the academic discipline that aims at the distinction, description and classification of species. These classifications may be based on morphological or on genetic characteristics of species, or on both. All subsequent ecological, physiological or evolutionary research needs as a fundamental basis the results of taxonomy.

Taxonomists are experts with many years of experience in the study of selected groups of organisms. A taxonomist usually works for 10 years or more before he is considered to be an expert on a certain group of organisms. The reasons for this are to be found in the research subjects' complexities. For example, there are more than 110,000 described species of bees and wasps (Hymenoptera) (Sharkey 2007: 521–48). Hence, it is impossible for a single scientist to know them all. Therefore, specialists either focus on a particular regional fauna (there are at least 11,000 European species!) or on one or very few families such as Andrenidae or Apidae (true bees), Ichneumonidae or Braconidae (parasitic wasps).

Some ecologists tend to disdain taxonomy, claiming that it is a pure descriptive activity, not a 'real' science. Themselves not being able to handle the diversity of species, many nevertheless tend to segregate these potential partners, mainly for reasons of use of financial resources, with the consequence that several so-called 'biodiversity studies' do not at all consider species diversity.[1]

In order to allow for a phylogenetically correct classification, taxonomists have to consider the evolutionary history (phylogenetic relationships) of 'their' species (Wägele 2005: 1–365). In order to find out about the defining characteristics, they need to know the extent of intraspecific variability (races are not species!). This implies that they need to keep track of a wide range of published data. While it might not be so in all other biological disciplines, in taxonomy it is a blunder to overlook already published data. Every naming of a species is based upon and comprises a particular scientific hypothesis (Wägele et al. 2011: 25). The quality of such a hypothesis depends on the experience of the scientist, the methods used, and the available data, and therefore it is not fundamentally different from any other scientific hypothesis.

What are ecosystem services?

Usually, the description of ecosystem services is based on an anthropocentric view. They provide answers for questions such as: What are ecosystems doing for man? What is the monetary value of these services? These questions are indeed relevant and arguments in favour of ecosystem conservation can be derived from such reflections.

The monetary value of ecosystem functions has been discussed especially during the last decade. Costanza et al. (1997: 253–60), for example, listed some major functions and tried to estimate their monetary value. Important ecosystem services are: The regulation of atmospherical chemical composition, climate regulation, ecosystem resilience in response to environmental fluctuations, water regulation and water supply, erosion control, soil formation, nutrient cycling, pollution and waste treatment, pollination, biological control of pest organisms, refugia for all sorts of organisms, raw material for industry and rural households, conservation of genetic resources, and opportunities for non-commercial activities (recreation, cultural activities) (cf. review in de Groot et al. 2002: 393–408). For oceans, the main services are food resources for millions of people and oxygen production. The authors estimated that for the entire biosphere the average value of 17 selected ecosystem services is about 33 trillion US$ per year.

Given the monetary value of these ecosystem services, obviously their destruction will imply considerable financial disadvantages. However, the exact estimation of the monetary value of effects of disturbances in natural ecosystems is extremely difficult. For example, deforestation may lead to a loss of water catchment. One of the consequences of loss of water catchment is that water supply for rural areas is reduced. Furthermore, the loss can have locally different effects on climate, evaporation, cloud formation, etc. Forests not only keep moisture, they also use water, with the consequence that rapidly growing trees may reduce flows in creeks and rivers (Calder 1998) instead of increasing the availability of water for farms. In other situations, due to deforestation, water is not retained in landscapes and quickly flows into the oceans. As the example shows, the calculation of this single ecosystem service requires careful study of the local hydrology for each geographic area. At a global scale, the loss of large forest areas has led to an increasing number of devastating floods. Between 1900 and 2000 nearly 100,000 persons were killed and several millions lost their homes (Bradshaw et al. 2007: 2379–95).

Note that studies of ecosystem services often do not consider species diversity and the function of single species in complex ecosystems. Therefore, research on ecosystem functions is not necessarily a contribution to biodiversity research.

Cost-benefit analyses and the nonmonetary value of species and landscapes

The high monetary value of products like firewood, timber, fish, medicine from plants, palm oil, etc. triggers a large-scale exploitation and destruction of natural habitats. Often, this exploitation is supported by development programmes, like the World Bank's investments in the palm oil industry in Southeast Asia (Roberts 2010: 1–6). Such programmes are increasingly being criticized by various Non-Governmental Organisations (NGOs) engaged in nature conservation, like the World Wide Fund For Nature, World Wildlife Fund (WWF), Fauna and Flora International, Earthwatch, and Say-No-To-Palm-Oil. These conflicts between industrial and conservational interests are more often resolved in favour of the more powerful industrial complexes than in favour of nature conservationists. Powerful industries, however, are not the only players interested in exploiting natural resources. In the case of subsistence agriculture, which still is the basis for the survival of large portions of the population in developing countries, population growth requires an increasing transformation of natural habitats into agricultural land and destruction of forests for fuelwood consumption (cf. for example Müller and Mburu 2008: 968–77) or to gain space for cattle grazing and for plantations. Even planting fuelwood in farms (e.g. Patel, Pinckney and Jaeger 1995: 516–30) can lead to a loss in biodiversity, because usually only few and often even alien tree species are being used. The original forest diversity is not restored.

'Preventing development projects because of their adverse impact on biodiversity may disproportionately affect the economies of less developed countries' (Spash and Hanley 1995: 129, 191–208). For decision makers, the solution of conflicts between economic development and biodiversity conservation might be a cost–benefit analysis. However, if cost–benefit analyses compare only the land-use strategies relevant for industry with the monetary value of products extracted from a near-natural forest, a good portion of the monetary value of biodiversity is ignored. As has been shown, these values are often difficult to calculate. They include the function of forests as water-capture areas, their role for climate mitigation, pollution assimilation, nutrient cycling, they are providers of pollinating bees, of seeds that are dispersed naturally, of pest-controlling organisms, of soil fauna and fungi that contribute to soil fertility. Natural habitats are sources for new genetic varieties that are currently not being used in agriculture and fisheries. These ecosystem functions are only partially studied and require more research. Further factors are loss of tourism, or the role of healthy landscapes for the health and well-being of people. In addition, the loss of genetic diversity and of organisms that might become important for future generations are incalculable parameters.

The impression often is that in the trade-off between current benefits and supposed disadvantages of biodiversity losses, the profits are only for a few. The losses, however, are for many and will even have to be dealt with by future generations.

Aside from monetary values, a number of different nonmonetary values should also be considered. Nonmonetary values include aesthetic and recreational benefits, the role of habitats and species for the local culture and for religious practices (e.g. Tiwari, Barik and Tripathi 1998: 1229–44) and for the identification of people with their home country. Especially for indigenous people, a healthy ecosystem is part of their essential needs (Alcorn 2010: 1–48). In Europe, a growing proportion of the human population appreciates the proximity to nature. 'In contrast when living in the country was synonymous with cultural backwardness and poverty, today many people desire to live in the country to enjoy less polluted environments and beautifully diverse countryside' (Negri 2005: 3–25).

Natural habitats are also sites for the evolutionary adaptation of species, where populations can survive without human effort. This seemingly intrinsic value is also interesting from an anthropocentric point of view, as it provides opportunities for intense biological research on some of the most valuable renewable resources our planet can offer. We also have to keep in mind that it has been shown that higher species diversity implies greater ecosystem stability and productivity (summary in Tilman 2000: 208–11), a function of diversity whose value cannot be estimated.

Since these cause–effect mechanisms are widely unknown, environmental education and information about the role of species in ecosystems must be implemented in conservation strategies. Given our considerable knowledge gaps and the yet unknown but not unlikely future relevance of species diversity for humanity, the question of whether biodiversity loss really matters should not be answered lightly.

Drivers of biodiversity loss

Without doubt, the growth of the human population is the only major factor that caused and still causes an unnatural rapid extinction of species and a large-scale transformation of landscapes (e.g. Anzagi and Bernard 1977: 63–8; Leakey and Lewin 1995: 1–288; Carpenter and Bishop 2009: 715–22). One of the resulting tragic conflicts is the battle for biodiversity,[2] which has led to deadly violence in some parts of Africa. The conflict between state interests (tourism, prestige, conservation of biological treasures) and poachers or impoverished peasants searching for food cannot be solved with 'shoot-on-sight' orders (Neumann 2004: 813–37).

To discuss the negative effects of population growth implies the necessity to discuss birth control – which is a social and political taboo in many developing countries. According to my own experiences, families in developing countries tend to take having numerous children to be a blessing (cf. also Caldwell and Caldwell 1990: 118–25; Bledsoe 2002: 1–416). The unavoidable consequence is increasing poverty and the reduction of farm size (e.g. Mehta and Kumari 1990: 153; Ahlburg, Kelley and Oppenheim Mason 1996: 360; Murton 1999: 37–46) in rural areas, a development that also threatens natural parks and other reserves for biodiversity. We need a new concept of an economy that can persist without growth of human populations and human exploitation of resources. Otherwise, we will be confronted by an irreversible impoverishment and sterility of our planet.

What are we losing?

We are losing the treasures that are unique for our planet, which cannot be found anywhere else in the universe. We are losing information, inventions of nature, patents for constructions (e.g. nanotubes, structural pigments, water-repellent surfaces, termite mould ventilation systems, the Velcro principle, etc.) for biochemical pathways, for antibacterial defences, etc. that required about 3 billion years for their development and testing in the real world. We are losing highly adaptive networks of ecosystem functions that survived – slowly changing and evolving – for millions of years, while we are still at the beginning of our understanding of these systems.

In April 2002, the parties of the Convention on Biological Diversity (CBD) decided 'to achieve by 2010 a significant reduction of the current rate of biodiversity loss' (United Nations 1992). 2010 was also declared the international year of biodiversity. It is clear today that we have failed to meet this goal. Although some progress has been made and the level of awareness in policy makers has been raised, biodiversity is still in decline (Stokstad 2010: 1272f.). The rate of loss of natural habitats is slowing down, in some cases endangered species (for those whose existence is known) are better protected, the number of protected areas has increased, but the growing human population is consuming more and more biological resources and the majority of biodiversity-rich unprotected habitats are threatened. A major impediment for a sound assessment of the situation is the lack of data. The estimation of the annual number of species that become extinct, for example, depends on assumptions about the total species numbers (cf. e.g. Dirzo and Raven 2003: 137–67). However, this number is unknown for the global biodiversity and for nearly all biotopes (cf. e.g. Pimm et al. 1995: 347–50).

The expansion of *Homo sapiens* is – on geological time scales – a fast catastrophic event comparable to a few mass extinction events that occurred during the past 500 million years. In the past, huge global environmental disturbances clearly were associated with mass extinctions. Extinctions also occur in more 'normal' times; however, there is a notable difference: as shown by the fossil record of marine invertebrates, during exceptional global disturbances the number of extinctions is higher and all sorts of organisms are affected, even those that have a wide distribution range (Payne and Finnegan 2007: 10511–60). In 'normal' times the extinction risk is higher for species with a small distribution range. We know that the global eradication of all sorts of organisms is also happening in our times (which geologists name the 'anthropocene').

Based on literature reviews, according to Smith et al. the extinction of 486 animal species and about 600 plant species since about 1600 has been documented (Smith et al. 1993: 375–8), but even these numbers are probably underestimated. Most of these are conspicuous species. It is well known that species existing in small areas such as islands are highly threatened by man. And it is not just modern civilisation or European colonial activities that destroyed island species like the famous dodo (*Raphus cucullatus*) from Mauritius or the giant north Atlantic auk (*Alca impennis*, eaten among others by hungry seamen and used to stuff feather beds). Rather the species of giant moas (ostrich-related birds) of New Zealand were already hunted down and eaten in prehistoric times. Fossils of bird species found on islands document that many species disappeared after the arrival of man (Olson 1989: 50–3). Polynesians arrived on remote Pacific islands within the last 1000 to 4000 years and exterminated more than 2000 birds (Pimm et al. 1995: 347–50, and literature cited therein). Freshwater species restricted to local rivers and lakes proved to be similarly vulnerable. Since 1900, the North American freshwater fauna has lost at least 123 species (Ricciardi and Rasmussen 1999: 1220–2). These extinctions are not a projection into the future; they have already happened.

Even formerly wide-spread and common species were eradicated in the past. A well-documented case is the North American passenger pigeon (*Ectopistes migratorius*), which in the nineteenth century was still one of the most common birds. Millions of these birds were sold as delicacies; the last bird disappeared in 1914 (Halliday 2003: 157–62). Formerly common European species like hares (*Lepus europaeus*) and sparrows (*Passer domesticus*) are becoming rare and measures to protect them have become necessary.

Among the better-studied animal groups are amphibians, whose species numbers are declining at an alarming rate. A comparison of extinction rates seen in the fossil record with present-day processes has shown that the actual extinction rate is estimated to be more than 200 times faster than in the past and is probably still accelerating (McCallum 2007: 483–91). Most extinctions of the past 500 years occurred after 1980. This suggests that in the near future we will witness a catastrophic loss of amphibian species. Major causes are probably habitat loss, negative effects of global warming, and especially the human-mediated transport of diseases (chytridiomycosis) (Kriger and Hero 2009: 6–10).

Studying the level of local populations of species, Hughes et al. (1997: 689–92) estimated that globally 16 million populations are being destroyed each year in tropical forests alone. When all populations of a species are destroyed, the species is extinct. Today, at a global scale many known species are highly threatened.[3] However, we must keep in mind that a large majority of species (probably about 90 per cent) remains unknown to science, that nobody observes these species, and that there are no data on the conservation status for these organisms and thus for most of our planet's biodiversity. It has been shown that extinction rates of understudied organisms are not lower than in groups for which better data exist. Marine invertebrates, for example, are rarely monitored. However, modelling using parameters from well-studied terrestrial organisms predicts that roughly between 240,000 and 2.4 million marine species are currently threatened (McKinney 1999: 1273–81). Considering all types of organisms, Pimm and co-workers estimated that recent extinction rates are 100 to 1000 times larger than in pre-human times, and that this rate will increase a further 10 times in the next 100 years. However, any absolute estimate of extinction numbers requires knowledge of the actual number of existing species. Unfortunately, this number is not known; it may vary between 10 and 100 million species (Pimm et al. 1995: 347–50). Therefore, we simply do not know what we have lost during the past 200 years, and how many thousands of species become extinct every year.

What we do know, however, is in which situations species might be endangered: small populations in small areas, slowly reproducing species, and species that evolved in regions where predators are absent are particularly vulnerable. At a larger scale, changing landscapes are the most important threat. A consequence of increasing land use is, for example, habitat fragmentation. Large forests disintegrate into small isolated forest areas and lose species. A study of a 4 ha forest fragment in the botanical garden of Singapore revealed that in 100 years about 50 per cent of the species originally present were lost (Turner et al. 1996: 1229–44). Observations of birds in tropical forests prove that fragmentation leads to local extinctions (Christiansen and Pitter 1997: 23–32; Turner 1996: 200–9).

An isolated area of 100 ha rainforest can lose half of the bird species in less than 15 years (Ferraz et al. 2003: 14069–73). Research efforts are necessary to discover which habitat size is required to sustain populations for a long time. Small habitats imply inbreeding and possible genetic 'degeneration' (susceptibility to deleterious mutations), scarcity of mating partners, sensibility for changes in temperature and rainfall, and abundance of parasites.

Predictions about the effects of habitat fragmentation are often based on models (e.g. Gu, Heikkilä and Hanski 2002: 699–710; Fahrig 2002: 346–53). Modelling, however, is hardly possible for species-rich areas, because for sound modelling, data are needed on life cycles, dispersal abilities, food requirements, etc. for each species. Furthermore, conditions in areas that separate suitable habitats are relevant (application of insecticides in agricultural areas, availability of migration corridors through hostile land, frequency of predators). These data are not available. Therefore, the extinction risk for populations in small isolated habitat fragments cannot be reliably estimated in most cases.

As expected, deforestation also leads to local or even total extinctions, and in a much more drastic way than fragmentation. Many animals and plants lose their habitat. For example, in a case study of a Brazilian region where between 1973 and 1993 forest was converted into pastures, of 326 plant species counted in the forest only 20 survived in pastures (Fujisaka et al. 1998: 17–26). Since in the tropical forest biodiversity-hotspots only 12 per cent of the original vegetation still exists, a large number of species are in immediate danger of extinction within the next 50 years (Pimm and Raven 2000: 843–5). For the next 100 years, even if deforestation comes to a halt, the momentum of the past will lead to massive losses of biodiversity (Brook et al. 2006: 302–5).

Another cause for anthropogenic extinctions, besides hunting and habitat destruction or overexploitation (fishes, trees, Steller's sea cow), is the accidental dispersal of animals by travelling people. A large portion of extinctions have been caused by rats, cats, dogs, and other animals, including insects (Clavero and García-Berthou 2005: 110). In Hawaii, some 60 species of endemic[4] forest birds disappeared, one factor being an avian malaria disease imported accidentally with a mosquito species (Sodhi, Brook and Bradshaw 2009: 514–20). The introduction of the Nile perch (*Lates niloticus*) into Lake Victoria with the goal to improve fisheries caused the extinction of several hundred unique fish species that were hunted or outcompeted by the alien predator (Witte et al. 1992: 1–28). These species existed only in this lake. They are gone forever.

The combination of local extinction of unique species and their replacement by ubiquitous species, usually dispersed by man, leads to a 'biotic homogenization' (McKinney and Lockwood 1999: 450–3) or – referring to the analogy of spreading fast-food chains – to a 'MacDonaldisation' of our planet. This process also takes place in agro-ecosystems: we are losing varieties of cultivated species at a global scale, which are being replaced by a few 'modern' races propagated by a powerful agro-industry (Negri 2005: 3–25). The agro-industry stresses the necessity to intensify food production, but never discusses the necessity to stop the growth of the human population, which would solve most problems.

A further factor which is little studied is the effect of habitat change due to global warming. We can expect that especially mountain fauna and flora will get into critical situations when highland forests warm up and have less rainfall.

Nature reserves do not guarantee species survival

The idea motivating the maintenance of nature reserves is that areas must be protected where a valuable diversity of life forms can survive. This implies not only the protection of objects (e.g. single elephants, old trees), but the reservation of space for natural processes. However, populations confined to a small or medium-sized area can only survive for a long time if (a) the population is large enough to endure fluctuations of numbers of individuals without running into a bottleneck. The latter is a situation where only very few reproducing individuals remain, with the consequence of loss of genetic diversity and the risk of absence of suitable mating partners; (b) the area must be large enough to allow for natural processes of regeneration and aging of vegetation, without temporarily losing habitat parameters essential for single species. To prevent this, a mosaic with different states of vegetation cycles is required. If areas are too small to maintain sufficient genetic diversity of a species, migration corridors to other areas can help to keep the species alive; and (c) climatic fluctuations can be buffered in regions that offer different microclimates, e.g. with moist valleys, dryer plains and habitats at different elevations within a continuous reserve.

Extinction in nature reserves is a fact. Theoretical studies have shown that stochastic extinction risks in protected areas are high when the areas are fragmented. In small fragments, population sizes reach the carrying capacity of the habitat much faster, and the probability that single patches become extinct is higher than in a continuous larger region (Burkey 1989: 75–81), even when poaching, invasive species, etc. do not exist. In practice, extinction probabilities usually are not estimated, among others, because inventories of fauna and flora of protected areas and monitoring programs are very incomplete or non-existent.

An interesting case study is that of butterflies living on protected calcareous grasslands in Western Germany (Wenzel et al. 2005: 542–52). Comparing two inventories, one from the year 1972, the second from 2001, an alarming > 50 per cent decline of autochthonous non-ubiquitous species was detected. Many of these are poor dispersers, need larger habitats or depend on specific host plants. The reserve size was relevant in this case, but also the structure and use of the surrounding landscape. To understand such phenomena, monitoring programmes are urgently needed. This case shows: it is not enough to put a fence around a valuable habitat.

How fast can lost biodiversity recover?

In view of ongoing extinctions, laypersons often state that nature can regenerate and therefore losses are not so tragic. What is usually ignored is that new faunas and floras never are the same as ancestral communities; the world changes distinctly after each mass extinction. Furthermore, regeneration of diversity in the sense of increase in species numbers and numbers of life forms takes geological time scales. Geochemical evidence suggests that marine productivity in open-ocean ecosystems did not fully recover for more than 3 million years after the Cretaceous-Tertiary mass extinction (D'Hont et al. 1998: 276–9). It may take several million years for the reappearance of lineages that vanished from the fossil record (Erwin 1998: 344–9), cases which suggest that only few individuals survived and populations recovered very slowly. After smaller crises during the Late Cambrian and early Ordovizian, recovery to diverse faunas took 3 million years (cf. review in Erwin 2001: 5399–403). Mass extinction may lead to a collapse of habitats with their complex networks, which have to grow anew. Recovery has been estimated to start after a delay of 5 million years in the case of the end-Permian extinction (Lehrmann et al. 2006: 1053–6). Even though each historical case is different, it is clear that evolution of new diversity takes not thousands, but millions of years.

Concerning single species, it has been estimated that single fossil species survive on average for a time period of 1 to 10 million years (May, Lawton and Stork 1995: 1–24). The estimated age of our own species varies, depending on the point in time we select to accept a fossil as being a genuine *Homo sapiens*. Our lineage diverged from the variety *Homo sapiens neandertalensis* about 600,000 years ago (the 'modern' variety being much younger) (Green et al. 2008: 416–26). For a global balance of species numbers, the rate of evolution of new species should be similar (between 1 and 10 million years). There are well-documented cases of speciation within a time span of 1 to 1.5 million years. A good example is flightless grasshoppers, which live on volcanic mountains in East Africa. Since these species are unable to fly and cannot cross hot savannah regions, they evolved in situ on young volcanoes after reaching them during periods of cooler climatic conditions. This allows for dating the time needed for the evolution of a new species. It seems that in several lineages of grasshoppers new species evolved on Mt. Kilimanjaro and Mt. Meru after the volcanoes rose over the savannah; that is, within 1–1.5 million years (Hemp et al. 2007: 85–96; Hemp et al. 2010: 581–95). However, despite this relatively long time span, related species are today still very similar to each other. They did not evolve from scratch but from an already specialized ancestor, and there is not much new diversity in the sense of new body constructions or niche adaptations that evolved during this time.

There are some cases of fast evolution of species. A well-studied example is the radiation of cichlid fishes in Lake Malawi (East African Rift system). There exist about 450–500 closely related species which evolved within the lake (Shaw et al. 2000: 2273–80). These species probably appeared after the last drought about 1 million years ago, which was survived by only a few species (review in Seehausen 2006: 1987–98). The speciation interval has been estimated to be 10,000 years. A similar radiation took place in Lake Tanganyika, but these radiations are older (2–5 Myr).

In contrast to the case of the East African grasshoppers, the young species of African cichlids show distinct differences in body shapes and several niche adaptations. However, the question remains whether this visible variation really represents biological and genetic diversity. The radiation of African cichlids can be compared with the rearing of dogs by humans. Selection of dog races has led to a large variety of shapes, sizes, and abilities to hunt prey. Nevertheless, all these variants are dogs. In the same way, the cichlid species flocks in African lakes are still closely related and genetically highly similar.

Therefore, all available evidence suggests that the evolution of biodiversity in the sense of diversity of body constructions, physiologies, ecological abilities and genetic equipment takes geological time scales. Within a few generations of humans it will not be possible to witness the recuperation of depauperate ecosystems. Experience from the past (paleontological data) suggests that 'today's anthropogenic extinctions will diminish biodiversity for millions of years to come' (Kirchner and Weil 2000: 177–80). Our descendants will have to live in a poorer world.

Which fields of biodiversity research need further support?

Biologists are currently not able to deliver the data needed for decision making to stop the loss of biodiversity. It is clear that this loss is dramatic and that data and methods are urgently needed. Ongoing processes will reduce the availability of regenerative biological resources, the resilience and functions of ecosystems, and the beauty of landscapes. Some of the most important requirements are:

- Better support for taxonomists in their effort to inventory biodiversity, because without identification of species the presence and loss of biodiversity cannot be documented;
- the development of technologies for a fast assessment of species diversity;
- the funding of well-designed monitoring programmes. Monitoring in this sense is the statistically sound long-term observation of the composition of fauna and flora in selected habitats;
- the analysis of monitoring data to better understand the causes of fluctuations and of extinction risks;
- the analysis of parameters that allow for the survival and reintroduction of species (e.g. size and structure of protected areas, properties of plant and animal communities).

Outlook

It is obvious that scientific research alone will not save our planet's biological treasures. We need a public understanding of ongoing processes of habitat destruction, and more awareness of the role of species diversity and the urgent necessity to stop the growth of humanity. The probability for favourable political decisions is much higher when many voters manifest their concern. To work towards this goal, it is not necessary to be a scientist. However, decision makers need reliable data and feasible recommendations. Awareness for climate change is currently underpinned by a bulk of long-term observations obtained with the help of weather stations and satellites. We need the same quality of data to develop arguments and recommendations for the conservation of the diversity of wildlife.

Notes

1. It would not be appropriate to name here ongoing ecological research projects; I recommend having a look at corresponding websites; particularly, examples of 'flux studies', 'productivity studies', and species lists for studied habitats.
2. Cf. e.g. <http://www.hutton.ac.uk/news/time-change-strategy-battle-biodiversity>, <http://www.isgtw.org/feature/whats-name-%E2%80%94-battle-save-brazils-biodiversity>, <http://ecomerge.blogspot.de/2011/05/borneo-battle-for-biodiversity.html> and <http://www.new-ag.info/en/developments/devItem.php?a=1585> (accessed 13 June 2013).
3. Cf. the IUCN Red List of threatened species. Online. Available <http://www.iucnredlist.org/> (accessed 13 June 2013).
4. Endemic = occurring only in one region.

References

Ahlburg, D.A., Kelley, A.C. and Oppenheim Mason, K. (eds) (1996) *The impact of population growth on well-being in developing countries*, Berlin: Springer.

Alcorn, J. (2010) *Indigenous people and conservation*, Macarthur Foundation Conservation white paper series. Online. Available HTTP: <http://production.macfound.org/media/files/CSD_Indigenous_Peoples_White_Paper.pdf> (accessed 12 June 2013).

Anzagi, S.M. and Bernard, F.E. (1977) 'Population pressure in rural Kenya', *Geoforum* 8: 63–8.

Bledsoe, C.H. (2002) *Contingent lives: fertility, time, and aging in West Africa*, Chicago, IL: University of Chicago Press.

Bradshaw, C.J.A., Sodhi, N.S., Peh, K.S.H. and Brook, B.W. (2007) 'Global evidence that deforestation amplifies flood risks and severity in the developing world', *Global Change Biology* 13: 2379–95.

Brook, B.W., Bradshaw, C.J.A., Koh, L.P. and Sodhi, N.S. (2006) 'Momentum drives the crash: Mass extinction in the tropics', *Biotropica* 38: 302–5.

Burkey, T.V. (1989) 'Extinction in nature reserves: the effect of fragmentation and the importance of migration between reserve fragments', *Oikos* 55: 75–81.

Calder, I.R. (1998) *Water-resource and land-use issues*, SWIM Paper 3, Colombo, Sri Lanka: International Irrigation Management Institute (IIMI).

Caldwell, J.C. and Caldwell, P. (1990) 'High fertility in sub-Saharan Africa', *Scientific American* 262: 118–25.

Carpenter, P.A. and Bishop, P.C. (2009) 'The seventh mass extinction: human-caused events contribute to a fatal consequence', *Futures* 41: 715–22.

CBD – United Nations (1992) *Convention on biological diversity*. Online. Available HTTP: < http://www.cbd.int/doc/legal/cbd-en.pdf> (accessed 12 June 2013).

Christiansen, M.B. and Pitter, E. (1997) 'Species loss in a forest bird community near Lagoa Santa in southeastern Brazil', *Biological Conservation* 80: 23–32.

Clavero, M. and García-Berthou, E. (2005) 'Invasive species are a leading cause of animal extinctions', *Trends in Ecology and Evolution* 20: 110.

Colwell, R.K. (2009) 'Biodiversity: Concepts, Patterns', in S.A. Levin, S.R. Carpenter, H.C. Godfray et al. (eds) *The Princeton Guide to Ecology*, Princeton, NJ: Princeton University Press, 257–63.

Colwell, R.K. and Coddington, J.A. (1995) 'Estimating terrestrial biodiversity through extrapolation', in D.L. Hawsksworth (ed.) *Biodiversity – measurement and estimation*, London: Chapman and Hall, 101–18.

Costanza, R., d'Arge, R., de Groot, R., Farber, S., Grasso, M., Hannon, B., Limburg, K., Naeem, S., O'Neill, R., Paruelo, J., Raskin, R.G., Sutton, P. and van den Belt, M. (1997) 'The value of the world's ecosystem services and natural capital', *Nature* 387: 253–60.

De Groot, R.S., Wilson, M.A. and Boumans, R.M.J. (2002) 'A typology for the classification, description and valuation of ecosystem functions, goods and services', *Ecological Economics* 41: 393–408.

D'Hont, S., Donaghay, P., Zachos, J.C., Luttenberg, D. and Lindinger, M. (1998) 'Organic carbon fluxes and ecological recovery from the Cretaceous-Tertiary mass extinctions', *Science* 282: 276–9.

Dirzo, R. and Raven, P.H. (2003) 'Global state of biodiversity and loss', *Annual Review of Environmental Resources* 28: 137–67.

Erwin, D.H. (1998) 'The end and the beginning: recoveries from mass extinctions', *Trends in Ecology and Evolution* 13: 344–9.

Erwin, D.H. (2001) 'Lessons from the past: biotic recoveries from mass extinctions', *Proceedings of the National Academy of Science* 98: 5399–403.

Fahrig, L. (2002) 'Effect of habitat fragmentation on the extinction threshold: a synthesis', *Ecological Applications* 12: 346–53.

Faith, D.P. (1995) 'Phylogenetic pattern and the quantification of organismal biodiversity', in D.L. Hawksworth (ed.) *Biodiversity – measurement and estimation*, London: Chapman and Hall, 45–8.

Ferraz, G., Russel, G.J., Stouffer, P.C., Bierregaard, R.O., Pimm, S.L. and Lovejoy, T.E. (2003) 'Rates of species loss from Amazonian forest fragments', *Proceedings of the National Academy of Science* 100: 114069–73.

Fujisaka, S., Castilla, C., Escobar, G., Rodrugues, V., Veneklaas, E.J., Thomas, R. and Fisher, M. (1998) 'The effects of forest conversion to annual crops and pastures: estimates of carbon emissions and plant species loss in a Brazilian Amazon colony', *Agriculture, Ecosystems and Environment* 69: 17–26.

Gray, J.S. (2001) 'Marine diversity: the paradigms in patterns of species richness examined', *Scientia Marina* 65: 41–56.

Green, R.E., Maalaspinas, A.S., Krause, J., Briggs, A.W., Johnson, P.L.F., Uhler, C., Meyer, M., Good, J.M., Maaricic, T., Stenzel, U., Prüfer, K., Siebauer, M., Burbano, H.A., Ronan, M., Rothberg, J.M., Egholm, M., Rudan, P., Brajkovic, D., Kucan, Z., Gusic, I., Wikström, M., Laakonen, L., Kelso, J., Slatkon, M. and Pääbo, S. (2008) 'A complete Neandertal mitochondrial genome sequence determined by high-throughput sequencing', *Cell* 134: 416–26.

Gu, W., Heikkilä, R. and Hanski, I. (2002) 'Estimating the consequences of habitat fragmentation on extinction risk in dynamic landscapes', *Landscape Ecology* 17: 699–710.

Halliday, T.R. (2003) 'The extinction of the passenger pigeon Ectopistes migratorius and its relevance to contemporary conservation', *Biological conservation* 17: 157–62.

Hellmann, J.J. and Fowler, G.W. (1999) 'Bias, precision, and accuracy of four measures of species richness', *Ecological Applications* 9: 824–34.

Hemp, C., Schultz, O., Hemp, A. and Wägele, J.-W. (2007) 'New Lentulidae species from East Africa (Orthoptera: Saltatoria)', *Journal of Orthoptera Research* 16: 85–96.

Hemp, C., Kehl, S., Heller, K.G., Wägele, J.-W. and Hemp, A. (2010) 'A new genus of African Karniellina (Orthoptera, Tettigoniidae, Conocephalinae, Conocephalini) integrating morphological, molecular and bioacoustical data', *Systematic Entomology* 35: 581–95.

Hughes, J.B., Daily, G.C. and Ehrlich, P.R. (1997) 'Population diversity: its extent and extinction', *Science* 278: 689–92.

Kirchner, J.W. and Weil, A. (2000) 'Delayed biological recovery from extinctions throughout the fossil record', *Nature* 404: 177–80.

Kriger, K.M. and Hero, J.M. (2009) 'Chytridiomycosis, amphibian extinctions, and lessons from the prevention of future panzootics', *Ecohealth* 6: 6–10.

Leakey, R. and Lewin, R. (1995) *The sixth extinction: patterns of life and the future of humankind*, New York: Anchor Books.

Lehrmann, D.J., Ramezani, J., Bowring, S.A., Martin, M.W., Montgomery, P., Enos, P., Payne, J.L., Orchard, M.J., Hongmei, W. and Jiayong, W. (2006) 'Timing of recovery from the end-Permian extinction: geochronologic and biostratigraphic constraints from south China', *Geology* 34: 1053–6.

May, R.M. (1995) 'Conceptual aspects of the quantification of the extent of biological diversity', in D.L. Hawksworth (ed.) *Biodiversity – measurement and estimation*, London: Chapman and Hall, 13–20.

May, R.M., Lawton, J.H. and Stork, N.E. (1995) 'Assessing extinction rates', in J.H. Lawton and R.M. May (eds) *Extinction rates*, Oxford: Oxford University Press, 1–24.

McCallum, M.L. (2007) 'Amphibian decline or extinction? Current declines dwarf background extinction rate', *Journal of Herpetology* 41: 483–91.

McKinney, M.L. (1999) 'High rates of extinction and threat in poorly studied taxa', *Conservation Biology* 13: 1273–81.

McKinney, M.L. and Lockwood, J.L. (1999) 'Biotic homogenization: a few winners replacing many losers in the next mass extinction', *Trends in Ecology and Evolution* 14: 450–3.

Mehta, P. and Kumari, A. (1990) *Poverty and farm size in India*, New Delhi: Mittal Publications.

Müller, D. and Mburu, J. (2008) 'Forecasting hotspots of forest clearing in Kakamega Forest, Western Kenya', *Forest Ecology and Management* 257: 968–77.

Murton, J. (1999) 'Population growth and poverty in Machakos District, Kenya', *The Geographical Journal* 165: 37–46.

Negri, V. (2005) 'Agro-biodiversity conservation in Europe: ethical issues', *Journal of Agricultural and Environmental Ethics* 18: 3–25.

Neumann, R.P. (2004) 'Moral and discursive geographies in the war for biodiversity in Africa', *Political Geography* 23: 813–37.

Olson, S.L. (1989) 'Extinction on islands: man as a catastrophe', in D. Western and M.C. Pearl (eds) *Conservation for the twenty-first century*, New York: Oxford University Press, 50–3.

Patel, S.H., Pinckney, T.C. and Jaeger, W.K. (1995) 'Smallholder wood production and population pressure in East Africa: evidence for an environmental Kuznets curve?', *Land Economics* 71: 516–30.

Payne, J.L. and Finnegan, S. (2007) 'The effect of geographic range on extinction risk during background and mass extinctions', *Proceedings of the National Academy of Science* 19: 10511–60.

Pimm, S.L. and Raven, P. (2000) 'Extinction by numbers', *Nature* 403: 843–5.

Pimm, S.L., Russell, G.J., Gittleman, J.L. and Brooks, T.M. (1995) 'The future of biodiversity', *Science* 269: 347–50.

Ricciardi, A. and Rasmussen, J.B. (1999) 'Extinction rates of North American freshwater fauna', *Conservation Biology* 13: 1220–2.

Roberts, J.M. (2010) 'World Bank's palm oil development strategy should focus on economic freedom', *Backgrounder* 2426. Online. Available HTTP: <http://thf_media.s3.amazonaws.com/2010/pdf/bg2426.pdf> (accessed 12 June 2013).

Schuh, R.T. and Brower, A.V.Z. (2009) *Biological systematics: principles and applications.* Ithaca, NY: Cornell University Press.

Seehausen, O. (2006) 'African cichlid fish: a model system in adaptive radiation', *Proceedings of the Royal Society London* B 273: 1987–98.

Sharkey, M.J. (2007) 'Phylogeny and classification of Hymenoptera', *Zootaxa* 1668: 521–48.

Shaw, P.W., Turner, G.W.F., Rizman Idid, M., Robinson, R.L. and Carvalho, G.R. (2000) 'Genetic population structure indicates sympatric speciation of Lake Malawi pelagic cichlids', *Proceedings of the Royal Society London* B 267: 2273–80.

Smith, F.D.M., May, R.M., Pellew, R., Johnson, T.H. and Walter, K.R. (1993) 'How much do we know about the current extinction rate?', *Trends in Ecology and Evolution* 8: 375–8.

Sodhi, N.S., Brook, B.W. and Bradshaw, C.A.J. (2009) 'Causes and consequences of species extinctions', in S.A. Levin (ed.) *Princeton Guide to Ecology*, Princeton, NJ: Princeton University Press, 514–20.

Spash, C.L. and Hanley, N. (1995) 'Preferences, information and biodiversity preservation', *Ecological Economics* 12: 191–208.

Stokstad, E. (2010) 'Despite progress, biodiversity declines', *Science* 329: 1272f.

Tilman, D. (2000) 'Causes, consequences and ethics of biodiversity', *Nature* 405: 208–11.

Tiwari, B.K., Barik, S.K. and Tripathi, R.S. (1998) 'Biodiversity value, status, and strategies for conservation of Sacred Groves of Meghalaya, India', *Ecosystem Health* 4: 20–32.

Turner, I.M. (1996) 'Species loss in fragments of tropical rain forest: a review of the evidence', *Journal of Applied Ecology* 33: 200–9.

Turner, I.M., Chua, K.S., Ong, J.S.Y., Soong, B.C. and Tan, H.T.W. (1996) 'A century of plant species loss from an isolated fragment of lowland tropical rain forest', *Conservation Biology* 10: 1229–44.

Wägele, J.-W. (2005) *Foundations of phylogenetic systematics*, München: Verlag Dr. F. Pfeil.

Wägele, H., Klussmann-Kolb, A., Kuhlmann, M., Haszprunar, G., Lindberg, D., Koch, A. and Wägele, J.-W. (2011) 'The taxonomist – an endangered race. A practical proposal for its survival', *Frontiers in Zoology* 8: 25. Online. Available HTTP: <http://www.frontiersinzoology.com/content/8/1/25> (accessed 12 June 2013).

Wenzel, M., Schmitt, T., Weitzel, M. and Seitz, A. (2005) 'The severe decline of butterflies on western German calcareous grasslands during the last 30 years: a conservation problem', *Biological Conservation* 128: 542–52.

Wilson, E.O. (1997) 'Introduction', in M. Reaka Kudla, D.E. Wilson and E.O. Wilson (eds) *Biodiversity II. Understanding and protecting our biological resources*, Washington, D.C.: Joseph Henry Press, 1–3.

Wilson, E.O. and Peters, F.M. (eds) (1988) *BioDiversity*, Washington, D.C.: National Academy Press.

Witte, F., Goldschmidt, T., Wanink, J., van Oijen, M., Goudswaard, K., Witte-Maas, E. and Bouton, N. (1992) 'The destruction of an endemic species flock: quantitative data on the decline of the haplochromine cichlids of Lake Victoria', *Environmental Biology of Fishes* 34: 1–28.

2 Biodiversity and the 'substitution problem'[1]

Dieter Birnbacher

Biodiversity – a 'hard case' for environmental philosophy

There is more than one reason why biodiversity is, for environmental ethics, a 'hard case'. One reason is that the concept easily defies definition. Like simplicity (a concept that plays a crucial role in the philosophy of science and scientific methodology), diversity seems intuitively easy to grasp but, on a closer look, proves to be a concept of considerable complexity. What does it mean to say that a natural system is 'diverse', absolutely or to a certain degree? What are the items on which biodiversity is predicated: biological species, alleles, or the various properties of natural systems? If properties, should only biological properties be included, or also non-biological ones such as aesthetic and symbolic properties, which play an important role in the psychology and the politics of conservation? Should the mere number of items count, or should diversity be defined by some qualitative measure, e.g. by the amount of difference between the items considered, provided these differences are in some way significant? Obviously, counting is not enough. Even a picture that is black all over can exhibit a great diversity of shades of black, but that would not be sufficient to make it particularly diverse in colour. The distinction, customary in ecology, between species diversity, ecosystem diversity and genetic diversity shows that the concept of biodiversity is in need of specification to make it commensurate with the findings and theories in various biological disciplines. The biological distinction is certainly helpful as a first step, but what is needed for an ethical analysis is an inclusive concept of biodiversity, covering all relevant dimensions.

Similar difficulties beset the attempt to spell out why biodiversity, however defined, is a value and why we should be concerned to preserve it. Is the value of biodiversity intrinsic, i.e. should it be pursued for its own sake? What is the point of preserving as much diversity as possible independently of human appreciation? Metaphysical systems such as Gottfried W. Leibniz' monadology were based, among others, on a 'principle of plenitude' to the effect that a more diverse world is to be preferred to a more homogeneous world. But this premise was embedded in a metaphysics that provided, at the same time, for a divine observer who was said to value diversity. In a modern metaphysics of diversity, such as Nicholas Rescher's (1980), devoid of a divine observer, the same principle seems no longer compelling. But even if diversity is taken as intrinsically valuable, the question arises whether the protection of biodiversity is a reasonable moral and political end in itself, or only in combination with other ends. Is it desirable to have a greater diversity of flu viruses than a lesser diversity of them (cf. Marggraf and Streb 1997: 238)? What is desirable in the great diversity of parasites newly evolved in the last centuries of human history? And would it be desirable to have a greater variety of exotic plants and animals in our vicinity, if that contributes to biodiversity world-wide, paralleling the increasing multiculturalism of many industrialised societies?

These questions arise not least because the ecological doctrine, formerly dominant, that diversity in ecosystems is a condition of stability has been falsified so that an argument for the protection of biodiversity, once taken for granted, no longer carries much weight. Ecosystems with a great number of different species such as the tropical rain forest have been shown to be much less resilient to external pressures than relatively simple ecosystems such as the much less diverse European beech forest (Remmert 1980: 255). Instead of a condition of stability, diversity seems rather to be a consequence of stability.

This acerbates the ethical question why the protection of biodiversity is important and why we should devote public resources to its preservation. At the same time, it shows that at least some of the arguments put forward by preservationists are, at best, incomplete. The mere existence, or rarity, of a biological species or allele cannot be sufficient to make it worthy of preservation. Otherwise, noxious bacteria and infectious agents would have the same right to protection as whales and elephants. If, however, the value of biodiversity is, at least partly, extrinsic, i.e. valued for its causal role in preserving or enhancing other values, the question arises what these values are and what their status is: are they *biocentric*, such as non-human nature's chance to evolve undisturbed by human intervention, or *anthropocentric*, such as aesthetics, education, scientific progress and long-term economic well-being?

In my view, the justification of the protection is, in fact, extrinsic. Its main justification lies in the interest we can suppose future generations will have in being given the chance to actually find the plenitude of natural beings in the actual world and not only to read of them in books. Another reason to preserve a great variety of biological species, ecosystems and landscapes is that we cannot predict what kind of nature later generations will be interested in. We can be sure, however, of two things: that preferences in this respect will undergo considerable change, just as they did in past history; and that the interest in experiencing nature untrammelled by civilisation will continue to grow as it becomes scarcer due to the expansion of human activities and the growth of human population. There are very few regions of the planet from which civilisation recedes. The general tendency is that increasing parts of nature are dominated by purposeful human activities, leading to a considerable simplification and impoverishment of natural systems. It is ironic that in the present industrialised world with its advanced agricultural technologies, the greatest biodiversity is no longer to be found in the countryside but in the scattered biotopes of densely populated agglomerations such as Berlin (cf. Reichholf 2007: 191).

The 'substitution problem'

In the following, attention is drawn to another problem surrounding biodiversity as a value concept: its indifference to *individuality*. Biodiversity, taken as a central value concept of nature ethics, confronts a problem identified by Eric Katz (1985) as the 'substitution problem'. The problem is that, in principle, every entity can be replaced by another entity possessing the same valuable properties. Provided that the overall value of nature is not diminished, or even increased, any replacement of one entity by another is legitimate, or even required. 'Overall value' means here the aggregate of the intrinsic and extrinsic values of the natural entity in question, ascribed on the basis of a certain underlying axiology and a prognosis of its short- and long-term effects.

The original context of Katz's diagnosis of the 'substitution problem' was that of ecological *holism*: according to holism, only natural wholes such as biotopes, ecosystems or biological species are recognised as bearers of natural value. This implies that individual natural entities have value only *qua* components of the wholes of which they are parts. Except for cases in which the value of a natural whole partly derives from the identity of some of its components (as with natural monuments, for example), every single component of the valuable whole is in principle replaceable by another individual with the same function or role. Thus, the value of a forest does not in general depend on the individuality of the trees of which it is made up. It follows that even massive changes in an ecosystem by human intervention can be justified given that the overall value of the relevant whole remains constant. Often, the overall value can even be increased by substitution, for example if substitution increases its diversity, complexity, interrelatedness or aesthetic attractiveness. If *function* or *role* is all that matters, any entity can in principle be replaced by another with the same function. If, as in Aldo Leopold's famous dictum, 'a thing is right when it tends to preserve the integrity, stability, and beauty of the biotic community' (Leopold 1949: 224f.), and 'integrity' is interpreted as the maintenance of the system in its *qualitative* aspects, any substitution of natural entities by human intervention is permitted that maintains or even adds to the integrity of the biotic community in this sense. It is not permitted, of course, if 'integrity' is interpreted as inviolability, or a strict prohibition of human interference.

The logical basis of substitutability in ecological holism is the ontological truth that the identity of a *whole* of some sort is in general compatible with the destruction, or loss, of one or the other of its individual components. While maintaining the integrity of an *individual* entity implies assuring its survival as an individual, maintaining the integrity of a *system*, *organic unity* or *whole* implies no more than the maintenance of its essential properties, its defining qualitative aspects.

Katz interpreted the 'substitution problem' as a problem peculiar to holistic conceptions of natural value. In fact, holism is only a special case. The problem is more general and arises in the same way for non-holistic conceptions. In *holistic* axiological models, individuals are substitutable under their *functions* for valuable wholes. In *individualistic* axiological models, individuals are substitutable under their value-conferring *properties*. Substitutability is not avoided by substituting holistic by individual value, as Katz seems to think when he postulates that 'an environmentally conscious moral decision maker cannot merely consider the instrumental value, the functional operation of the entities in the ecological system; he must also consider the integrity and identity of the entities in the system, i.e. their intrinsic value' (Katz 1985: 255). If 'intrinsic value' in this quotation is interpreted in the usual way, i.e. as a value ascribed to a thing in virtue of its own properties (in contrast to extrinsic value ascribed in virtue of its effects), individual intrinsic value (exceptions apart) fares no better, with respect to substitutability, than holistic intrinsic value. However intrinsic value is conceived, as beauty, richness, complexity or whatever, it allows for substitution of individuals in the same way for ecological individualism as for ecological holism. Each of these values is preserved by substituting one individual by another with the same value. Even in the *community* model proposed by Katz, a model that is to take into account *both* the instrumental value of the individual entity for a larger organic whole *and* the intrinsic value of the individual, substitutability is preserved. If I substitute, in my garden, a beautiful forsythia by an even more beautiful jessamine, without thereby jeopardising the ecological functioning of my garden as a biotope, I certainly *increase* rather than *diminish* the overall value of my garden, as far as its intrinsic value is concerned.

The 'substitution problem' in axiological nature ethics and in the ethics of biodiversity

Axiological or value ethics bases human obligation on the real or potential existence of entities exhibiting certain value properties. The basic moral obligation postulated by axiological ethics is either to assure the maintenance of a given level of value-instantiation (in a certain domain), or to increase the level of value-instantiation (in a certain domain), depending on whether the ethical system takes a more conservative or a more progressive perspective. (An essentially *conservative* ethic of maintaining a certain level of value-instantiation is part and parcel of the idea of *sustainability*.)

Axiological approaches continue to dominate nature ethics, but there are also a few rival theories. For a long time, virtue ethical positions played a more prominent role in the German-speaking world than in the Anglo-Saxon world, such as Albert Schweitzer's ethic of *reverence of life*, and Kantian approaches basing obligations toward animals and other components of non-human nature on the concept of *human dignity*. In the meantime, virtue ethical approaches to the ethics of nature and particularly to animal ethics are widespread also in the Anglo-Saxon sphere (cf. Hursthouse 2006). Another interesting, but little regarded, alternative to axiological approaches is the *genealogical* approach suggested by biologist Ulrich

Kattmann (1997). According to Kattmann, obligations in respect of the natural world do not arise from the *values* embodied in nature but from our *position* within it. Responsibility for nature is based not on any qualitative properties of natural individuals or systems but on relations of descent and kinship. We are responsible for nature not in the way a custodian is responsible for treasures valued for their beauty, rarity or old age, but in the way a father is responsible for his children, i.e. unconditionally and irrespective of any specific value attributed to them by himself or others. Ecological responsibility is dictated by fact and in so far beyond dispute and human discretion.

Some of these rival approaches to nature ethics are inspired by the very same observation underlying the 'substitution problem': that axiological approaches primarily attach value to general properties (qualities, functions, relations) exhibited by natural entities and not to individuals as individuals. They attach value to individual natural entities *qua* bearers of certain properties but are ultimately indifferent to the identity of these individuals. In this way, they are at odds with one of the central impulses of nature conservation, the impulse to preserve natural entities for their own sake, in view of their singular individuality, or, to use a rather old-fashioned expression, their *haecceitas*.

Values are treated by axiological approaches as *supervenient* on certain descriptive properties. If an individual natural entity is intrinsically worthy of protection, this is, according to axiological theories, in virtue of some general property the entity has. It may, for example, be worthy of protection because it adds to the overall level of beauty, complexity, and stability of an ecosystem. However, most values for which it is protected under the axiological scheme have no relation to its concrete individuality.

All this applies to the value of biodiversity as it does to other natural values, and even to a heightened degree. If a natural entity is protected because it adds to the overall level of some general natural property, the value for which it is protected is related to one or many of its individual properties. This relation is no longer present if a natural entity is valued only for its contribution to biodiversity. The only property by which it contributes to biodiversity is that of being sufficiently *different*, in genotype, phenotype or some other relevant property, from other natural entities. The reason is that biodiversity is, first, essentially a holistic value (a value attaching to wholes) that does in no way imply the sacrosanctness of any individual (genome, ecosystem, or species) contributing to it; and that it is, secondly, a structural or second order value independent of the identity of the qualities on whose differences biodiversity is based. Like a differential in mathematics, it characterises the structure of a function but does not specify its identity. Since, for biodiversity, only this general structure matters, not only is any individual in principle replaceable by another individual, also their qualities are replaceable by other qualities. If, in practice, restrictions of substitution are in place to preserve biodiversity, these are due to *contingent* factors, such as the slowness of biological evolution. As Kristin M. Jakobson and Andrew K. Dragun state, 'the genetic impoverishment caused by present extinction rates will not be replenished for about 5 million years' (Jakobson and Dragun 1996: 60). This fact may well afford a reason to protect each species individually to secure species diversity in the long run. From a practical point of view, Nicholas Rescher is certainly right in pointing to the fact that the preservation of endangered species is a much safer strategy to maintain biodiversity than the hope to substitute lost species by new ones synthesised by genetic recombination (cf. Rescher 1980: 90ff.). But the *theoretical* position underlying this warning is the position that overall value would be preserved by such a substitution, given that the overall level of existence and the level of diversity are maintained. No doubt, this position is perfectly consistent given the Leibnizian perspective Rescher takes on biodiversity; however, it will come as a shock to devoted environmentalists.

The example of species biodiversity shows that if there is a problem with substitutability, it does not only come in degrees but also on different levels of the hierarchy of natural properties. The extent to which substitution is possible under an axiology of natural values depends, first, on the particular axiology chosen and on how this axiology is organised: whether it is *monistic* or *pluralistic* and whether these values are *commensurate* or not. It depends, second, on what level of *generality* the values of the axiology are meant to function, allowing or not allowing for an aggregation of different kinds of value into one overall value. If the values of a pluralistic axiology are taken to be commensurate, we get substitutability among values in addition to substitutability of individuals under one single value. Trade-offs become possible between the degrees in which different values are realised by a natural entity, so that entity e_1 which realises values v_1, v_2 and v_3 to degrees d_1, d_2 and d_3 respectively will not only be replaceable by an entity e_2 that realises the same values to the same degree, but also by an entity e_3 that realises values v_4, v_5 and v_6 to degrees d_4, d_5 and d_6 respectively. Even if jessamine, in my personal axiology, is less beautiful than forsythia, jessamine may make up for this by attracting, and supporting, a far greater variety of butterflies, thus realising the same overall value. Not only is the beauty of e_1 traded off against the beauty of e_2, but beauty is traded off against ecological function. If the values concerned are incommensurate, the range of substitutability among values, and thereby of entities bearing different kinds of value, is narrowed. If values v_1 to v_5 are lexicographically ordered with v_1 (beauty, say) at the top, substitution of jessamine for forsythia would diminish the overall value of my garden, no matter what other relevant differences there are, and would be incompatible with even the weakest normative principle of an axiologic ecological ethics, the principle of value preservation.

It might be questioned whether talk of substitution and substitutability makes sense at this high level of aggregation. That e_1 is replaceable by e_2 usually means that e_1 is replaceable by e_2 *in a certain respect r*, with r functioning as a specified value aspect, a 'covering value' (Chang 1997: 5ff.). Can r be an abstract value without any specific content? It makes perfect sense to say that e_1 is replaceable by e_2 with respect to beauty or ecological function (which is to say that e_1 is more beautiful, or has a more substantial ecological function, than e_2). But does it make sense to say that e_1 is substitutable by e_2 with respect to value or valuableness, without specifying the concrete nature of this value?

There seems to me no good reason why the abstractness of the value aspect in this case is incompatible with meaningful talk of substitution and substitutability. Even in cases in which the overall value of a state of affairs is no specific value, but the aggregate of all value aspects present (for which we have no specific label apart perhaps from 'desirability'), it seems a perfectly legitimate aspect with respect to which two states of affairs can be equivalent or non-equivalent, substitutable or non-substitutable. The situation may be compared to cases in which we judge that a certain distribution is *just*, in an overall sense, where justice is the aggregate of a number of partial aspects of justice (like fairness, equality, merit, need, etc.). We do compare and evaluate distributions in terms of overall justice, despite the fact that this measure cannot be explicated but by reference to a diversity of criteria or principles of justice. An example is the measures of overall justice by which the various schemes of distribution of social goods or social burdens are justified, such as taxes, social services or scarce organ transplants. In the latter case, usually a set of highly diverse principles of distributional and compensatory justice are taken account of, along with other rights-based and consequentialist considerations, and integrated into one complex algorithm.

Why is substitution a 'problem'?

How serious a problem is the 'substitution problem'? It must be said that this is fundamentally unclear. It is even unclear whether substitutability is a problem at all. Whether or not it is a problem is a persistent issue in the controversy between mainstream economists on the one hand, and mainstream environmental ethicists on the other. For economists, substitution tends to be a normality. For a significant fraction of environmental ethicists, and for many practising environmentalists, substitution is a provocation.

Explanations are not far to find. Economists typically favour *anthropocentric* axiologies, whereas environmentalists typically do not. 'Efficiency' in economics usually means the maximisation, or satisficing, of *human* preferences. Since in an anthropocentric axiology the value of non-human natural entities can only be *extrinsic* value, i.e. *resource* value in the widest possible sense (including aesthetic, scientific, spiritual and other kinds of 'inherent' anthropocentric value in William K. Frankena's (1979: 13) sense), it is evident that bearers of extrinsic values are exchangeable by their very nature. A mere means is always substitutable by functional equivalents. If natural entities are viewed as means for human ends, or as raw material to be transformed, in one way or other, into human goods, the relation between non-human nature and value is contingent. Nothing depends on the exact nature of the natural inputs into the human states and activities in which intrinsic value resides. In the controversy about 'weak' or 'strong' sustainability, economists, as a rule, are on the side of the weak variant.

Another reason is that quite a number of economists subscribe to some variant of *welfarism* (such as utilitarianism) or to some other *monistic* value system, whereas environmentalists typically favour pluralistic value systems, often with lexicographic orderings. Value monism implies substitutability, whereas value pluralism does not.

Thirdly, it is obvious that the resources of economics are much better suited to problems allowing unlimited substitution. There is a natural convergence of what the mathematical machinery of economics is most apt to deal with and the intellectual predilections of those who wield this machinery. Environmentalists, on the other hand, typically tend to have an almost instinctive dislike of formal approaches sometimes misleading them into a full-blown rejection of formal methods of systematising values, establishing priorities and calculating costs and benefits. This corresponds to the fact that environmentalists typically tend to reject substitutability of natural entities.

It is far from obvious, however, whether they have a clear theory of what they are doing. As we have seen, the fact that the majority of environmentalists are in favour of non-anthropocentric and non-welfarist axiologies and subscribe to some form of axiological pluralism is not sufficient to explain the rejection of substitutability. Even granting that there are problems in any *practical* substitution of natural entities, e.g. because of possible side-effects, risks of irreversibility, and limited possibilities of creating new species or alleles, all this cannot justify the strictness with which substitution is rejected in some strands of ecological ethics even as a thought experiment. David Ehrenfeld, for example, writes as if each natural value is in principle incommensurate with any other so that the condition of equivalence basic to substitutability is never fulfilled:

> The … hazard is that formal ranking is likely to set Nature against Nature in an unacceptable and totally unnecessary way. Will we one day be asked to choose between the Big Thicket of Texas and the Palo Verde Canyon on bases of relative point totals? The need to conserve a particular community or species must be judged independently of the need to conserve anything else. Limited resources may force us to make choices against our wills. But ranking systems encourage and rationalise the making of choices.
>
> (Ehrenfeld 1978: 203ff.)

Edward O. Wilson expresses a similar conviction by saying in a newspaper article that 'each biological species is a masterwork of evolution and irreplaceable' (Wilson 1995: 33). Though the exact meaning of 'irreplaceable' is far from clear in this statement, it is, I think, obvious that its intention is to warn us not to deal with natural entities (and especially natural species) in the calculating and actuarial spirit characteristic of economics.

Is it possible to give a rational account of this rejection of substitution and substitutability? What is the reasoning behind this wholesale opposition to any principled weighing, comparing and prioritising of values, the part and parcel not only of economics but of axiological thinking in general?

One potential answer is that the assumption of substitutability is inherently *reductionist*. It might be taken to ignore the individuality and specificity of each individual entity by subsuming it under some general value category, thereby making it comparable to others. This answer is, however, as unclear as it is popular. To criticise some way of thinking as 'reductionist' implies that one or more aspect of a thing is neglected which is held to be particularly important, thus leading to false or at least distorted ways of thinking about it. An axiological approach to natural entities, however, is not reductionist in any of these ways. It is true that axiology implies comparability, especially with non-lexicographic orderings of values. This does not imply, however, that the specificity of each single constellation of values present in a natural entity is thereby somehow eliminated or denied. Even if the Big Thicket of Texas and the Palo Verde Canyon get a certain ranking, or a point value, in the valuation scheme of some nature protection agency, this abstract aggregate value is the result of a number of more specific and more fine-grained valuations taking into account a plurality of value aspects. (Think, again, of the diversity of factors that enter into the judgement of the overall 'justice' of a certain distribution scheme, for example in the allocation of scarce organ transplants.)

Karl Marx, in his critique of money in 'The Capital', famously called money the 'radical Leveller of all differences', in which 'all qualitative differences of commodities are wiped out', thus drawing an analogy to G.W.F. Hegel's concept of *Entfremdung*, the reification of an inherently complex and dynamic entity with its own history into a seemingly homogeneous, inert and unhistoric object (Marx 1965: 145f.). A sum of money no longer shows 'what has been transformed into it'. Likewise, the resultant number of points in a priority list of natural sites worthy of preservation does not show what kinds of concrete value have gone into it. But is that a case of reduction? Aggregation amounts to a *reduction* only if not all relevant value aspects have been taken into account. In this respect, the comparison with monetary value is inadequate. While the market value of a thing does not in general give an adequate picture of the value ascribed to it (the consumer usually pays less for it than he would be prepared to pay on the basis of his preferences), this need not be the case with a more elaborate and adequate scheme of valuation. A systematic axiological approach to ecological choices is not at all subject to the same criticism as the attitude of the snob who, according to Oscar Wilde, knows the price of everything and the value of nothing.

A second potential answer is that what is at stake is not commensurateness but the fact that by subjecting natural entities to any valuation scheme whatsoever we are exercising *control* over it. We impose, as it were, our own valuation scheme on something to which human valuation is alien, subjecting nature to some kind of *heteronomy*. The psychological basis of this feeling seems to be this: man has arrived late on the earth. It is not adequate that man as a late product of evolution arrogates to himself the role to decide what in nature should exist and what not. It is simply arrogance to judge on what we have not created and could not have created.

But this idea does not lead us very far either. Talk of heteronomy makes sense only where talk of autonomy makes sense. Since most parts of nature are incapable of valuations of their own, the terminology of heteronomy seems to be out of place. There would be 'arrogance', to take up Ehrenfeld's title *The arrogance of humanism*, in valuing nature and thereby subjecting it to human standards only if we could confidently say that nature has a will of its own.

If there is no 'arrogance' in human valuation as such, isn't there 'arrogance' in the concrete *act* of changing nature according to these valuations, i.e. in the act of actively substituting given natural entities by others, thus changing, modelling and re-modelling biotopes, ecosystems and landscapes?

In response to this it may be asked why there is a problem in actively changing nature according to human values when nature itself is a process in constant flux, involving the replacement of individuals, biotopes, ecosystems and species, though at a much slower pace. What is particularly problematic in *anthropogenic* changes in contrast to *spontaneous* changes? Katz in his article of 1985 obviously thinks substitution problematic only if effected by human intervention, not if effected spontaneously, e.g. by one biological species driving another out of the system and taking its place. But what, it might be asked, is special about anthropogenic substitution? It cannot be the case that, as Katz seems to imply, in the case of anthropogenic substitution some kind of intrinsic value is lost which is not lost in the case of spontaneous substitution. In both cases, intrinsic value is lost (or gained), depending on what is substituted for what. Apart from possible side-effects, the axiological overall value of a natural system is not affected by whether substitution comes about spontaneously or by human intervention. If the intrinsic value of each natural item is bound up with its 'integrity and identity' (Katz 1985: 255), this intrinsic value is also lost by natural elimination and extinction.

The reason Katz gives for this asymmetry is that only humans are moral subjects and are therefore responsible for the loss of intrinsic value involved in substitution. But this seems to misplace the asymmetry. Humans are responsible not only for active interventions but also for omissions, and often, humans are able to alter the course of events in such a way that a spontaneous process of substitution is prevented. For Katz, however, 'natural' substitutions are never problematic independently from whether they can be prevented by human interference. If weeds grow in my garden and drive out my flowers, though I could prevent them from doing so this would, according to Katz, constitute no comparable problem.

What is seen as problematic about substitution, then, is *neither* whether the sum total of intrinsic value is diminished, preserved or increased in the process, *nor* whether it falls within the range of human responsibility, but whether it involves any *active destruction or transformation* of given natural entities. The distinction between *activity* and *passivity*, *agency* and *non-agency* and not that between these other factors seems to be the central variable.

The theoretical problem with this variable, however, is that it has no place in a purely axiological approach, at least not in axiologies of the non-moral sort we have hitherto considered. The distinction between active and passive is a characteristically *deontological* distinction. This explains the difficulties in integrating the values of 'integrity and identity' into a purely axiological scheme. Seen from a purely axiological point of view, integrity and identity are values, if they are values, of a general sort, not tied to any particular individual. If e_1 is replaced by e_2, the integrity and identity of e_1 is replaced by the integrity and identity of e_2, just as other value properties of e_1, say beauty, complexity, diversity or productivity, are replaced by those of e_2. The fact that substitution is effected by the *infringement* both of the identity and the integrity of e_1 has no effect on the value balance.

What underlies Katz's concern about substitution seems, then, a deontological principle prohibiting not so much substitution as such but the acts of active harming and transforming natural entities involved in many of them. As an action-related rather than an end-state principle, this principle resists integration into a purely axiological approach.

This is, however, not the end of the matter. There remain two problems with Katz's way of stating his position, a meta-ethical one and a normative one.

The meta-ethical problem is that Katz's formulation tends to make what is at bottom a deontological norm of human behaviour appear as some kind of intrinsic value of the things that are the object of this behaviour. The wrongness of actively destroying and transforming natural entities is expressed in axiological language as the value of the integrity of these natural entities. What is misleading about this is that the special logical role of a value of integrity as opposed to other values is left implicit. A similar problem attaches to Immanuel Kant's concept of human dignity, which in fact has led to analogous misunderstandings. Human dignity should be understood as a deontological norm prohibiting treating a person only as a means and not also as an end in itself. In Kant's theory, however, this norm is given the logical status of a value of the object of the forbidden instrumentalisation. Human dignity is described as a *property* of human beings. This way of construing human dignity, however, invites the question why it should be contrary to human dignity to substitute one human being for another (for example by selective abortion and subsequent birth of another possible less ill-fated child) since the sum total of human dignity is kept constant in the process.

How plausible is a deontological constraint on human substitution of natural entities? Stated as a principle rigidly prohibiting any destruction of natural items, this principle seems hardly acceptable even on a narrow interpretation of 'natural items' excluding natural entities that owe their very existence to human intervention. Nevertheless, a principle of this kind *does* seem to underlie Ehrenfeld's and Katz's reservations about the irreversible elimination of variola, the smallpox virus. Destruction of a pest is held by these authors to be a problem even if keeping it alive implies risks of destruction on an enormously greater scale.

A more plausible strategy would be to soften the rigidity of an appropriate deontological constraint by integrating it into a mixed deontological-axiological value scheme as *process-values* which contribute to the overall value of a state of nature without dominating it. An *extended* axiological framework on these lines seems more suited to incorporate the specific negative valuation put by many environmentalists on processes of substitution involving destructive human interventions. At the same time it would not prejudge the issue of what exact weight to give to these values.

An example of such an extended axiological framework is provided by Paul W. Taylor's 'ethics of respect for nature' (Taylor 1986). Though Taylor's nature ethics starts from an axiological postulate, the moral rules constructed on this basis are of an explicitly deontological kind. The only intrinsic value in nature recognised by Taylor's axiology is the teleonomic structure of biological organisms. This by itself leaves ample room for substitutions between individual organisms since the value of the teleonomic structure of individuals is assumed to be the same for 'higher' and 'lower' organisms. Substitution is restricted, however, by a rule of non-maleficence as well as (following Tom Regan 1983: 297ff.) by a 'principle of minimum wrong'. Both rules prohibit harming a biological organism even if the net sum of the underlying value is thereby increased. The beings that are benefited or harmed by an act of substitution are not regarded 'as so many "containers" of intrinsic value or disvalue' but as beings 'to which are owed prima facie duties' (Taylor 1986: 172, 283f.). As with Katz, and differently from Robin Attfield (1983), 'harm' in this context means *active* harm; not, for example, negligence or the failure to prevent harm coming from other sources (Taylor 1986: 172). In contrast to Katz, however, the rule of not harming is stated by Taylor as a prima facie rule which allows for trade-offs with other deontological constraints as well as with the values specific to the human sphere ('respect for persons'). In this way, it accords much better with widely shared intuitions.

Limits to substitution in an axiological framework: historical values

Are there limits to substitution even in an unextended axiological framework? It might be thought that substitution can be satisfactorily excluded by giving some items of an ontology *infinite* value. But this strategy does not preclude substitution provided that what is substituted is also of infinite value. What we have to look for in order to limit substitution is a value property that necessarily vanishes, or deteriorates, with substitution, like genuineness of the original of a work of art vanishes with substitution by a perfect copy.

Plausible candidates for axiological values by which substitution is limited are *historical* values such as *age* and the fact that something is of *natural* origin (cf. Elliot 1982, Attfield 1994). Age, be it the ontogenetic age of an individual plant, the phylogenetic age of a biological species, or the geological age of a landscape, is an important criterion for quite a number of environmentalists. For example, Eric Ashby thinks that it 'would … be vandalism, and therefore immoral to destroy unnecessarily something which we cannot create and which is the expression (and not the end-product) of millennia of evolution' (Ashby 1980: 28f.). Though substitution of younger entities by older ones is conceivable, substitution in the reverse direction tends to be the rule, especially where natural systems are harvested for human use, including the slaughter of domestic animals. *Naturalness* and *wildness* understood in a genetic sense are historical values of equal importance, at least where substantial areas of wilderness persist. In Central Europe, where even so-called 'virgin forests' are in general no older than a few centuries, efforts at conservation tend to concentrate on naturalness and wildness in a purely *phenomenal* sense, referring not to the *history* of natural entities but to their *appearance*.

As far as historical interest is directed at singular objects with a certain genetic or causal role, objects of historical value like manuscripts, documents, and relics seem irreplaceable. However, historical value can in turn be regarded as an axiological value so that historically valuable items can at least be substituted for each other. This kind of value has recently been termed 'origin value' (Basl 2010: 136). Historical values, however, cover only a small part of what we value in nature. Historical values like age and authenticity are relevant to natural monuments (a concept that was coined, not surprisingly, in the period of historicism) but it may be doubted whether they carry much weight, for example, with biological species. Is the spider, as Albert Schweitzer once suggested to one of his guests in Lambarene who was about to kill the arachnid in his room, a worthy object of protection because it is, as a species, so much *older* than man? Historical value has its legitimate place in a museum, but it is doubtful if the museum perspective should dominate our practical dealings with nature.

In general, historical authenticity is a value of limited scope. We are interested in authenticity where it goes together with additional values, such as causal importance and artistic quality, or with relational values like the role a work of literature has had in one's personal biography (even when it has long lost its personal value as a work of art). It does not seem to make much of a difference how old a component of nature is if it is devoid of all further qualities that recommend it for preservation, or which, like the smallpox virus, strongly recommend it for elimination.

Limits to substitution in an axiological framework: relational values

We have so far only mentioned in passing the probably strongest reservations against substitution: *personal* values rooted in affective ties like love and friendship. Personal ties transcend axiological value because their objects are not general *properties* but individual *objects*. The objects of love and friendship are inexchangeable precisely because these emotions are directed at individuals *qua* individuals and not *qua* possessors of general properties. That is why love, more clearly than friendship, is an unsuited object for why-questions.

An immortal apotheosis of the strict individuality of love is Heinrich von Kleist's comedy *Amphitryon*: Alkmene proves unable to distinguish Jupiter who has transformed himself into a 'perfect copy' of her husband from her real husband Amphitryon. Nevertheless she is adamant in maintaining that she loves only one of the copies, the authentic Amphitryon. In her insistence on identity and authenticity she is not alone, however. Identity is not only valued by mortals. Even Jupiter, in spite of his divinity, insists on his irreplaceability. He prides himself that he, though indistinguishable from Amphitryon in order not to betray himself, is slightly more perfect than a perfect copy in surpassing Alkmene's legitimate husband in terms of passionate love-making.

The absence of substitutability in affective relations is, of course, strictly *relative*. An object of love or friendship is irreplaceable not in the impersonal sense of a philosophical axiology but only with respect to the individual emotionally related to it. This holds not only for persons and pets but also for natural objects. A part of nature may be greatly important for an individual, the population of a certain geographical area or a whole generation, and have no or very little importance for others living at different places and times. This value essentially depends on the relations humans build up towards their natural environment. If there is axiological value in these relations (there certainly is), it must not be sought in the value that is attributed to the objects of these relations but in their contribution to these intrinsically valuable relations, either as causes and causal factors (instrumental value) or as intentional objects (inherent value).

As far as natural entities are concerned, these relations are highly diverse in quality, intensity and extent. The intentional objects of love, awe, admiration, and sentimental attachment range from single animals, plants or mountains to the whole of nature, creating, in every single case, a certain resistance against change and substitution. We cling to what we know and have grown familiar with, an effect studied under the names of *endowment effect* (Thaler), *status quo bias* (Samuelson / Zeckhauser) or *loss aversion* (Kahneman / Tversky). People in general demand much more to give up an object than they would be willing to pay to acquire it (cf. Kahneman et al. 1991). The status quo is looked upon as a kind of *possession*, with a corresponding reluctance to replace what exists with what will exist in the future even if it is certain that the substitute will be a perfect equivalent.

Substitution within the constraints of biodiversity protection

The issue of relational values provides an opportunity to reprise a point made at the beginning: that it is mainly the *variability* of what is loved and liked in nature between cultures and historical periods that is one of the strongest reasons for the protection of biodiversity. Relational value is ultimately a variant of *extrinsic* and not of *intrinsic* value. But that does not diminish its significance and its legitimacy. The variability of the relations people are able to establish to natural objects, structures and qualities, documented by environmental psychology (cf. Oechsle 1988: 68), is a strong reason for the protection of biodiversity in nature on the social and political level. At the same time, there is a strong cause for biodiversity protection not only in a *conservative* sense, but also for an active *furtherance* of the evolution of diversity wherever this is possible without jeopardising other important ecological and non-ecological goals.

Within the constraints of securing a high level of biodiversity as an ecological and potentially economic safety margin for future generations, however, anthropogenic substitutions of natural individuals and natural systems seem unobjectionable. Substitution may even be ecologically productive, as by counteracting tendencies to ecological simplification, actively supporting endangered species, or replacing common varieties by rare varieties and frequent ones by endangered ones. Nature is not sacrosanct. We should feel free to improve it. Biodiversity might, after all, be a much less conservative value than it appears at first sight.

Note

1 An earlier version of this chapter appeared under the title 'Limits to substitutability in nature conservation' in Oksanen 2004: 180–95.

References

Ashby, E. (1980) 'The search for an environmental ethic', in S. McMurrin (ed.) *The Tanner lectures on human values*, vol. I, Cambridge: Cambridge University Press, 1–47.

Attfield, R. (1983) *The ethics of environmental concern*, Oxford: Blackwell.

Attfield, R. (1994) 'Rehabilitating nature and making nature habitable', in R. Attfield and A. Belsey (eds) *Philosophy and the natural environment*, Cambridge: Cambridge University Press, 45–58.

Basl, J. (2010) 'Restitutive restoration: New motivation for ecological restoration', *Environmental Ethics* 32: 135–47.

Chang, R. (1997) 'Introduction', in R. Chang (ed.) *Incommensurability, incomparability, and practical reason*, Cambridge, MA: Harvard University Press, 1–34.

Ehrenfeld, D. (1978) *The arrogance of humanism*, New York: Oxford University Press.

Elliot, R. (1982) 'Faking nature', *Inquiry* 25: 81–93.

Frankena, W.K. (1979) 'Ethics and the environment', in K.E. Goodpaster and K.M. Sayre (eds) *Ethics and problems of the 21st century*, Notre Dame, IN: University of Notre Dame Press, 3–20.

Hursthouse, R. (2006) 'Applying virtue ethics to our treatment of the other animals', in J. Welchman (ed.) *The practice of virtue ethics: Classic and contemporary readings in virtue ethics*, Indianapolis, IN: Hackett Publishing, 136–55.

Jakobson, K.M. and Dragun, A.K. (1996) *Contingent valuation and endangered species. Methodological issues and applications*, Cheltenham: Edward Elgar Publishing.

Kahneman, D., Knetsch, J.L. and Thaler, R.H. (1991) 'Endowment effect, loss aversion and status quo bias', *Journal of Economic Perspectives* 5: 193–206.

Kattmann, U. (1997) 'Der Mensch in der Natur: Die Doppelnatur des Menschen als Schlüssel für Tier- und Umweltethik', *Ethik und Sozialwissenschaft* 8: 123–31.

Katz, E. (1985) 'Organism, community, and the substitution problem', *Environmental Ethics* 7: 241–56.

Leopold, A. (1949) 'The land ethic', in A. Leopold (ed.) *A Sand County almanac and Sketches here and there*, New York: Oxford University Press, 201–26.

Marggraf, R. and Streb, S. (1997) *Ökonomische Bewertung der natürlichen Umwelt. Theorie, politische Bedeutung, ethische Diskussion*, Heidelberg: Spektrum.

Marx, K. (1965) *Das Kapital*, vol. 1, Berlin: Dietz Verlag.

Oechsle, M. (1988) *Der ökologische Naturalismus. Zum Verhältnis von Natur und Gesellschaft im ökologischen Diskurs*, Frankfurt am Main: Campus.

Oksanen, M. and Pietarinen, J. (eds) (2004) *Philosophy and biodiversity*, Cambridge: Cambridge University Press.

Regan, T. (1983) *The case for animal rights*, Berkeley, CA: University of California Press.

Reichholf, J.H. (2007) *Eine kurze Naturgeschichte des letzten Jahrtausends*, Frankfurt am Main: Fischer.

Remmert, H. (1980) *Ökologie. Ein Lehrbuch*, 2nd edn, Berlin: Springer.

Rescher, N. (1980) 'Why preserve endangered species?', in N. Rescher (ed.) *Unpopular essays on technological progress*, Pittsburgh, PA: University of Pittsburgh Press, 79–92.

Taylor, P.W. (1986) *Respect for nature. A theory of environmental ethics*, Princeton, NJ: Princeton University Press.

Wilson, E.O. (1995) 'Jede Art ein Meisterwerk', *Die Zeit* June 23 1995: 33.

3 Biodiversity

A methodological reconstruction of some fundamental misperception

Mathias Gutmann

Introduction

Biodiversity is a biological concept – this assumption does not come as a surprise, save the objection that a term which is used in context A (say, biology) is not only on grounds of this specific application 'by nature' a concept that exclusively belongs to context A. The use of the same term in context B (say, engineering) will generate no misunderstanding, as far as we assume that term 'a' now represents concept 'b' (and not concept 'a' anymore). To give an example, the term 'field' ('a') may be understood as a biochemical description of factors (a), which are thought to hold a determinable role in the process of morphogenesis (A). In terms of electrodynamics (B), however, the same term refers to a bundle of organised forces, generated between charges of specific qualities (b). The term obviously designates two different concepts, which are explicable unanimously, *given* the respective context of theory and laboratory practices.[1]

Keeping this caveat in mind, we may easily argue for the intuition that 'biodiversity' is coined at least in biological terms, an assumption which gains some plausibility by citing a well-known and most comprehensive definition:

> "Biological diversity" means the variability among living organisms from all sources including, inter alia, terrestrial, marine and other aquatic systems and the ecological complexes of which they are part; this includes diversity within species, between species and of ecosystem.
>
> (Harper and Hawksworth 1995: 6)

Very obviously, the definition is of a biological nature and nothing leads to the conclusion that the term 'biodiversity' might be used in a metaphorical way. However, taking a closer look at the definition and its constituting parts, some doubts are reasonable. The development of these doubts on grounds of a methodological analysis of the underlying biological concepts, their necessary (but mostly implicit) premises and anticipation and the tools, which are applied in order to deal with the respective pieces of nature, is the aim of this chapter.

Is 'being divers' a property of a particular?

The definition given above seems to be as comprehensive as it is explicit but it shows at least one central systematic problem, which is tightly connected with the referent of the term 'variability'. We might be tempted to answer that it is – very obviously – the variability found 'among living organisms' that we are seeking and thus we only needed to explicate this term 'living organism' in order to know what 'biodiversity' refers to. And again, there seem to be no further obstacles that we might easily apply one or the other parameterisation-tools, e.g. the Shannon-, the Simpson- or the Brillouins-Index in order to finally reach a more or less agreeable estimation of the diversity of the respective organisms (cf. Back and Türkay 2002).

The parameterisation itself seems at first glance to be just a formal procedure, which allows us to determine the diversity of any given manifold of elements. And this can be done *irrespective* of the specific nature of the elements under scrutiny, as long as they are characterised by certain features, such as being a car, being a natural number or being an insect. The example of the Shannon-Index helps to clarify this point[2] (cf. Back and Türkay 2002: 237–42):

$$H'(p_1, p_2, \ldots, p_s) = -\sum_{i=1}^{s} p_i \log p_i$$

with

$$p_i = \frac{N_i}{N}$$

The index itself is obviously without any biological implications; it was indeed derived without relevant reference to biological knowledge at all, namely within the context of information theory. If, however, the index is to be applied in order to generate *biological* knowledge, we have to explicate the variables in reference to biological concepts. From this point of view, e.g. the 'belonging to a species' provides – or should provide – the standard for the estimation of the living organisms' diversity. This brings us back to the question of what it means to belong to a species. It is a well-considered result of the thorough discussion on the species-problem during the last four decades that the term 'species' is used in (at least) two fundamentally different ways, which are often confused (cf. Gutmann 1996). On the one hand, we call species a formal entity, introduced by abstraction, referring to a *relation of equivalence*. Thus 'white-hairedness' can be explicated by referring to something that shared its 'having white hairs' with something else. In doing so, we introduce the species 'white hairs' by constructing a class via the underlying predicate. In consequence, we are not stating that two tokens of 'having white hairs' are identical,[3] but that they represent the same class (designated as 'white hairs'). In biology we usually refer to species of this kind in terms of taxonomic features, e.g. 'having six/eight/ten arthropodia', 'being homoiothermal/poikilothermal' or 'having gonads in retroperitoneal position'. By stating that A is 'the same' as B, we are referring to a class-building predicate, and thus the relation between A and B is a three-termed relation (because *A* resembles *B* in *reference to x*), which allows us to deal with A in the same way as with B (in reference to x, etc.). This relation of equivalence is reflexive (A equals A), symmetrical (if A equals B then B equals A, etc.), and transitive (if A equals B, and B equals C, then A equals C).

On the other hand, the term 'species' refers to some other *type of relation between two things*, thus stating, e.g., that A is the brother of B, that both are the *descendents* of an honourable family or that dogs are more closely related to wolves than to cats. Given the adequacy of the statements we can derive that, for example, A and B belong to the class of 'brothers', and that if C is the brother of D, they belong to the same class as C and D. So far, there is no difference to our example above concerning the sameness of colour. The difference of the underlying logical grammar becomes observable, however, when we consider the decomposition of the following statements: 'A and B are white-haired' and 'A and B are brothers': obviously the first statement can be expressed as the conjunction of two statements, namely 'A is white-haired' and 'B is white-haired'. This decomposition is a problem in the second case, at least under the rule of salva veritate, because 'A is a brother' and 'B is a brother' means that A and B belong to the same class (namely referring to 'being a brother') whereas the statement could be read as 'A is the brother of B' (which makes all the difference to C or D, but the identity to 'C and D'). On a more abstract level, we may refer to those relations as functional relations of different types. In biology this includes – beside the intuitively relevant reproductive relations – all types of functional ascriptions (e.g. in technical terms biology, physiology, ecology, etc.). In terms of biodiversity research we are dealing with both types of relations, whereas the second context is the (biologically) most relevant. Accordingly, 'being (bio-)divers' is not (just) a property of (just) a particular.[4]

Is 'being a member of a species' a property of a living entity?

Taking into consideration the homonymy of 'species', we are confronted with the – methodologically – most interesting problem of biodiversity research, namely to provide an answer to the question of how exactly we know about the 'speciousness' of particulars – in our case of 'living organisms'. The discussion of how to determine the species an animal or plant belongs to is as old as philosophy and sciences themselves and is best summarised by Darwin's statement concerning the status of the term 'species':

> In short: we shall have to treat species in the same manner, as those naturalists treat genera, who admit that genera are merely artificial combinations made for convenience. This may not be a cheering prospect, but we shall at least be freed from the vain search for the undiscovered and undiscoverable essence of the term species.
>
> (Darwin 1897: II, 301)

The main point here is not Darwin's opposition to the allegedly unchangeable character of populations (!): this undoubtedly eminent and important dispute of contemporary sciences has been decided conclusively and definitively since then. Far more important is the fact that Darwin emphasises the difference between a natural process on the one hand and the means that are applied in order to describe and conceptualise it on the other. Setting aside Darwin's own solution to this problem, the (methodological) consequence of the insight that the *means of description* should not be confused with the very *objects of the description* is best expressed in Darwin's words again:

> The endless dispute whether or not some fifty species of British brambles are good species will cease. Systematists will only have to decide (not that this will be easy) whether any form be sufficiently constant and distinct from other forms, to be capable of definition; and if definable, whether the differences be sufficiently important to deserve a specific name.
>
> (Darwin 1897: II, 300)

Hence, before we can identify species and determine their number, their structure and the relations between their respective members and between each other, we need to determine the criterion we are going to use. However, by doing so we are confronted with the startling fact that there is not just one, but a couple of differing species definitions that do not necessary result in co-extensive groupings (i.e. classes of equivalences; cf. Gutmann and Janich 1998, 2002 a, b). Referring to the respective definientia, we can discern between, e.g., phonetic,[5] genetic, physiological, morphological, evolutionary, ecological or phylogenetic[6] species definitions, each facing specific methodological as well as pragmatic problems. Let us start with the so-called Biological Species Concept (BSC), according to which a species is defined as follows (for further reading cf. Gutmann and Janich 1998):

> A species consists of a group of populations which replace each other geographically or ecologically and of which the neighboring ones intergrade or interbreed wherever they are in contact or which are potentially capable of doing so (with one or more of the populations) in those cases where contact is prevented by geographical or ecological barriers. Or shorter: Species are groups of actually or potentially interbreeding natural populations, which are reproductively isolated from other such groups.
>
> (Mayr 1942: 120)

The BSC actually provides the core of the oldest definitorical schema. In a certain sense it goes back as far as to Aristotle – at least according to the empirical criterion Aristotle refers to (in the sense of kai anthropos anthropon genna: Metaphysics XII, 1070b). This represents a logical structure of 'species', which refers to the generative relation of particulars and universals, i.e. the logical problem we dealt with in the section above.

Unfortunately, in contrast to its presumed simplicity, it poses some serious problems. Firstly, reproductive isolation (or conversely interbreeding) between living entities is no transitive relation (at least not formaliter – materialiter it may prove to be so), the projectability and thus the explanatory power of a related statement on a group of not interbreeding (or conversely interbreeding) living entities is rather restricted.[7] Secondly, it does only apply to certain types of living entities in restricted spatio-temporal regions. Thirdly, reproductive isolation (or interbreeding) between organisms may be sex-related, i.e., the criterion may not even be symmetric (cf. Laven 1953, Neumann 1971; further reading cf. Gutmann 1996; Gutmann and Janich 2002b). Mayr considers at least some of these limitations himself and adds a rather interesting aspect:

> The application of a biological species definition is possible only in well-studied taxonomic groups, since it is based on a rather exact knowledge of geographical distribution and on the certainty of the absence of interbreeding with other similar species.
>
> (Mayr 1942: 121)

The demand for the groups under consideration to be well studied is by no means trivial: it implies the validity of some biological knowledge, which, by definition, must be independent of the identification of a group as a species in the first place and which – in the second place – enables us to determine the (species-defining) characters of the entities under consideration. We obviously run into a dilemma, because we

1 either have to know already whether reproductive isolation exists (because the group is well studied), in which case we apply some knowledge that should have been provided by applying the BSC, or
2 we apply the BSC without the (supposedly) necessary knowledge, in which case the application does not provide us with valid knowledge on the species status of the groups.

In both cases, it is not the BSC (alone) that allows us to determine species in nature, an insight that leads Mayr himself towards an astonishing remark:

> As can be seen from these definitions, it is necessary in the cases of interrupted distribution to leave it to the judgment of the individual systematist, whether or not he considers two particular forms as "potentially capable" of interbreeding – in other words, whether he considers them subspecies or species.
>
> (Mayr 1942: 120)

The consequences of this characterisation of the systematists relevance for identifying species 'in nature' are well known as the cynics species concept (cf. below). However, considering some of the problems connected with the BSC and in slight contrast to it, the Genetic Species Concept (GSC) restricts the characters or traits, which are assumed to indicate the membership of a species. For example, the sequence-similarity allows the estimation of genealogical distance. However, GSC faces the same methodological problem, as it is also assuming that 'belonging-to-a-species' is a property of particular living entities: the degree of similarity is by no means defined 'by nature' (whether it should be less than 5, 3 or 1 per cent sequence-difference; for further reading cf. Ammann and Rossello-Mora 2001). The underlying atomism which presupposes the belonging-to-a-species as a property of particulars generates similar limitation problems we had to face in the case of the BSC. There are some further methodological difficulties and even contradictions which result from the use of different types of characters or different distance measures, as the example given by Anthony G. O'Donnell shows:

> However, while all available 16S rRNA sequences of legionellae show marked sequence similarities, DNA-DNA hybridization studies of the same species may show less than 5% homology with the type strain of the genus, Legionellae pneumophila ... Similarly, Fox et al. (1992) have shown that although strains of Bacillus globisporus and B. psychrophilus share greater than 99.5 % similarity in 16Sr RNA sequence homology they show less than 50% reassoziation in DNA pairing experiments.
>
> (O'Donnell, Goodfellow and Hawksworth 1995)

There are similar problems even if an explicitly *evolutionary* point of view is taken. For example, the Phylogenetic Species Concept (PSC) takes into consideration not just populations and their reproductive behaviour but the entire evolutionary lineage:

> A species is a single lineage of ancestral descendant populations of organisms which maintains its identity from other such lineage and which has its own evolutionary tendencies and historical fate.
>
> (Wiley 1992: 80)

The idea behind defining species via their evolutionary descent is probably the most comprehensive approach conceivable in biology (we will come to a functionalistic alternative in the next section) and as such it has to face the same type of objections as other holistic approaches. In this case we have not only to assume the *potential* stability or instability of inbreeding-relations and their mutuality and symmetry, but the non-trivial independence of evolutionary statements of the species-category; and, to make things worse, we have to determine the *biological* meaning of certain procedures (such as parsimony, maximum likelihood, bootstrapping, etc.) which allows us to shift from mere clades to actual taxonomic entities (such as species, genera or phyla; for further reading cf. Back and Türkay 2002; Gutmann and Janich 2002b).

Methodologically, the crucial point here is not the pure number of alternative definitions of the term species (which sometimes even are mutually exclusive); rather it is the very fact that the determination of species ('in nature') is not an unambiguous, formal procedure of counting entities, but a process of *judgement*. Thus, the parameterisation is actually guided by specific epistemic purposes. Assuming that living entities are per se belonging to species, we might indeed be bound to accept Philip Kitcher's diagnosis:

> The most accurate definition of "species" is the cynic's. Species are those groups of organisms, which are recognized as species by competent taxonomists. Competent taxonomists, of course, are those who can recognize the true species.
>
> (Kitcher 1992: 317)

This circularity was to a certain extent already realised by Darwin. From a methodological point of view, it comes as a surprise only if we overlook that describing a particular living entity *as a member* of a species is not just a statement of fact – but a specific biological structuralisation of living entities in terms of functional relations.

Functional structures and functional replaceability

We have exclusively focused on the term species thus far in order to gain more clarity for the line of argument.[8] However, even this seemingly simple term revealed an intrinsic ambivalence, serving as a designator of class-element relations on the one hand, and as a designator of more complex relations (such as the genealogical or evolutionary) on the other. Those relations were paradigmatically explicated in terms of genealogy – which was, of course, the main focus[9] of Darwin's attempts to formulate the foundation of evolutionary theory. However, even genealogical relations are just *one* token of a type of relations which provide the methodological framework for biological objects. By providing biological descriptions of living entities we actually structure objects of everyday *lifeworld*[10] in terms of functional relations. In doing so, we are constructing a *standardised* scientific language which refers to this pre-scientific basis – such as breeding animals and plants, using them in terms of labour and work, etc. (for further reading cf. Gutmann 1996). According to this methodological reconstruction of the semantics of biological theory[11] we are leaving the specifically restricted validity of the ordinary language game by dealing with *organisms* instead of *living entities*. The term organism then designates the result of the functional structuring of living entities as functional units. In contrast to organisms, living entities are *not* just functional units, comprising organs or tissues – although they can be described as such.[12]

Biologists are interested in learning more about the relations between those functionally structured entities, these relations themselves being of a functional nature. Functionalism, however,comes at a price as it leads to some counterintuitive results. The functions we are ascribing to parts, individuals or groups of living entities are well defined only within the respective theoretical framework, i.e. in reference to a (prior order implicated) structuring. Accordingly, there usually are no one-to-one relations between structures and functions. From a biological point of view, the function of, e.g. the eyes, is not merely to serve as light sensors – even though they (often) certainly do have this function.

This insight is usually expressed by the caveat that one and the same structure can play several functional roles at once. The lungs in higher developed mammalia (such as *homo sapiens*), for example, are of central importance for gas-exchange, but they are functionally involved in thermoregulation and important for the capability to carry heavy weights; the eyes in vertebrates are used in sensory activities, but they play a crucial role for the mechanical constitution of the skull during embryogenesis. What is true for (virtually) any part of a living entity holds true for ascending and descending of further steps in the respective scales of descriptions as well: organs, tissues, cells, molecules, or populations and ecosystems, respectively, are to be understood as functionally arranged units within a complex, multi-levelled framework of a multitude of relations. Hence, ascribing functions to parts of particulars, to the respective particular itself, or to groups of particulars does not lead to an atomistic but to a relationistic construal of such units;[13] they are not defined 'as such' but 'in relation to' each other.

According to this functionalistic reading, the term species does not just refer to 'a group of living entities' whose members stand in certain relation to each other, but to *organisms*. Consequently, they are to be understood as an element of a (functional) system. This allows us to introduce some type of functional Ecological Species Concept (ESC):[14]

> The intensive definition required may be obtained by considering a hyperspace, every coordinate (X1, X2, X3 …) of which corresponds to a relevant variable in the life of a species of organism. A hypervolume can therefore be constructed, every point of which corresponds to a set of values of the variables permitting the organism to exist.
>
> (Hutchinson 1965: 32)

This approach results in a system which is literally composed of organisms and their respective relations. The parameter hypervolume represents these relations of the resulting system; it is the (logical) form of a biological system. According to this functionalistic reading, the boundaries of the system, the respective functions (which then are considered to be instantiations of the relations), and the functional units (of which organisms are considered as instantiations) are not to be understood as parts of nature. They are elements of a formal structure depending on the respective epistemic purpose and the means of description. Such 'ecosystems' are not to be found in nature, but refer to the activities of biologists, as they are describing and structuralising pieces of nature *as systems*.[15]

Consequently, a particular living entity can be part of several different systems at once and, conversely, one and the same system can refer to a number of different types of organismic structures. The components of such systems stand as mutual relations of functional equivalence. That is, they are exchangeable in respect to their external or internal functional relations.[16] Accordingly, in one description – defined via the parameter hypervolume of the ESC – two living entities could belong to different species, whereas in another description – i.e. in terms of taxonomy or systematics – they would belong to a single species.

Different types of (hierarchical) systems are based upon different biological descriptions which in turn are neither co-extensional concerning the tokens of the respective levels nor do they necessarily represent the same concept (even if the same term applies) concerning the correlation of the levels themselves (from Gutmann and Neumann-Held 2000):

Table 3.1 An overview of several hierarchies that can be found. Note the differing structure of these hierarchies as well as the different levels that are mentioned

Ecological (Odum 1991: 39)	*Physiological (Odum 1991: 39)*	*Genealogical hierarchy (Eldredge 1985: 187)*	*Ecological hierarchy (Eldredge 1985: 187)*
Biosphere	Individual	Monophyletic taxa	Regional biotas
Biogeographical region	Organ system	Species	Communities
Biom	Organ	Demes	Populations
Landscape	Tissue	Organisms	Organisms
Ecosystem	Cell	Chromosomes	Cells
Living community	Organelle	Genes	Molecules
Population	Molecule		
Organism			

Table 3.2 This table continues the overview of Table 3.1.

Bio-levels (Mahner & Bunge 1997: 178)	*Vertical aspects (Solbrig 1994: 43)*	*Horizontal aspects (Solbrig 1994: 43)*
B1 = elementary cells = set of all elementary cells	Molecules	Populations, species, living communities
B2 = composite cell level = the set of all composite cells	Cells	
B3 = organ level = the set of all (multicellular) organs	Organs	
B4 = multicellular organismic level = the set of all multicellular organisms		
B5 = population level = the set of all bio-populations		
B6 = community level = the set of all communities (biocoenoses)		
B7 = biosphere level = the set of all biospheres		

Corresponding hierarchies may be constructed by referring to specific sub-levels of these levels,[17] e.g. by referring to specific functional systems such as the immune-, endocrine- or neuro-system.

The functionalistic reading of biological units (such as species or ecosystems) has severe consequences for the parameterisation of biodiversity. Following this line of reasoning, being a member of species A does not necessarily equal 'not being a member of species B'. Furthermore, insofar as species are nothing but components of systems, the same multi-relational structure is potentially to be found on all levels of the description.

Hierarchies and the ambivalence of numbers

Aside from the fact that species are just one component of biodiversity, and considering the challenges described above, if we are to understand the peculiarities of the parameterisation of biodiversity, we are confronted by an even more delicate problem, namely the problem of comparability. 'Being a member of a species' is as little the (sole) property of a particular living entity as 'being a component of a system' is the (sole) property of an organism. It is not just the pure number of species or their internal and external relations that matter in attributing diversity values to a given system. The even more startling problem consists in the fact that the degree of diversity varies in the methods and the formal structure of measurement applied.

The methodological problem here is the different standards of weighing characteristic for different diversity indices. In order to understand these difficulties, it should be kept in mind that species are but one – namely the lowest – type of taxonomic unit within a hierarchically ordered system. A rough sketch of the hierarchical-enkaptic system we are referring to in terms of *taxonomically* characterised systems (which are usually called communities – a term with misleading anthropomorphic connotations) can be presented with Ax (1984: 244) as follows:

Phylum	Chordata
Classis	Mammalia
Ordo	Primates
Familia	Hominidae
Genus	Homo
Species	Homo sapiens

Of course, much more fine-grained representations are possible, e.g. by considering inter-taxa differentiation, such as (Ax 1984: 88; Storch and Welsch 1997: 695ff.):

Ordo	Primates
Sub-order	Haplorhini
Infra-order	Simiformes Catarrhini

Super-family	Hominoidea
Family	Hominidae
Sub-family	Homininae
Tribe	Hominini
Genus	Homo
Species	Homo sapiens

As this taxonomic sketch already reveals, the respective levels are of different cardinality depending on the higher groups under which species are subsumed. Now, if we assume three groups of species – two families and one order and one phylum in the first group, two orders and one phylum in the second group, and three orders and three phyla in the third group – the diversity may differ in correlation to the index used (main differences refer to diversity, equity, dispersion, etc.; for further reading cf. Back and Türkay 2002). However, even the formal structure of the measures influences the evaluation of the various pieces of nature, as it depends on the use of criteria such as dispersion, distance, taxon-richness or species-richness. So, for example, the monotony and continuity in species or distances are not necessarily preserved to the same extent by applying different measures (cf. Back and Türkay 2002). In consequence, even the same number as an expression of the diversity of some parts of nature, described as ecosystems, may represent differently structured biodiversities (!).

Why should we parameterise biodiversity?

Provided that the parameterisation of biodiversity is necessarily guided by purposes, we may wonder *why* we *should* parameterise it at all. And this question is the more pressing the less we assume species to be natural components of a given environment. Given there is no 'natural quantity' of species, we can only answer this question by referring to explicable or already explicated purposes that underlie the procedures used in order to structure, classify and organise biotic matters.

There are at least two types of purposes that can be discerned according to the epistemic status of the resulting statements:

a The respective scientific interest of enquiry determines the selection of the procedures used to discern groups of organisms as tokens of species. In this context it can be useful to 'borrow' methods from other sciences such as chemistry, physics and informatics. These may be applied, e.g. to determine growth rates, the abundance and composition of certain chemical substances (such as alcaloids) or the genealogical structure of some putatively monophyletic taxa. However, the resulting taxonomies (and their respective species) are tightly connected with the guiding scientific purpose and, not surprisingly, the validity of the resulting system[18] is of an *alethic* nature.

b In a certain contrast to scientific purposes the realm of non-scientific purposes seems not to show any determinable limitations. But even if we do not aim at scientific explanations, prognosis or explications we can – and must – give specific criteria in order to determine whether scientific knowledge is applied successfully in terms of know-how; these criteria (the aptness of application), however, will be of a *pragmatic* and not an alethic nature. This implies that even if some knowledge is regarded as (scientifically) insufficient or inaccurate, we will not deny its applicability in terms of successful action.

Construing species as *scientific* (here: biological) units on the one hand and as *pragmatic* units on the other, we can explicate biodiversity itself as a scientific and as a pragmatic concept, whereby in the latter case the term 'biodiversity' – although referring to explicable or explicate biological knowledge – becomes a kind of metaphorical phrase.

Biodiversity as a metaphor

Biodiversity explicated as a biological concept solely referring to species reflects some characteristics of scientific concepts in general: universally valid statements ought to be valid irrespective of the context of utterance, at least in principle.

By understanding universality as being of a prescriptive (instead of a purely descriptive) nature, we are not committed to assume that a given statement 'is' universally valid but that a given statement *ought to be* universally valid. The difference is of some methodological importance, because in the first case we are confronted with the – probably impossible – task of actually guaranteeing the context invariance of a given statement, whereas in the second case even the falsification of a given statement does not imply the impossibility of universality.

This situation changes fundamentally when we are shifting from a scientific context to a non-scientific context of application: in this case, we are not asking for a scientific explanation (for example, on the evolutionary mechanism of population transformation) and, consequently, the definienda of the term 'biodiversity' cease to be of a (purely) scientific nature. 'Species' is now regularly connected with non-scientific purposes, e.g. horti- and agricultural production.

The relevance of concepts such as 'biodiversity' or 'species' cannot be underestimated within our modern industrialised production of breeding lines. They provide valuable knowledge about the kinds and types of animals or plants that are used for breeding and cultivation (cf. Gutmann and Weingarten 2004).

Considering modern biochemical production, we may add pharmacological taxonomies, dealing with the chemical compounds of plants used for medical purposes, etc., or the utilisation of living entities for the maintenance of the (ecological) status of parts of nature (e.g. for oxygen production, the filtering of water, the preservation of availability and accessibility of ground water, etc.). In summary, we can present a – by no means complete – list of useful 'services' parts that nature can 'deliver':

Economic considerations. In contrast, biological resources represent a significant contribution to economic activity and – provided they are managed prudently – therefore to sustainable development. Prescott-Allen & Prescott-Allen (1986) produced the first analysis of the importance of wild species to the United States economy.

Agriculture and pest management. Among the uses of biodiversity for economic activity which ordinarily escape mainstream economic calculus is the use of genetic traits from wild relatives of domestic crop species. The international centres for various crops such as rice or wheat are continually turning to wild relatives for disease and pest resistance.

Pharmaceuticals. Very often production of a pharmaceutical product initially requires a lot of material harvested in nature to extract the active ingredient but this is often superseded by the ability to synthesize. Although this means in one sense the biological source is no longer necessary, it is important to recognize that it derives from the original inspiration, that is, the template provided by a wild species.

Environmental applications. The value of biodiversity to waste management and environmental clean-up problems, through a technique known as bioremediation, is increasing rapidly. The discovery of microorganisms with odd metabolisms and appetites can greatly facilitate solving such problems. One of the more intriguing is the bacterium found in the sediments of the Potomac river which has the ability to break down the ozone destroying chemicals known as chlorofluorocarbons.

Molecular-level benefits. Genetic engineering now makes it possible to introduce desirable genetic traits from one species into another which is not closely related. Pest resistant genes from Bacillus thuringiensis have been transferred to a variety of crop species. A freeze resistant strain of tobacco has been produced by inclusion of a gene from winter flounder. The development of the "Flavr Savr" tomato about which there has been so much controversy in the United States actually involves only the manipulation of tomato genes to delay softening which normally comes with ripening so that ripe tomatoes can be shipped long distances without rapid spoiling.

(Lovejoy 1997: 82ff.)

In all these cases we apply scientifically based taxonomies that allow us to identify organisms which show the necessary features; additionally, we might even be able to recognise the conditions of their optimal reproduction. Consequently, the (ultimate) purpose of their application is not the generation of a scientifically sound system of living entities or the corroboration of scientific theory; rather the successful utilisation itself shall be enhanced, assessed and maintained. 'Utilisation' here refers to direct consumption of parts of nature or their indirect contribution to the reproduction of conditions of societal reproduction itself. Accordingly, in each of these cases 'biodiversity' is but a metaphor that indicates practical, not epistemic, criteria of application.

Reproduction of reproductivity

According to this reconstruction, parts of nature are (first of all) means for achieving certain non-scientific, non-epistemic purposes. In this respect, the aim of applying concepts such as biodiversity or species may be the preservation and development of conditions under which human societies reproduce. In this context, the term 'reproduction' is not to be reduced to its biological meaning: reproduction of human societies designates all kinds of human activities that are considered to be relevant not only for a society to be viably preserved but to be capable of reproduction, including all types of symbolic reproduction. This implies that even the sources of our knowledge on nature, namely sciences, become means of this comprehensive type of reproduction (cf. e.g. Cassirer 1987, 1988, 1990).

However, the metaphorical use of the term biodiversity provides us with a guideline for policy in order to shape and to structure parts of nature for societal reproduction. This guideline is easily gained by considering the fact that each and every tool becomes worn out during usage. Consequently, a certain effort is necessary in order to reproduce the tools themselves – this 'reproductive labour' then reduces the maximally possible productive capacity.

Joachim Spangenberg (1999) discerns two fundamental cases, the first referring to sampling of biogenic products, extending from hunting and gathering to extensive forms of agri- and horticulture, and the second case referring to intensive agri- and horticulture. This second case includes the (possibly industrial) use of by-products of parts of nature: the production of oxygen, the stabilisation of ground water, the utilisation of river banks for the sake of water filtering, etc. (cf. Spangenberg 1999). In this description, the components of biodiversity (here: species of animals, plants or microorganisms) play the role of productive means[19] and it is exactly this position which allows us a non-naturalistic interpretation of the biodiversity metaphor: whereas the first case represents the simple utilisation of parts of nature, i.e. the use of its products, the second case represents the investment of work into the reproduction of the productive means. However different the indicator for the success of our actions may be (e.g. the increase of bio-mass in relation to the space needed, the substance output per square metre, etc.), the general purpose of dealing with biodiversity (as a politically explicated biological metaphor) refers to the utilisation of pieces of nature in terms of societal reproduction. According to Gutmann and Weingarten (2004), this utilisation of pieces of nature should avoid purely consumptive use; the ultimate aim of utilisation is the development of options and potentials of pieces of nature for further use (the 'utilisation of utilisation'; cf. Gutmann and Weingarten 2004). But we should add a second, more creative aspect here, namely the possibility of re-interpreting given means-end relations as a starting point for the transformation of the respective utilisation practise itself (cf. Gutmann and Weingarten 2004).

Conclusion

The term biodiversity and the use of biological terms can have a non-trivial metaphorical function in the socio-cultural context. In this case, those terms do not aim solely at alethic validity. The term 'ecosystem' may provide a framework for economic action – it indicates that certain descriptions of parts of nature, of species or other aspects of an environment point to the use of the respective parts of nature for the sake of societal production and reproduction. The purposes aimed at by using biodiversity and biological terminology are not of a scientific nature, i.e. they do not consist of pure description, explanation or prognosis. Predominantly they serve as a source of means for non-scientific purposes. Scientific descriptions are irreplaceable and indispensable tools for societal purposes insofar as they do not simply depict nature as such, but enable us to understand the reproduction of the parts of nature as tools for further societal production and reproduction. Consequently, by reconstructing the term biodiversity as a metaphor for some aspects of the reproduction of human societies, we should understand the so-called 'biodiversity crisis' neither as deviance from nor convergence towards a presumed 'natural' equilibrium, disturbed by the activity of human societies – it is rather a crisis of productive and reproductive relations within and of human societies.

Notes

1 This does not exclude the possibility of family resemblance in historical and even in conceptual terms, e.g. by assuming the transfer and retransfer of expressions between distinct but contemporary scientific theories.

2 This observation is independent of the fact that there are several types of parameterisation tools in use which differ in terms of their formal nature (cf. Back and Türkay 2002). The authors discuss the intrinsic problems of the respective underlying concepts and their respective application in greater detail.

3 This indeed would be a self-contradicting statement, considering the fact that we are actually dealing with *two* tokens! Hence, we are confronted with the intricate dialectics of the principium identitatis indiscernibilium (for further discussion cf. Tugendhat and Wolf 2004, Kamlah and Lorenzen 1973, Lorenzen 1987).

4 From an evolutionary point of view, assuming that the features or traits of organisms are the result of an evolutionary process, the first type of relations can be included. The *grade* of similarity between some organisms then is thought to express the relative evolutionary distance between them and may become the measure of genealogical relations. However, the logical difference between the two types remains the same.

5 The term 'phenetic' refers to all characters or traits of an organism that build its respective phenotype (i.e. those characters which – presumably – make a 'genetic' difference).

6 'Phylogenesis' and 'evolution' are by no means synonymous expressions; for further reading cf. Mayr 1997 and Gudo et al. 2007.

7 The relevance of these restrictions increases by actually considering the aspect of potentiality, which is indeed substantial and necessary, because otherwise the projectability of the species category would collapse completely.

8 The methodological reconstruction of the two other relevant terms (gen and ecosystem) leads to corresponding results. The semantics of biological theories and descriptions are always (at least to a certain extent) a matter of normative (but not

necessarily ethical) construction which refers to explicit or at least explicable epistemic purposes and their underlying scientific practices (for further reading cf. Gutmann and Janich 2002a).

9 Even though it was obviously not the only one, as we can learn from his extensive considerations of functional, embryological and even 'ecological' aspects of organismic constitution (for further reading cf. Gutmann 2002, Hertler 2001).

10 Following a constructivist extension of Husserl's concept, we are not just referring to ordinary language but to know-how and knowledge in terms of (often implicit and non-scientific) practices, which provide a starting point for the introduction of scientific knowledge and know-how (such as the use of numbers, measurement devices, the building of houses, the processing of ore, the cultivation of plants, etc.; cf. Lorenzen 1987; Janich 1992, 1997).

11 This is not only theory-laden, but based upon pre-scientific and scientific (but not biological) know-how such as physical and chemical procedures (cf. Gutmann 1996).

12 As already stated, living entities are modelled *as* organisms. This means that they are described and structured *as if they were* functional units. For a detailed reconstruction of the model procedures applied in order to gain 'organismic descriptions' of this type cf. Gutmann 2002.

13 This philosophical statement has immediate practical consequences: properties refer to particulars *qua* particulars in a certain context. For example, the invasiveness or the nativeness of a species cannot be its exclusive property. Thus, trying to single out invasiveness-defining criteria can be as vain as asking for the species-defining criterion itself (cf. below; for further reading on this problem cf. Williamson 1993, Sukopp and Sukopp 1993, Noble 1989).

14 The ESC is just one example for functional structuring in biology.

15 Alfred Lotka's introduction of ecosystems on the background of the metaphor of a world engine may serve as an example here; accordingly, living entities are to be understood *as if they were* energy transformers (for a critical reconstruction of this approach cf. Gutmann and Janich 2002b).

16 Of course regarding the specific logical structure of the relation itself at the same time.

17 From a constructivist perspective, there are indeed no 'levels in nature'; the term 'level' represents elliptically the results of the respective underlying structuring procedures, their guiding epistemic questions, etc. (cf. Gutmann and Janich 2002a, b).

18 For the difference between taxonomy and systematics cf. Gutmann 2002.

19 These are 'produktionsmittel' in Marx's terminology – we should keep in mind that our considerations are not of an economic nature. They are part of some socio-theoretical assumptions which probably underlie the economic theory itself. From a methodological point of view, then, economic theory needs a basis in everyday lifeworlds as much as biology.

References

Amann, R. and Rossello-Mora, R. (2001) 'Mikrobiologische Aspekte der Biodiversität', in P. Janich, M. Gutmann and K. Prieß (eds) *Bodiversität. Wissenschaftliche Grundlagen und gesellschaftliche Relevanz*, Berlin: Springer, 161–80.

Aristotle (2002) *Metaphysics*, trans. J. Sachs, 2nd edn, Santa Fe, NM: Green Lion.

Ax, P. (1984) *Das phylogenetische System*, Stuttgart: Gustav Fischer.

Back, G. and Türkay, M. (2002) 'Quantifizierungsmöglichkeiten der Biodiversität', in P. Janich, M. Gutmann and K. Prieß (eds) *Bodiversität. Wissenschaftliche Grundlagen und gesellschaftliche Relevanz*, Berlin: Springer, 235–80.

Cassirer, E. (1987) *Philosophie der symbolischen Formen. Band 2: Das Mythische Denken (1924)*, Darmstadt: Wissenschaftliche Buchgesellschaft.

Cassirer, E. (1988) *Philosophie der symbolischen Formen. Band 1: Die Sprache (1923)*, Darmstadt: Wissenschaftliche Buchgesellschaft.
Cassirer, E. (1990) *Philosophie der symbolischen Formen. Band 3: Phänomenologie der Erkenntnis (1929)*, Darmstadt: Wissenschaftliche Buchgesellschaft.
Darwin, C. (1897) *Origin of Species*, vols I and II, New York: AMS Press.
Eldredge, N. (1985) *Unfinished Synthesis*, New York: Oxford University Press.
Gudo, M., Gutmann, M. and Syed, T. (2007) 'Ana- und Kladogenese, Mikro- und Makroevolution – Einige Ausführungen zum Problem der Benennung', *Denisia* 20: 23–36.
Gutmann, M. (1996) *Die Evolutionstheorie und ihr Gegenstand. Beitrag der Methodischen Philosophie zu einer konstruktiven Theorie der Evolution*, Berlin: VWB.
Gutmann, M. (2002) 'Aspects of Crustacean Evolution. The Relevance of Morphology for Evolutionary Reconstruction', in M. Gudo, M. Gutmann and J. Scholz (eds) *Concepts of Functional, Engineering and Constructional Morphology: Biomechanical Approaches on Fossil and Recent Organisms. Senckenbergiana lethaea* 82 (1): 237–66.
Gutmann, M. and Janich, P. (1998) 'Species as Cultural Kinds. Towards a Culturalist Theory of Rational Taxonomy', *Theory in Biosciences* 117: 237–88.
Gutmann, M. and Janich, P. (2002a) 'Überblick zu den methodischen Grundproblemen der Biodiversität', in P. Janich, M. Gutmann and K. Prieß (eds) *Bodiversität. Wissenschaftliche Grundlagen und gesellschaftliche Relevanz*, Berlin: Springer, 3–27.
Gutmann, M. and Janich, P. (2002b) 'Methodologische Grundlagen der Biodiversität', in P. Janich, M. Gutmann and K. Prieß (eds) *Bodiversität. Wissenschaftliche Grundlagen und gesellschaftliche Relevanz*, Berlin: Springer, 281–353.
Gutmann, M. and Neumann-Held, E. (2000) 'The Theory of Organism and the Culturalist Foundation of Biology', *Theory in Biosciences* 119: 276–317.
Gutmann, M. and Weingarten, M. (2004) 'Preludes to a Reconstructive "Environmental Science"', *Poiesis and Praxis* 3 (1–2): 37–61.
Harper, J.L. and Hawksworth, D.L. (1995) 'Preface', in D.L. Hawksworth (ed.) *Biodiversity. Measurement and Estimation*, London: Chapman and Hall, 5–12.
Hertler, C. (2001) *Morphologische Methoden in der Evolutionsforschung*, Berlin: VWB.
Hutchinson, G.E. (1965) *The Ecological Theatre and the Evolutionary Play*, New Haven, CT: Yale University Press.
Janich, P. (1992) *Grenzen der Naturwissenschaften*, München: Beck.
Janich, P. (1997) *Kleine Philosophie der Naturwissenschaften*, München: Beck.
Kamlah, W. and Lorenzen, P. (1973) *Logische Propädeutik. Vorschule des vernünftigen Redens*, Mannheim: Bibliographisches Institut.
Kitcher, P (1992) 'Species', in M. Ereshefsky (ed.) *The Units of Evolution*, Cambridge, MA: MIT Press, 317–41.
Laven, H. (1953) 'Reziprok unterschiedliche Kreuzbarkeit von Stechmücken und ihre Deutung als plasmatische Vererbung', *Zeitsch. f. indukt. Abstammungs- und Vererbungslehre* 85: 118–36.
Lorenzen, P. (1987) *Lehrbuch der konstruktiven Wissenschaftstheorie*, Mannheim: BI.
Lovejoy, T.E. (1997) 'Biodiversity: What is it?', in M. Reaka Kudla, D.E. Wilson and E.O. Wilson (eds) *Biodiversity II. Understanding and Protecting our Biological Resources*, Washington, D.C.: Joseph Henry Press, 7–14.
Mahner, M. and Bunge, M. (1997) *Foundations of Biophilosophy*, Berlin: Springer.
Mayr, E. (1942) *Systematics and the Origin of Species*, New York: Columbia Univiversity Press.
Mayr, E. (1997) *This is Biology*, Cambridge, MA: Belknap Press.
Neumann, D. (1971) 'Eine nicht-reziproke Kreuzungssterilität zwischen ökologischen Rassen der Mücke Clunio marinus', *Oecologia* 8: 1–20.

Noble, I.R. (1989) 'Attributes of Invaders and the Invading Process: Terrestrial and Vascular Plants', in J.A. Drake, H.A. Mooney, F. di Castri, R.H. Groves, F.J. Kruger, M. Rejmánek and M. Williamson (eds) *Biological Invasions. A Global Perspective. SCOPE 37*, Chichester: Wiley, 301–13.

O'Donnell, A.G., Goodfellow, M. and Hawksworth, D.L. (1995) 'Theoretical and Practical Aspects of the Quantification of Biodiversity Among Microorganisms', in D.L. Hawksworth (ed.) *Biodiversity. Measurement and Estimation*, London: Chapman and Hall, 65–74.

Odum, E.P. (1991) *Prinzipien der Ökologie*, Heidelberg: Spektrum.

Prescott-Allen, C. and Prescott-Allen, R. (1986) *The First Resource: Wild Species in the North American Economy*, New Haven, CT: Yale University Press.

Solbrig, O.T. (1994) *Biodiversität*, Bonn: Rheinischer Landwirtschaftsverlag.

Spangenberg, J. (1999) 'Indikatoren für biologische Vielfalt', in C. Görg, C. Hertler, E. Schramm and M. Weingarten (eds) *Zugänge zur Biodiversität*, Marburg: Metropolis.

Storch, V. and Welsch, U. (1997) *Systematische Zoologie*, Stuttgart: Gustav Fischer.

Sukopp, H. and Sukopp, U. (1993) 'Ecological Long-Term Effects of Cultigens Becoming Feral and of Naturalization of Non-native Species', *Experientia* 49: 210–18.

Tugendhat, E. and Wolf, U. (2004) *Logisch-semantische Propädeutik*, Stuttgart: Reclam.

Wiley, E. (1992) 'The Evolutionary Species Concept Reconsidered', in M. Ereshefsky (ed.) *The Units of Evolution*, Cambridge, MA: MIT Press, 79–92.

Williamson, M. (1993) 'Invaders, Weeds and the Risk from Genetically Manipulated Organisms', *Experientia* 49: 219–24.

4 Framing biodiversity

The case of 'invasive alien species'

Ulrich Heink and Kurt Jax

Introduction

Framing is a ubiquitous phenomenon that pervades science, policy and the mass media. The concept of frames was appropriated in the 1970s by cognitive psychology (Minsky 1974), sociology (Goffman 1974) and linguistics (Fillmore 1975). Frames are schemata of interpretation for concepts. They help organise complex concepts by embedding them into a story line or connecting them stereotypically to certain actions or events. For example, the frame for a children's birthday party can be characterised by presents, a birthday cake, lemonade and games like pot hitting or sack racing. A single word activates its defining frame, and also much of the system its defining frame is in (Lakoff 2010).

Importantly, frames do not only exist in people's minds; they are also susceptible to external framing in communication, as explained by Robert Entman (1993: 52): 'Framing essentially involves selection and salience. To frame is to select some aspects of a perceived reality and make them more salient in a communicating text, in such a way as to promote a particular problem definition, causal interpretation, moral evaluation, and/or treatment recommendation for the item described.' And certainly, people and organisations make use of the possibility of framing. For example, 'semantic battles' and the power of naming play a major role in the political struggle for getting allies and votes (Busse 1991). Frames that are often used are mentally more quickly available than others and hence suppress other possible frames (Lakoff 2010). Further, they are often directly connected to emotions (van Dijk 1980) which makes them very persistent. For these reasons it is hard to reframe existing frames.

Framing is most effective when situations are complex, facts are uncertain and values contested (Benford and Snow 2000). This can be well illustrated by the issue of climate change. In 2003, Frank Luntz advised the Bush administration in a memo on 'cleaner, safer, healthier America' on the right use of language in global warming debate. Under the title 'Winning the global warming debate' and the subtitle 'Language that works' he argues:

> Unnecessary environmental regulations hurt moms and dads, grandmas and grandpas. They hurt senior citizens on fixed incomes. They take an enormous

> swipe at miners, loggers, truckers, farmers – anyone who has any work in energy intensive professions. They mean less income for families struggling to survive and educate their children.
>
> (Luntz 2003: 139)

Further, he advises talking about 'climate change' instead of 'global warming', because '"Climate change" is less frightening than "global warming"' (Luntz 2003: 142). Moreover, 'change' avoids insinuating human agency.

Pielke (2005) also criticises 'misdefining "climatic change"' and the emphasis on human agency in the climate change debate, but for different reasons. He maintains that the focus on the causes of climate change creates a bias for greenhouse gas reduction and against adaptation policies towards climate change.

And last, according to Crist (2007), framing of climate change as the most urgent environmental problem of our time causes dangers which exceed those of climate change for two reasons: first and similar to Pielke (2005), the urgency frame induces stereotypically a call for a techno-fix of the problem. And second, the climate issue detracts attention from other environmental issues, especially the biodiversity issue.

The example of framing climate change is instructive for several reasons. First, the framing of an issue can lead to different actions. This is remarkable, as frames often are not overtly evaluative or prescriptive and hence do not suggest decisions for or against a certain action. Second, politics and scientists meanwhile are much more aware of the importance of framing, so that a great effort is made to direct the framing of issues. And last, framing is pervasive not only in different fields of science and policy but also on different levels of discourse: from the definition of terms, over semantic relations between terms of interest and other terms and the embedding in different discourses.

Biodiversity and especially invasive alien species are fields associated with great uncertainties and often a clash of values. It is still hardly possible to predict species invasions (Williamson 1999; Angert et al. 2011). While David Wilcove et al. (1998) claim that invasions are after habitat destruction the greatest threat to biodiversity, this claim produced some doubt by others (Gurevitch and Padilla 2004; Latimer et al. 2004). Mark Sagoff (2005) maintains that one cannot really judge if non-native species do harm to the natural environment because the concept of 'harm to the natural environment' is nebulous and undefined. In a reply, Daniel Simberloff (2005) insists that non-native species threaten the natural environment. These different opinions suggest that both the concept of 'harm to the natural environment' and the topic of invasive alien species are very much prone to framing.

In the following we depict attempts to frame and reframe the biodiversity concept and the concept of 'invasive alien species'.[1] These play an important role in the description, evaluation and management of biodiversity. How to deal with invasive alien species is a very controversial topic in conservation biology originating in differing worldviews on biodiversity. We show how framing of the issue of invasive alien species is connected to different perceptions of biodiversity.

We investigate how biodiversity and invasive alien species are framed in definitions and how values are assigned to them. Further, we shed light on metaphors which are used in the context of invasive alien species and normative concepts for biodiversity. We assume that an understanding of framing puts argumentation on the value of biodiversity and invasive alien species on a more solid ground and avoids manipulation by certain interest groups.

Framing biodiversity

Biodiversity can be framed on different levels. It is explicitly framed by definitions. In general, these definitions are descriptive in that they clarify what objects fall in the category 'biodiversity'. However, it can also be framed in a normative way. Motives for conservation give an idea which attributes contribute to the value of biodiversity and may thus help to separate desired from less desired parts of biodiversity. Such motives are often embedded in normative concepts. These normative concepts often implicitly refer to individual and cultural 'meanings' of biodiversity which are the basis for its protection. In the following sections we give an overview of definitions, motives and normative concepts.

Defining biodiversity

According to Article 2 of the Convention on Biological Diversity (CBD), '"Biological diversity" means the variability among living organisms from all sources including, inter alia, terrestrial, marine and other aquatic ecosystems and the ecological complexes of which they are part; this includes diversity within species, between species and of ecosystems.' Thus, the concept is comprehensive in two respects: it comprises different levels of ecological organisation (genes, species, ecosystems) and includes compositional, structural and functional diversity.

This definition is widely accepted among both policy-makers and scientists. Some criticism arose because of the buzzword character of biodiversity and related terms like 'ecosystem function' (Noss 1995; Goldstein 1999; Hull et al. 2003; cf. Lovett 1994 for an opposite view on this subject), and the fact of biodiversity being a construct which is not measurable. There were also some protests against including the diversity of abiotic components into the definition of biodiversity (DeLong 1996) because, strictly speaking, the term 'biologic' only refers to biotic components. If abiotic components are included, it was recommended to refer to such a concept as 'ecological diversity' (DeLong 1996).

Some more serious attacks on the biodiversity concept originate from a conservation point of view. Sarkar and Margules (2002) maintain that a definition of biodiversity should acknowledge the purpose to prioritise places. Therefore they offer a 'relative' biodiversity concept, which serves to determine which one of two places has a higher biodiversity against a background set of places Π: 'place A contributes more (same/less) biodiversity than B relative to Π if and only if, its list of features has more (less/the same number of) entries not in the list of Π than the feature list of B' (Sarkar and Margules 2002: 303). They add that this definition 'can be used in a straightforward way to generate additions to a system of reserves' and point to the complementarity principle in reserve selection.

Paul Angermeier (1994) complains that current conceptions of biodiversity fall short of management needs because they fail to distinguish between native and artificial diversity. In a similar vein, the definition of biodiversity by Osvaldo Sala et al. (2000: 1770) 'excludes exotic organisms that have been introduced and communities such as agricultural fields that are maintained by regular human intervention'. According to these opinions, alien organisms or cultivated ecosystems can never increase biodiversity (cf. also Sagoff 2005).

Especially before the coming into force of the CBD, biodiversity was understood as being limited to elements and composition of biodiversity (OTA 1987; Angermeier and Karr 1994; Callicott, Crowder and Mumford 1999). Diversity thus excludes ecological structures and processes and can simply be expressed as a 'number of kinds of items' (Angermeier and Karr 1994, 692). This concept was widely criticised. Noss (1990b) argues that processes, such as interspecific interactions, natural disturbances and nutrient cycles, are crucial to maintaining biodiversity. Therefore, it should not be restricted to species richness. The relation of biodiversity to 'structure, processes, functions and interactions' was explicitly referred to by the 'ecosystem approach' in decision 5/23 of the fifth meeting of the Conference of the Parties to the CBD (UNEP 2000: 104).

These attempts to frame the biodiversity concept were made mainly because it should match strategies to protect biodiversity (in principle reaching back to the 'invention' of biodiversity in the 1980s; cf. below). If one regards the definition of biodiversity as common ground, apart from the inclusion of ecological structures and ecosystem functions into the biodiversity concept, these attempts have not been very successful. Overtly normative concepts for biodiversity (e.g. Angermeier and Karr 1994) probably were not acceptable for the scientific community, as this would question their independence from policy and the paradigm of scientific objectivity.

However, framing of biodiversity as compositional, structural and functional diversity does not necessarily resonate with dominating mental frames. In the biodiversity context, this means that although the term is explicitly defined as compositional, structural and functional diversity, it may activate mainly frames connected to species richness. The use of the biodiversity term in different contexts gives some evidence for that. Heink and Kowarik (2010) found out that a majority of papers which use biodiversity indicators refer to species richness as a measure for biodiversity. Similarly, Balvanera et al. (2006) point out that 393 out of 446 ecosystem property measurements in 103 publications on the effects of biodiversity on ecosystem functioning and services referred to species richness. Moreover, 'biodiversity indicator' is sometimes defined as a variable that is correlated with species richness (Ekschmitt et al. 2003). A book on species concepts in systematics is titled 'Species – the Units of Biodiversity' (Claridge et al. 1997). Also the title 'Biodiversity – a Biology on Numbers and Difference' (Gaston 1996) does not immediately suggest ecosystem functions as part of biodiversity. In 2002, the Conference of the Parties to the Convention on Biological Diversity (COP) committed to achieving a significant reduction in the current rate of biodiversity loss at the global, regional and national levels (2010 target, UNEP 2002b). Such an anticipation of biodiversity loss is generally staged as a race against the clock for preventing species extinction (Collins and Kephart 1995).

The different definitions show that there is a great tension between attempts to define biodiversity for scientific purposes and those for supporting biodiversity protection. This ambiguity is inherent in the term from the very beginning of the debates. David Takacs (1996: 2) notes:

> In the name of biodiversity, biologists hope to increase their say in policy decisions, to accrue resources for research, gain a pivotal position in shaping our view of nature, and, ultimately, stem the rampant destruction of the natural world. To accomplish all this, biologists speak for an array of values that go far beyond what one might think of as falling within the ambit of their expertise. Their factual, political, emotional, aesthetic, ethical, and spiritual feelings about the natural world are embodied in the concept of *biodiversity*; so packaged, *biodiversity* is used to shape public perceptions of, feelings about, and actions toward the world.

Thus, from the very start, biodiversity was a term with the hybrid function of describing biodiversity components and for evaluating them (Potthast 2007).

Framing of biodiversity by conservation motives

We have shown so far that biodiversity can be defined in different ways and that these definitions are regularly influenced by ideas to protect biodiversity. In contrast to a scientific definition of biodiversity, biodiversity policy is overtly based on values which cannot be derived from ecology. Decisions have to be made concerning which parts of biodiversity should be prioritised. Conservation motives determine the background against which biodiversity is evaluated. These motives are the source for deriving evaluation criteria and can be found in different normative concepts (cf. Callicott, Crowder and Mumford 1999). In contrast to conservation motives, normative concepts do not only give a baseline for evaluation but include a line of reasoning for biodiversity conservation and a way of conceiving biodiversity (e.g. as a functioning machine, a cybernetic system or a harmonious, holistic entity). Below, we describe the three most important conservation motives – item conservation, conservation of states and conservation of processes – and show how they are related to prevalent normative concepts for biodiversity.

Item conservation

The 'itemising approach' (O'Neill et al. 2008, 167ff.) is characterised by developing a list of items (e.g. species, habitats) which are worth protecting. The value of each item increases according to its importance for completing the list and scores are assigned to different items to calculate the value of a site. The criteria used to operationalise the item conservation approach and calculate such scores are: rarity, threat and species richness. As threatened species or habitats are the first that might get lost, they are the ones that can most probably render the list of items incomplete.

Item conservation is probably the most prevalent biodiversity conservation motive. This importance is reflected by the CBD. As the CBD aims at the conservation of biodiversity, there is a conservation focus on rare and threatened species and habitats as those are most likely to be extinguished and thereby reduce biodiversity. Thus, the criteria 'rarity' and 'threat' can directly be derived from the aims of the CBD. Next to them, 'species richness' is one of the most popular criteria in conservation evaluation (Prendergast et al. 1993). This demonstrates the importance of the itemising approach.

Conservation of states

The idea of 'conservation of states' is to preserve a defined composition or structure of biodiversity. The reference state which should be achieved is in general a former natural state or a historical state. In both cases, the desired state serves as an original from a certain point of time, which should be restored. However, this need not necessarily be the case. The intended state can in principle also be a target vision ('Leitbild', Potschin et al. 2010). But even such visions will most probably refer to some historical or natural state.

In general, naturalness can be defined as the absence of human influence (McIsaac and Brün 1999). In the conservative naturalness concept[2] which is referred to here, naturalness of biodiversity is determined by a comparison to a former or – rarely – a present state of biodiversity considered not to be influenced by human culture. The criterion which is encompassed by historical naturalness concepts is here called naturalness in a strict sense or pristineness.

Naturalness is one of the fundamental motives in biodiversity conservation. The goal of Article 8 d) of the CBD is to promote the protection of ecosystems, natural habitats and the maintenance of viable populations of species in natural surroundings. The European Habitats Directive (European Union 1992) aims at 'the conservation of natural habitats and of wild fauna and flora' as is indicated by its title.

Within the conservative cultural value concept the state of biodiversity is measured against the reference of the original historical state. Biodiversity of a certain location is regarded as a cultural heritage site or a monument.

The importance of species and habitats for cultural history depends on two properties: their connection to a bygone time period and their distinctiveness. The historical connection is given when species or habitats appeared or developed under former socio-economic conditions which nowadays have ceased to exist. Landscapes which serve for the demonstration of either historical conditions or regeneration from previous impacts (landscapes of cultural heritage) have cultural importance. Distinctiveness is the extent to which species or habitats contribute to the character, identity and uniqueness of a landscape (Heink 2009). Loss of distinctiveness brings about a loss of identification with a formerly familiar surrounding and a deprivation of home without moving from the place of residence.

In the preamble to the CBD, the 'cultural, recreational and aesthetic values of biological diversity and its components' are highlighted. The cultural value, however, is quite underrepresented in the actual text of the CBD. It is worth mentioning, though, that domesticated and cultivated species are explicitly addressed in the CBD. Moreover, cultural services have become important in the assessment of ecosystem services.

Process conservation

In the 1980s the protection of processes became an important topic in conservation biology (e.g. White and Bratton 1980; Pickett and White 1985; for Germany: Scherzinger 1990). One reason for its advent was that the linear relationship between species diversity and ecological stability in its general form proved false. Another reason was the growing awareness that many species could not be saved from extinction with a static conservation approach. Many ecological processes which shape habitats (like windbreakage or inundation) cannot be adequately matched by conservation management. Hence, the understanding of nature as dynamic and evolving in a sometimes accidental and erratic manner is regarded as a major change for conceptualising biodiversity conservation. The discussion of process conservation is mainly restricted to natural processes while human-induced processes are widely ignored.

Conservation of natural processes focuses on a free development of nature without the influence of humans and does not address preserving the remnants of history. The criterion which is encompassed by this ahistorical naturalness concept is here called wildness. We deliberately chose the term 'wildness' here which is distinguished from 'wilderness' (Cole 2000; Jax 2010). While the latter refers to a 'pristine' area that has not been used by humans, 'wildness' refers to the absence of human control, to 'free-running' of natural processes. Wildness is thus determined by the magnitude of effects of past and present human activities acting on biodiversity of a given site. The concept is related to the concept of hemeroby which is defined as an 'integral expression of the sum of those effects of past and present human activities on the current site conditions or vegetation, which prevent the development to a final stage' (Kowarik 1990: 58). Hence, this definition determines naturalness as the ability of a system's self-regulation (cf. Kowarik 1999). But self-regulation (without human interference) is only one aspect of natural processes. It does not consider the processes leading to this state of self-regulation. For example, tree plantations can reach more quickly a state of self-regulation than ecosystems developing by natural succession. Further, the paradigm of natural processes assumes that a final stage of ecosystem does not exist, as processes keep the ecosystem in flux. Therefore, we maintain that self-regulation does not necessarily lead to a final stage. The protection of natural processes can be legitimised by eudaimonistic arguments. It is the experience of untamed nature which produces awe and admiration and which gives wildness an importance for human lives.

Beside natural processes cultural processes can also be appreciated, as landscapes which balance economic, ecological, aesthetic and cultural features (cultural landscapes in a strict sense, Wöbse 2002). Not any human-induced process is regarded as valuable, but only those based on a balanced consideration of these societal concerns. Sustainability in a narrow sense, i.e. a consumption of resources that does not exceed reproduction in the long run, can therefore be used as a benchmark for the evaluation of present cultural processes and instrumentally of the biodiversity which is produced through these processes (e.g. a certain species composition).

Feelings of awe and admiration are not as well developed for cultural processes as for natural processes. However, careful management of limited resources certainly deserves respect, as they testify human achievements in adapting to societal and environmental challenges. Kowarik (2008) thus advocates being open to new dynamic processes: if former land use changes had been impeded by arguments of a static preservation approach, the spectrum of valued habitats today would be more restricted.

Process conservation can be regarded as an end in itself (as described above); however, it can also be understood instrumentally for the conservation of species or ecological communities (Soulé 1996; Potthast 2000; Piechocki et al. 2004). If the conservation of processes is understood in an instrumental sense, it is determined in advance which state of biodiversity is aimed at as a result of dynamic processes. For example, the process of ageing and decay of forest trees may be pursued to develop habitats for the great capricorn beetle (*Cerambyx cerdo*) which is listed in Annex II of the Habitats Directive. For cultural processes, limits for land use intensity may be determined to keep a minimum standard of species richness in intensively cultivated agricultural landscapes (Matzdorf et al. 2008).

It is noteworthy that item conservation, conservation of states and process conservation are rarely applied in pure form. For example, the maximisation of items by adding genetically modified organism is generally rejected. Here, the itemising approach is connected with an approach of protecting natural or cultural states. Further, it is not only important to protect species, habitats, etc., but also how this is done. Maintaining biodiversity by ex-situ conservation in botanical gardens and zoos cannot replace in-situ conservation in the wild. Here the process of conservation is important for item conservation.

Normative concepts in conservation: connecting worldviews with conservation motives

In conservation biology, conservation motives are often combined with normative concepts as a yardstick for evaluating biodiversity. These normative concepts do not only contain conservation motives but also – sometimes quite implicitly – worldviews and ideas of biodiversity which give a rationale for biodiversity protection. These concepts are often framed by the use of illustrative metaphors. Here, we focus on four prominent normative concepts: biological integrity, ecosystem health, ecosystem services, and variety. We give a short overview of how biodiversity is normatively framed by these concepts.

Biological integrity

Biological integrity has been defined as 'the capability of supporting and maintaining a balanced, integrated adaptive community of organisms having a species composition, diversity, and functional organisation comparable to that of the natural habitat of the region' (Karr and Dudley 1981: 56). Thus, biological integrity considers both the biotic components of the environment and the ecological processes that generate and maintain the described components (Angermeier and Karr 1994; Callicott, Crowder and Mumford 1999).

The integrity metaphor is derived from a cultural context. As Sagoff (1995, 162) writes, 'integrity is a virtue of character and makes one more likely to receive advantages such as trust and love. As a virtue, integrity is intrinsically good'. It suggests that ecological communities ideally are intact or complete.

Naturalness provides the benchmarks to evaluate biological integrity (Callicott, Crowder and Mumford 1999; Karr 2000). Any change of a pristine natural state or in natural species composition affects integrity negatively. The concept of biological integrity thus clearly reverts to historical nature concepts.

Ecosystem health

'An ecological system is healthy and free from "distress syndrome" if it is stable and sustainable—that is, if it is active and maintains its organization and autonomy over time and is resilient to stress' (Haskell et al. 1992: 9).[3]

The health metaphor suggests an organismic view of ecosystems. Ecosystems show symptoms of health (vigor, resilience, etc.; e.g. Mageau et al. 1995). Along this line, ecosystem properties can be analysed like organisms in terms of function and malfunction (Holland 1995). However, there is no empirical evidence to support an organismic view of ecosystems (Norton 1993; Shrader-Frechette and McCoy 1994; Norton 1995). In contrast to health from an organismic perspective, ecosystems do not possess an inherent optimum state.

Instead of naturalness as the yardstick for the assessment of biological integrity, ecosystem health focuses on 'normality' (Callicott, Crowder and Mumford 1999; Hull et al. 2003). As there is no reason to use statistical normality (e.g. mean values of measures for certain ecosystem functions) in a normative way, and organismic normality in terms of function and malfunction is not defensible for ecosystems, 'normal' ecosystem function is understood to mean ecological processes occurring as they have occurred historically (Callicott, Crowder and Mumford 1999). The idea is therefore to maintain natural and cultural processes, which are regarded as beneficial if they orient themselves to historical processes.

Ecosystem services

Ecosystem services 'are the aspects of ecosystems utilized (actively or passively) to produce human well-being. … Defined this way, ecosystem services include ecosystem organization or structure as well as process and/or functions if they are consumed or utilized by humanity either directly or indirectly' (Fisher, Turner and Morling 2009: 645). Ecosystem services include provisioning services as food and fuel, regulating services as air purification, erosion control, cultural services as cognitive development and aesthetic enrichment, and supporting services that are necessary for the production of all other ecosystem services, such as primary production or soil formation (Hassan, Scholes and Ash 2005).

The 'service' metaphor implies that biodiversity offers goods and activities for the benefit of somebody who is served. Such services are rarely for free and if they are, they are included in some more comprehensive service which has to be paid for. The 'service' metaphor thus has some crucial connotations. Those who are served are certainly humans. The aim to achieve human well-being is therefore already implied in the term. Moreover, the frame of monetisation and commodification resonates in the term 'service', although it is normally not included in the definition. The term further suggests that biodiversity or ecosystems are agents, which readily supply services as long as they are treated properly.

As the ultimate criterion for evaluating ecosystem services is human well-being, only those parts of biodiversity are of interest that enhance human well-being by producing ecosystem services. This is crucial, as these often depend, for example, on biomass production. However, an increase in biomass often leads to a decrease in other components of biodiversity, e.g. species richness. Hence the ecosystem service approach can conflict with other biodiversity goals.

The benchmark for evaluating provisioning services and regulating services is sustainability. For cultural services furthermore cultural heritage is of importance. As supporting services only act indirectly for providing human benefits, it is difficult to assign an evaluation criterion to these services.[4]

Variety

As Sarkar and Margules (2002: 302) note, 'Diversity, including biodiversity, connotes "difference" and "unlikeness" – the amount of variety. Richness does not capture this connotation.' For them, variety is so important that it is included in their definition of biodiversity (cf. above). The detrimental process counteracting particularity is biotic homogenisation. It is the process by which the genetic, taxonomic or functional similarities of regional biotas increase over time (Olden and Rooney 2006).

While the term 'variety' is quite neutral, with a slight positive connotation (cf. Lakoff and Johnson 1980), 'homogenisation' often is coupled with boring uniformity. This is even more the case if homogenisation is combined with a critique of globalisation and cosmopolitanisation: 'the trend is toward a globalisation of flora and fauna that threatens to homogenise the world's ecological assemblages into one giant mongrel ecology' (Hettinger 2001: 216).

Gábor Lövei (1997: 627) calls to 'stop this Macdonaldization' and warns that we 'may well find that we need variety not just as the spice of life – but for life itself'.

The baseline for evaluating variety is primarily historical naturalness which is often extended to cultural heritage value. In principle it could also be understood as an ahistorical itemising approach. The requirement then would simply be that two sites or regions differ in their biodiversity without considering how that biodiversity evolved. However, according to our knowledge such an approach has not been at issue so far.

With the normative concepts of biological integrity, ecosystem health, ecosystem services and variety, different criteria for the evaluation of biodiversity (e.g. naturalness, threat by humans) are covered. However, these normative concepts go far beyond bare evaluation criteria. Often implicit, they convey a certain idea of biodiversity (e.g. as an organism or as a person whose integrity can be threatened, etc.) and imply value judgements. These assumptions are framed by a language in which metaphors play a crucial role. Metaphors make ideas more illustrative and comprehensible (Lakoff and Johnson 1980). However, instead of clarifying worldviews they rather obscure the rationale for choosing a normative concept which is then hardly accessible in a discourse.

Framing invasive alien species

Alien species are discussed in two biodiversity conservation contexts. On the one hand, alien species are part of biodiversity, but it is often contested that they are worthy of conservation measures in case they are threatened. On the other hand, if alien species are invasive, they can have a great and sometimes adverse impact on other parts of biodiversity which is referred to in Article 8h of the CBD. But still there is a great uncertainty on the magnitude of effects on biodiversity by invasive alien species and how to evaluate these effects (representatively the debate between Sagoff 2005 and Simberloff 2005). Different points of view on that issue are expressed on several levels of discourse. Here we investigate how metaphors frame the discourse on invasive alien species, how different approaches for defining invasive alien species compete for acceptance, and last, how the perception of alien species is framed by different biodiversity conservation motives.

Metaphors in the discourse on invasive alien species

Our conceptual system is fundamentally metaphorical in nature. As Lakoff and Johnson (1980: 3) state, 'the way we think, what we experience and what we do every day is very much a matter of metaphor'. Both the terms 'invasion' and 'alien' are metaphors. 'Invasion' is a military metaphor which was already used in Charles Elton's classic book on invasion biology. The importance of invasion metaphor for the Briton Elton was linked to 'his country's psyche' after World War II (Davis, Thompson and Grime 2001: 99). Brendon Larson (2005) emphasises two properties of militaristic metaphors. First, in a war there are always two opposing sides. By this, invasive species are framed as our opponents. Second, one side is the good one and the other is evil. And certainly, it is not us who are on the evil side. The discourse on alien invasive species is replete with militaristic terminology as Larson, Nerlich and Wallis (2005) have shown for media coverage on invasive alien species (like 'war' or 'battle' against invasives). Further, invasive species are often characterised by personifications (e.g. 'aggressive') and as 'killers' (Larson, Nerlich and Wallis 2005).

The term 'alien' origins from the migration discourse. 'Alien' is connoted with persons not being well integrated and who do not belong to a certain place. Further associations with aliens from Mars are quite obvious ('The aliens have landed', Subramaniam 2001). Because of this normative loading, Charles Warren (2007) recommends shunning 'alien' in favour of the less pejorative 'introduced'.

Pejorative attributes for introduced species can be found for different stages of the invasion process (Eser 1999; Körner 2000). These stages are the introduction to a new region with subsequent emergence of non-established populations, the establishment of self-replacing populations, and finally the growth of populations and range extension. The first stage of invasion is referred to as 'nature out of place' (van Driesche and van Driesche 2000). Introduced species here are regarded as trespassers, which cross geographical but also cultural borders (Eser 1999) and enter a terrain for which they are supposed not to have access rights. In the next stage it becomes clear that they are here to stay. The crossing of cultural borders becomes even more distinct when species are characterised as competitive, persistent and intolerant. This competitive ability is expressed by designations like 'gap grabbers' and 'swampers' (Newsome and Noble 1986). By this, alien species are framed as behaving in an 'uncultivated' way. At the final stage these species get out of control (Larson, Nerlich and Wallis 2005). Attributes of unlimited fertility, prolific reproduction and dominance are attributed to them. Now as the 'invaders sweep in' (Enserink 1999), the question arises if they 'are taking over' (Hulme et al. 2010).

It can be seen from these examples that even in scientific discourses on alien species, value-laden metaphors are used. Militaristic, migration and mass reproduction metaphors serve to downgrade alien species. This does not need necessarily to be the case, as language also provides metaphors for coexistence by which the discourse might be framed (Larson 2005).

The discourse on defining invasive alien species

There is a large number of definitions for both alien and invasive species. While metaphors implicitly convey meaning and values to invasive alien species, definitions may serve specific purposes. In the following, we will discuss several definitions on alien and invasive species and refer to the purposes which motivate the development of these definitions.

Native and alien species

The first condition for a species to be referred to as invasive is that it is alien. Alien in general is equated with exotic, non-native or nonindigenous. According to the definitions by the CBD and David Richardson et al. (2000), two premises have to be fulfilled to qualify for being alien: first, a species is outside its original range and second, this range expansion has to take place as a result of human activity. Importantly, in these definitions naturalised species are still regarded as alien (cf. Table 4.1).

However, there are contrasting approaches for defining native or alien. Tina Heger's definition of non-nativeness (2004) is exceptional as it does so without reference to human agency. The defining criteria are a range expansion and the disappearance of dispersal barriers, regardless of the question if that disappearance is caused naturally or by humans.

Table 4.1 Definitions of alien and native species

'"Alien species" refers to a species, subspecies or lower taxon, introduced outside its natural past or present distribution; includes any part, gametes, seeds, eggs, or propagules of such species that might survive and subsequently reproduce.' (UNEP 2002a: 257)
Alien plants: 'Plant taxa in a given area whose presence there is due to intentional or accidental introduction as a result of human activity (synonyms: exotic plants, alien plants; nonindigenous plants).' (Richardson et al. 2000: 98)
Native species: 'A feral animal or plant species which a) has or had in historic times its distribution or migratory range within the borders of the country b) or which naturally expands these ranges into the country Native are also animal and plant species, if animals or plants which became feral or which became established by human intervention possess self-sustainable populations in the wild over several generations without human intervention.' (§ 10 lit.2 no. 7 BNatSchG, transl. by the authors)
'To become native, an exotic species must not only naturalise ecologically (i.e., adapt with local species and to the local environment), but it must also naturalise evaluatively. This means that for an exotic to become a native, human influence, if any, in the exotic's presence in an assemblage must have sufficiently washed away for us to judge that species to be a natural member of that assemblage.' (Hettinger 2001: 208) 'Exotics are species that have not significantly adapted with the local ecological assemblage. Once a species has significantly adapted (ecologically naturalised), it is no longer exotic. But such a species might still not be native.' (Hettinger 2001: 213)
'A non-native species is any plant species that occurs at a location outside its area of origin; the occurrence of the species must have been prevented in the past by a barrier to dispersal, and not by the conditions in the new habitat.' (Heger 2004: 12, transl. by the authors)

Ned Hettinger (2001; cf. Table 4.1) rejects criteria referring to biogeography and human agency and instead advocates 'ecological' and 'evaluative' criteria for defining native. According to this, 'exotic species' can also become native.

The ecological criterion for becoming native is adaptation with local species and to the local environment. Hettinger (2001) makes clear that by adaptation he does not mean fitting into a balanced, harmonious natural system. A species has for example adapted 'when it has changed its behavior, capacities, or gene frequencies in response to other species or local abiota' (Hettinger 2001: 198) or when it moves into a type of ecological assemblage that is already present in its home range.

'Evaluative naturalisation' is a quite complicated concept, which is not clearly defined. However, several aspects are mentioned which affect 'washing away human influence'. Among them are the magnitude of human influence (e.g. for domesticated taxa the influence is higher than for feral ones, if exotic species take on keystone roles or extirpate keystone species the impact is high), the 'temporal distance from human influence' (Hettinger 2001: 212) or the extent of natural processes wearing off human influence.

According to § 7 lit. 2 no. 7 of the Federal German Nature Conservation Law, a wild living species becomes native when it has naturalised, i.e., if a population of an introduced species 'survives in the wild for several generations without human intervention'. In contrast to Hettinger, criteria for becoming native are the sustainable reproduction of populations and missing human support.

Invasive species

As for 'alien species', there are different definitions for 'invasive species' (Table 4.2). Richardson et al. (2000) mention two criteria to define invasive species: they are naturalised (i.e., they possess self-reproducing populations), and spread over a great distance. In contrast, according to Davis and Thompson (2000), invasive species are 'novel' and have a large impact on the new environment. A coloniser is 'novel' if it expands its range into the region of interest or if it establishes 'in a new type of environment, without an extension of the range's latitude, longitude, or altitude' (Davis and Thompson 2000: 228). While Davis and Thompson (2001: 206) concede that impacts of invasive species are usually undesirable, they make clear that 'at least in principle, impact can be defined quite objectively'. In terms of impacts addressed, the COP definition is more specific in determining invasive species as those alien species which threaten biodiversity. In this context, this definition is clearly normative as in Art. 8h, CBD, measures against these species need to be taken. By this, it is made clear that invasive species are undesired. The policy-oriented definition of invasive species is sometimes adopted by ecologists (cf. Kowarik 2003b). For example, the Ecological Society of America (2004) defines invasive species as 'the subset of introduced species that persist, proliferate, and cause economic or environmental harm, or harm to human health'.

In summary, these definitions differ in their inclusion of the impact criterion. While Richardson et al. (2000) do not refer to impact as a defining criterion, Davis and Thompson (2000) allow for impacts as value-neutral effects, and the COP definition addresses only alien species which have adverse effects on biodiversity.

Table 4.2 Definitions of invasive species

Definition
'"Invasive alien species" means an alien species whose introduction and/or spread threaten biological diversity.' (UNEP 2002a: 257)
Invasive plants: 'Naturalized plants that produce reproductive offspring, often in very large numbers, at considerable distances from parent plants (approximate scales: > 100 m; < 50 years for taxa spreading by seeds and other propagules > 6 m/3 years for taxa spreading by roots, rhizomes, stolons, or creeping stems), and thus have the potential to spread over a considerable area.' (Richardson et al. 2000: 98)
This definition is based on the definition of naturalised plants. These are: 'Alien plants that reproduce consistently (cf. casual alien plants) and sustain populations over many life cycles without direct intervention by humans (or in spite of human intervention); they often recruit offspring freely, usually close to adult plants, and do not necessarily invade natural, seminatural or human-made ecosystems.' (Richardson et al. 2000: 98)
Invaders '... are novel and have a large impact, usually undesirable, on the new environment.' (Davis and Thompson 2000: 229).
Invasive species: 'A species which outside its natural range potentially causes a considerable threat to naturally occurring ecosystems, biotopes and species.' (§ 10 lit.2 no. 9 BNatSchG, transl. by the authors)
'Invasive plant species are non-native species, which spread into a new region. Thereby it does not matter to what extent or at what rate this spread occurs; further it is irrelevant, if the spread entails adverse effects on biocoenoses from a cultural or economic point of view.' (Heger 2004: 12, transl. by the authors)

Struggling for the right definition and evaluation of invasive alien species

As Davis and Thompson (2001) argue, to support understanding and effective management efforts, only one definition on invasive species should be used. For them, this definition should be based on the impact criterion because 'outside of the discipline of ecology, "invasive species" are usually explicitly defined on the basis of their impact' (Davis and Thompson 2001: 206). Further, they note that the term 'invader' has pejorative connotations and therefore should only be used for colonisers that have a large impact on the new environment. These arguments effected a response by Marcel Rejmánek et al. (2002) who claim that there is some historical continuity in ecology to use the 'invasion' term in a biogeographic sense and give some evidence for its application without reference to impacts.

In spite of the obvious dissent between Davis and Thompson (2000, 2001, 2002) and Rejmánek et al. (2002) concerning the understanding of 'invasive species', there is also some common ground. They silently seem to agree that if invasive species were regarded as detrimental it would be because of their impacts on biodiversity. Thus, the debate is narrowed down to the impact frame, without considering other objections towards invasive species.

Roughly ten years after this debate on defining invasive alien species, Davis et al. (2011) launched a *Science* paper where the evaluation of alien species is framed by the title 'Don't judge species on their origins'. In this paper they oppose the vilification of alien species and 'urge conservationists and land managers to organize priorities around whether species are producing benefits or harm to biodiversity, human health, ecological services and economies' (Davis et al. 2011: 154). They back up their arguments by claiming that 'invaders do not represent a major extinction threat to most species in most environments'. Again, there was a reply. Daniel Simberloff et al. (with 141 scientists, 2011) denied that scientists oppose alien species per se; quite the contrary, they acknowledge the benefits of alien species. However, they claim that 'Davis and colleagues downplay the severe impact of alien species' (Simberloff et al. 2011: 36) and cite a case of an alien species causing harm to biodiversity. Interestingly, there is no obvious dissent on the adversity of impacts of alien species between Davis et al. (2011) and Simberloff et al. (2011). Both sides agree that there are many alien species that do not do any harm, but there are some that do. However, they draw different conclusions. While Davis et al. (2011) maintain that if there are harmless alien species, the concept 'origin' should not be used for judging harm done to biodiversity, Simberloff et al. (2011) argue that if there are harmful alien species, origin is certainly a good indicator for harm.

The earlier and the later debates differ in that the first relates to the notion of invasive as having an impact on biodiversity while the latter deals with the question of whether harm to biodiversity can be judged by the geographic origin of species. By addressing the geographic origin of species, this debate initially seems to go beyond the question of whether impact alone is a criterion for judging adversity of alien species. However, in the further course of the dispute, the discussion is again reduced to the question of impact. The question arises whether there is another possibility for framing adversity of alien species other than in terms of biodiversity impacts.

One possibility is blunt xenophobia. However, although discussed elsewhere (Peretti 1998; Gröning and Wolschke-Bulmahn 2003), scientists in general are cautious to raise the charge of xenophobia. It was also avoided in the discussion between Davis et al. (2011) and Simberloff et al. (2011). However, the presence of alien species themselves can be regarded as adverse without referring to xenophobic stereotypes or to interactions with resident biodiversity. There are two strands of argumentation supporting this view. First, alien species are detrimental to the naturalness of an area. Alien species by definition do not belong to the natural flora and fauna, and as we have shown, naturalness is an important evaluation criterion in nature conservation. Second, alien species contribute to homogenisation of different parts of the Earth. That means, if alien species spread, different parts of the world are becoming more and more similar.

It is quite surprising that the first strand of argumentation was ignored. This was probably because the problem was framed by discussing extensively if it is justified to eradicate alien species with adverse impacts (Davis et al. 2011; Simberloff et al. 2011). However, if the question had been if, for a naturalised alien species on the brink of extinction in its novel range, conservation measures should be taken or if there are cases when the extirpation of a species in its native range is justified when it has an adverse impact on biodiversity, the discussion on the judgement of alien species might have taken a different course. Here, naturalness would probably play a major role in argumentation.

The second argument gets much support from Hettinger (2001: 216): 'The objection biological nativists can have to exotic species as exotics – at least in the current context – is that although they immediately add to the species count of the local assemblage and increase biodiversity in that way, the widespread movement of exotic species impoverishes global and regional biodiversity by decreasing the diversity between types of ecological assemblages on the planet. For example, adding a dandelion (*Taraxacum officinale*) to a wilderness area where it previously was absent diminishes the biodiversity of the planet by making this place more like everyplace else'. Although biotic homogenisation is sometimes discussed in ecology in context with invasive alien species, Hettinger's attribution of homogenisation to an increase of species is rather exceptional. In the context of alien invasive species it is rather narrowed down to a loss of species and is thus related to the item conservation motive. A reason for that might be the resonance of biodiversity with the species loss frame which suppresses the 'difference' frame.

Invasive alien species at the science-policy interface

Threat to biodiversity is a criterion for defining alien invasive species according to COP decision VI/23 (UNEP 2002a). Davis and Thompson (2001) favour this inclusion of impacts on biodiversity in the definition of invasive species for practical reasons which they already make clear in the title of their comment: 'Should ecologists define their terms different than others? No, not if they want to be of any help!' Rejmánek et al. (2002: 131) fervently defend the 'scientific' invasion term based mainly on reproduction and spread of an alien species against influences from policy: 'A recent attempt to make the long established ecological term "invasion" concordant with the definition in an Executive Order of one country's president (Davis and Thompson 2001) probably has no precedent'. Colautti and MacIsaac (2004: 137) support Rejmánek et al. (2001) by stating that it is imperative to 'provide clear, objective definitions and models to managers and other officials charged with protection of native biodiversity' and advocate a 'neutral' terminology to define invasive species.

Interestingly, the debate here shifts from conceptual aspects of invasiveness to the delimitation of science from policy. While Rejmánek et al. (2002) and Colautti and MacIsaac (2004) think that policy interests should not invade science, Davis and Thompson (2001) argue for using a definition which is directly applicable in a policy against harmful invasive species. While we will not go into the details of what should be the proper role of science in policy-making (extensively on this topic cf. Pielke 2007), we think that a definition of alien or invasive based on the criterion of human agency in long-distance dispersal is not defensible without referring to arguments from outside biogeography and population ecology. The dichotomy between human and natural agency is certainly a cultural one and not based in natural science. Thus, we suspect that the definition of invasive species focusing on human agency may be influenced to a greater extent by policy than Rejmánek et al. are aware of. Besides, influence on terms used in ecology by policy interest certainly has some precedent as we have expanded on the biodiversity term.

Invasive alien species in the biodiversity context

The acceptance of alien species depends heavily on the underlying conservation motive and derived evaluation criteria. The criteria 'species richness' and 'rarity and threat' derived from the itemising approach can be applied both to native and alien species. However, trade-offs are necessary when an alien species causes threat to native species (or vice versa!). The presence of alien species enhances species richness. However, if they outcompete other species, they may also reduce species richness. Thus, the evaluation of alien species according to the itemising approach depends on the species and the geographical scale of observation of richness, threat, etc.

Wildness can also be achieved with alien and native species, as in the process conservation approach ecological processes are crucial but the composition and structure of ecosystem components are irrelevant.

As long as alien species do not interfere with or support a sustainable use of biodiversity, they are also in line with the conservation of present human-induced processes. However, if they impede the use of benefits, e.g. from provisioning or regulating services, they are certainly regarded as detrimental. This applies for example to agricultural weeds which may lead to a loss in crop production, or the dieback of the annual Himalayan Balsam (*Impatiens glandulifera*) in winter which leaves river banks bare and exposed to erosion.

However, alien species are in general not accepted, if the goal is to preserve an 'original' natural state, e.g. the natural state after species returned from their glacial refuges. All alien species are introduced after this point of time. Similarly, the acceptance of alien species depends on the reference point of time selected to evaluate the importance for cultural heritage. For example, most of the agricultural weeds introduced before 1500 A.D. are greatly appreciated in Europe in contrast to species introduced after that point of time (Kowarik 2003a).

Different views of nativeness match well the dichotomy of process and state conservation. While for the definition of 'alien' by Richardson et al. (2000) the historically original range of a species is important, Davis and Thompson (2000: 228) state 'Like it or not, these species are here and they are not going back. Continuing to refer to them at this point as alien invaders, or exotics, or even novel plant species, is beginning to make little ecological [sic!] sense'. Thus, similar to Hettinger (2001) it depends on ecological process in which introduced species are involved if they become native.

However, the debate on invasive alien species is rarely related to conservation goals like conservation of natural or cultural states, or processes and item conservation. In contrast they are often embedded in normative concepts of biodiversity. Angermeier (1994) makes clear that 'artificial biodiversity' like alien species is not included in the concept of biological integrity. For Noss (1990a: 243) a sign for integrity is when a community is dominated by native species and he warns 'to abandon concern for species composition, and permit exotics and other species that thrive on anthropogenic disturbance to run roughshod over more sensitive, native species – the outcome can only be biotic impoverishment.'

The determination of homogenisation, the contrary of variety, is at times based on alien species: 'homogenization generally refers to the replacement of local biotas with nonindigenous species, usually introduced by humans' (McKinney and Lockwood 1999: 450; cf. also Olden et al. 2004). We have already shown the importance of the homogenisation concept for the evaluation of alien species: in principle they can be evaluated negatively without contributing to a loss of species and habitats with the aid of this concept.

Ecosystem health is not very often related to alien species, although the University of Georgia harbours a Center for Invasive Species and Ecosystem Health.[6] Recently, alien species are more often related to ecosystem services. While mainly a disruption of ecosystem services in terms of monetary costs was examined (e.g. Born, Rauschmayer and Bräuer 2005), there are also studies showing positive effects on ecosystem services (e.g. Lugo 2004; Gleditsch and Carlo 2011). This is in line with our claim that the ecosystem services concept with its focus on sustainable use allows for such a differentiated view in contrast, for example, to the concept of biological integrity.

Conclusions

We have shown that both biodiversity and invasive alien species issues can be differently framed. The reason for differences in framing is grounded in the use of these concepts in natural science and in policy. Each of these domains is wrestling to make concepts as fruitful as possible for their purpose. While science tries to describe components of biodiversity and the population biology and spread of certain species, policy-makers, environmentalists and concerned conservation biologists are interested in finding out what components of biodiversity should be protected and which species should be managed because of adverse effects on biodiversity.

It would be naïve to assume that the meaning of terms used in scientific publications are not influenced by policy. The biodiversity term was generated and disseminated from the very start to bring about a change in viewing and evaluating nature. As the alien-native dichotomy is not an ecological one, there is strong suspicion that it is politically motivated. And even if invasiveness is not defined by the impact of a species, as it depends on being alien, it still is a policy-oriented concept.

It is astonishing to what extent the understanding of alien invasive species is guided by normative considerations and on ideas of (the right sort of) biodiversity. The term 'native' can be conceptualised by means of a historic state of biodiversity or by ecological processes in which the introduced species is involved. These conceptualisations clearly reflect different ideas of what kind of biodiversity is worth conservation efforts. The term 'invasive' can be defined either by spread or by the impact of species on biodiversity. In the latter sense, the concept of 'invasive' is more useful for policy. However, this concept is also adopted by some ecologists.

We have given examples for framing invasive alien species on different levels. The most conspicuous term on the metaphorical level is the invasion term, which is derived from militaristic language. Although 'invasion' for ecologists might have a mere technical meaning they should be aware that for nonscientists it will still resonate with its original meaning (Larson 2005).

We think that the reframing of concepts by metaphors and of issues is possible and for invasive alien species in many cases desirable. For conceptual framing this probably takes a lot of time and effort (Lakoff 2010). However, the biodiversity term is a successful example of reframing existing concepts like wildlife, flora and fauna or simply nature (Takacs 1996). Therefore, in principle a reframing can also work for invasive alien species. We do not think that a long history of use justifies the reinforcement of the 'invasion' term. Accordingly the debate of problems which are caused by alien invasive species needs to be reframed. An investigation of underlying values which support general acceptance or rejection of alien species would be very helpful (as an example for such value elicitation, Schüttler, Rozzi and Jax 2011).

On the level of framing concerns which are raised against invasive alien species, we have shown that the debate on (detrimental) impacts on biodiversity diverts from the question of whether alien species are in principle an object of biodiversity conservation efforts. We think that although there is much debate on problems by invasive alien species, a serious dispute of general acceptance of alien species has hardly begun. Is it desirable to take measures for threatened alien species? Or are they as important for nature conservation as greenhouse plants, pets or genetically modified organisms?

The question is still open which concept of invasive alien species should be pursued in ecological science. Above, we discussed the models of the 'pure' scientist represented by Rejmánek et al. (2002) who hold that scientific terminology should not be influenced by policy, and the 'applied' scientist represented by Davis and Thompson (2000) who hold that it should. As Pielke (2007) notes, both models have their justification. The use of policy terms is widespread in science and it is crucial for applied ecology to use such terms to allow for policy relevance.

Being relevant, however, should not be confused with being prescriptive. Hence, scientific methodology should not be compromised by value systems external to science. For example, it would be highly problematic if scientific results were shaped by policy considerations. However, Pielke (2007) warns that pure scientists can become 'stealth issue advocates'. Stealth issue advocacy refers to situations in which a presumed pure scientist reduces the scope of choice available to decision-makers. We suspect the concept of invasive species favoured by Rejmánek et al.

(2002) of possessing features of stealth issue advocacy. On the one hand, they accept the pejorative resonance of the term 'invasive' and its ambiguous use in science and policy. As a result, for example, a 'scientific' list of invasive species which consists of alien species spreading rapidly outside their natural range can be adopted in policy, although policy measures are meant to be taken for a different subset of alien species which have an adverse impact on biodiversity. On the other hand, as invasive species is still a subset of alien species, the scope of species scientists refer to is already reduced. However, reducing the scope of species for political action is certainly not a scientific task.

Notes

1 'Alien' refers to species occurring outside their natural range. In a broad understanding 'invasive species' are the subgroup of alien species which further spread into their new environment. The meanings of the terms 'alien' and 'invasive' are extensively discussed below. We use the term 'invasive alien species' here as an umbrella term which means either alien or invasive or both. We refer specifically to 'alien' and 'invasive' species if we want to address only one of these groups. Although we dislike the expressions 'alien' and 'invasive' we stick to them, because most of the authors we quote here use these terms and they might give different meanings to terms we regard as synonymous to 'alien' and 'invasive'.

2 This nature conservation approach is sometimes called 'preservationism' (Minteer and Corley 2007).

3 Other authors maintain that ecosystem health is determined mainly by societal values (e.g. Okey 1996; Xu and Mage 2001). Thus, the concept of ecosystem health is certainly an ambiguous one. We do not discuss this meaning of ecosystem health here, as the 'health' metaphor suggests value baselines intrinsic to ecosystems. However, cf. e.g. Jax 2010: 153ff.

4 In fact, many recent classifications of ecosystem services meanwhile exclude 'supporting services' as a category because they emphasise the character of services as benefits, with ecological processes like primary production not leading to a clear benefit as such (e.g. Wallace 2007). Including them also brings about the danger of 'double counting' in assessment of ecosystem services. This happens, e.g., when primary production is counted as well as the apples or grain derived from it.

5 Heger (2004) refers to 'gebietsfremde Pflanzenarten' ('plant species alien to a region') which probably has the same meaning as non-native plant species.

6 Cf. <http://www.bugwood.org> (accessed June 13 2013).

References

Angermeier, P.L. (1994) 'Does biodiversity include artificial diversity?', *Conservation Biology* 8: 600–2.

Angermeier, P.L. and Karr, J.R. (1994) 'Biological integrity versus biological diversity as policy directives – protecting biotic resources', *Bioscience* 44: 690–7.

Angert, A.L., Crozier, L.G., Rissler, L.J., Gilman, S.E., Tewksbury, J.J. and Chunco, A.J. (2011) 'Do species' traits predict recent shifts at expanding range edges?', *Ecology Letters* 14: 677–89.

Balvanera, P., Pfisterer A.B., Buchmann, N., He, J.-S., Nakashizuka, T., Raffaelli, D. and Schmid, B. (2006) 'Quantifying the evidence for biodiversity effects on ecosystem functioning and services', *Ecology Letters* 9: 1146–56.

Benford, R.D. and Snow, D.A. (2000) 'Framing processes and social movements: an overview and assessment', *Annual Review of Sociology* 26: 611–39.

BNatSchG – *Gesetz über Naturschutz und Landschaftspflege (Bundesnaturschutzgesetz – BGBl. I S. 148)*, Berlin: Deutscher Bundestag, last revised 29 July 2009, in force since 1 October 2010.

Born, W., Rauschmayer, F. and Bräuer, I. (2005) 'Economic evaluation of biological invasions – a survey', *Ecological Economics* 55: 321–36.

Busse, D. (1991) 'Angewandte Semantik. Bedeutung als praktisches Problem in didaktischer Perspektive', *Der Deutschunterricht* 43: 42–61.

Callicott, J.B., Crowder, L.B. and Mumford, K. (1999) 'Current normative concepts in conservation', *Conservation Biology* 13: 22–35.

CBD – United Nations (1992) *Convention on biological diversity*. Online. Available HTTP: < http://www.cbd.int/doc/legal/cbd-en.pdf> (accessed 12 June 2013).

Claridge, M.F., Dawah, A.H. and Wilson, M.R. (eds) (1997) *Species: The units of biodiversity*, London: Chapman and Hall.

Colautti, R.I. and MacIsaac, H.J. (2004) 'A neutral terminology to define "invasive" species', *Diversity and Distributions* 10: 135–41.

Cole, D.N. (2000) 'Paradox of the primeval: ecological restoration in wilderness', *Ecological Restoration* 18: 77–86.

Collins, C.A. and Kephart, S.R. (1995) 'Science as news: The emergence and framing of biodiversity', *Mass Comm Review* 22: 21–45.

Crist, E. (2007) 'Beyond the climate crisis: a critique of climate change discourse', *Telos* 141: 29–55.

Davis, M.A. and Thompson, K. (2000) 'Eight ways to be a colonizer; two ways to be an invader: a proposed nomenclature scheme for invasion ecology', *ESA Bulletin* 81: 226–30.

Davis, M.A. and Thompson, K. (2001) 'Invasion terminology: should ecologists define their terms differently than others? No, not if we want to be of any help!', *ESA Bulletin* 82: 206.

Davis, M.A. and Thompson, K. (2002) '"Newcomers" invade the field of invasion ecology: question the field's future', *ESA Bulletin* 83: 196f.

Davis, M.A., Thompson, K. and Grime, J.P. (2001) 'Charles S. Elton and the dissociation of invasion ecology from the rest of ecology', *Diversity and Distributions* 7: 97–102.

Davis, M.A., Chew, M.K., Hobbs, R.J. et al. (2011) 'Don't judge species on their origins', *Nature* 474: 153f.

Delong, D.C. (1996) 'Defining biodiversity', *Wildlife Society Bulletin* 24: 738–49.

Ecological Society of America (2004) *Invasion!*, Washington, D.C.: Leaflet prepared by the Ecological Society of America.

Ekschmitt, K., Stierhof, T., Dauber, J., Kreimes, K. and Wolters, V. (2003) 'On the quality of soil biodiversity indicators: abiotic and biotic parameters as predictors of soil faunal richness at different spatial scales', *Agriculture Ecosystems and Environment* 98: 273–83.

Enserink, M. (1999) 'Predicting invasions: biological invaders sweep in', *Science* 285: 1834–6.

Entman, R.M. (1993) 'Framing – toward clarification of a fractured paradigm', *Journal of Communication* 43: 51–8.

Eser, U. (1999) *Der Naturschutz und das Fremde. Ökologische und normative Grundlagen der Umweltethik*, Frankfurt am Main: Campus.

European Union (1992) *Council Directive 92/43/EEC on the conservation of natural habitats and of wild fauna and flora, L 206, 22 July 1992.*

Fillmore, C.J. (1975) 'An alternative to checklist theories of meaning', *Proceedings of the First Annual Meeting of the Berkeley Linguistics*: 123–31.

Fisher, B., Turner, R.K. and Morling, P. (2009) 'Defining and classifying ecosystem services for decision making', *Ecological Economics* 68: 643–53.

Gaston, K.J. (ed.) (1996) *Biodiversity: A biology of numbers and difference,* Oxford: Blackwell Science.

Gleditsch, J.M. and Carlo, T.A. (2011) 'Fruit quantity of invasive shrubs predicts the abundance of common native avian frugivores in central Pennsylvania', *Diversity and Distributions* 17: 244–53.

Goffman, E. (1974) *Frame analysis. An essay on the organization of experience,* Boston, MA: Northeastern University Press.

Goldstein, P.Z. (1999) 'Functional ecosystems and biodiversity buzzwords', *Conservation Biology* 13: 247–55.

Gröning, G. and Wolschke-Bulmahn, J. (2003) 'The native plant enthusiasm: ecological panacea or xenophobia', *Landscape Research,* 28: 75–88.

Gurevitch, J. and Padilla, D.K. (2004) 'Are invasive species a major cause of extinctions?', *Trends in Ecology and Evolution* 19: 470–4.

Haskell, B.D., Norton, B.G. and Costanza, R. (1992) 'What is ecosystem health and why should we worry about it?', in R. Costanza, B.G. Norton, and B.D. Haskell (eds) *Ecosystem health: New goals for environmental management,* Washington, D.C.: Island Press, 1–18.

Hassan, R., Scholes, R. and Ash, N. (eds) (2005) *Ecosystems and human well-being,* Washington, D.C.: Island Press.

Heger, T. (2004) *Zur Vorhersagbarkeit biologischer Invasionen. Entwicklung und Anwendung eines Modells zur Analyse der Invasion gebietsfremder Pflanzen,* Berlin: Lentz.

Heink, U. (2009) 'Representativeness – an appropriate criterion for evaluation in nature conservation?', *Gaia* 18: 322–30.

Heink, U. and Kowarik, I. (2010) 'What criteria should be used to select biodiversity indicators?', *Biodiversity and Conservation* 19: 3769–97.

Hettinger, N. (2001) 'Exotic species, naturalisation, and biological nativism', *Environmental Values* 10: 193–224.

Holland, A. (1995) 'The use and abuse of ecological concepts in environmental ethics', *Biodiversity and Conservation* 4: 812–26.

Hull, R.B., Richert, D., Seekamp, E., Robertson, D. and Buhyoff, G.J. (2003) 'Understandings of environmental quality: ambiguities and values held by environmental professionals', *Environmental Management* 31: 1–13.

Hulme, P.E., Nentwig, W., Pyšek, P. and Vilà, M. (2010) 'Are the aliens taking over? Invasive species and their increasing impact on biodiversity', in J. Settele, L. Penev, T. Georgiev, R. Grabaum, V. Grobelnik, V. Hammen, S. Klotz, M. Kotarac and I. Kühn (eds) *Atlas of biodiversity risk,* Sofia: Pensoft, 132f.

Jax, K. (2010) *Ecosystem functioning,* Cambridge: Cambridge University Press.

Karr, J.R. (2000) 'Health, integrity, and biological assessment: the importance of measuring whole things', in D. Pimentel, L. Westra and R.F. Noss (eds) *Ecological integrity. Integrating environment, conservation, and health,* Washington, D.C.: Island Press, 209–26.

Karr, J.R. and Dudley, D.R. (1981) 'Ecological perspective on water quality goals', *Environmental Management* 5: 55–68.

Körner, S. (2000) *Das Heimische und das Fremde. Die Werte Vielfalt, Eigenart und Schönheit in der konservativen und in der liberal-progressiven Naturschutzauffassung,* Münster: Lit-Verlag.

Kowarik, I. (1990) 'Some responses of flora and vegetation to urbanization in Central Europe', in H. Sukopp, S. Hejny and I. Kowarik (eds) *Plants and plant communities in the urban environment,* The Hague: SPB Academic Publishing, 45–74.

Kowarik, I. (1999) 'Natürlichkeit, Naturnähe und Hemerobie als Bewertungskriterien', in W. Konold, R. Böcker and U. Hampicke (eds) *Handbuch Naturschutz und Landschaftspflege*, Landsberg: ecomed, 1–18.

Kowarik, I. (2003a) *Biologische Invasionen. Neophyten und Neozoen in Mitteleuropa*, Stuttgart: Ulmer.

Kowarik, I. (2003b) 'Human agency in biological invasions: secondary releases foster naturalisation and population expansion of alien plant species', *Biological Invasions* 5: 293–312.

Kowarik, I. (2008) 'Bewertung gebietsfremder Arten vor dem Hintergrund unterschiedlicher Naturschutzkonzepte', *Natur und Landschaft* 83: 402–6.

Lakoff, G. (2010) 'Why it matters how we frame the environment', *Environmental Communication – a Journal of Nature and Culture*, 4: 70–81.

Lakoff, G. and Johnson, M. (1980) *Metaphors we live by*, Chicago, IL: University of Chicago Press.

Larson, B.M.H. (2005) 'The war of the roses: demilitarizing invasion biology', *Frontiers in Ecology and the Environment* 3: 495–500.

Larson, B.M.H., Nerlich, B. and Wallis, P. (2005) 'Metaphors and biorisks – the war on infectious diseases and invasive species', *Science Communication*, 26: 243–68.

Latimer, A.M., Silander, J.A., Gelfand, A.E., Rebelo, A.G. and Richardson, D.M. (2004) 'Quantifying threats to biodiversity from invasive alien plants and other factors: a case study from the Cape Floristic Region', *South African Journal of Science* 100: 81–6.

Lövei, G.L. (1997) 'Biodiversity – global change through invasion', *Nature* 388: 627f.

Lovett, J. (1994) 'Biodiversity – not just a buzzword', *Search* 25: 146.

Lugo, A.E. (2004) 'The outcome of alien tree invasions in Puerto Rico', *Frontiers in Ecology and the Environment* 2: 265–73.

Luntz, F. (2003) 'The environment: a cleaner, safer, healthier America. The Luntz Research companies – straight talk', unpublished memo.

Mageau, M.T., Costanza, R. and Ulanowicz, R.E. (1995) 'The development and initial testing of a quantitative assessment of ecosystem health', *Ecosystem Health* 1: 201–13.

Matzdorf, B., Kaiser, T. and Rohner, M.-S. (2008) 'Developing biodiversity indicator to design efficient agri-environmental schemes for extensively used grassland', *Ecological Indicators* 8: 256–69.

McIsaac, G.F. and Brün, M. (1999) 'Natural environments and human culture: defining terms and understanding worldviews', *Journal of Environmental Quality* 28: 1.

McKinney, M.L. and Lockwood, J.L. (1999) 'Biotic homogenization: a few winners replacing many losers in the next mass extinction', *Trends in Ecology and Evolution* 14: 450–3.

Minsky, M. (1974) *A framework for representing knowledge. Artificial Intelligence, Memo No. 306*, Cambridge, MA: Artificial Intelligence Laboratory of the Massachusetts Institute of Technology.

Minteer, B.A. and Corley, E.A. (2007) 'Conservation or preservation? A qualitative study of the conceptual foundations of natural resource management', *Journal of Agricultural and Environmental Ethics* 20: 307–33.

Newsome, A.E. and Noble, I.R. (1986) 'Ecological and physiological characters of invading species', in R.H. Groves and J.J. Burton (eds) *Ecology of biological invasions*, Cambridge: Cambridge University Press, 1–20.

Norton, B.G. (1993) 'Should environmentalists be organicists?', *Topoi* 12: 21–30.

Norton, B.G. (1995) 'Objectivity, intrinsicality and sustainability: comment on Nelson's "Health and disease as 'thick' concepts in ecosystemic context"', *Environmental Values* 4: 323–32.

Noss, R.F. (1990a) 'Can we maintain biological and ecological integrity?', *Conservation Biology* 4: 241.

Noss, R.F. (1990b) 'Indicators for monitoring biodiversity: a hierarchical approach', *Conservation Biology* 4: 355.

Noss, R.F. (1995) 'Ecological integrity and sustainability: buzzwords in conflict?', in L. Westra and J. Lemons (eds) *Perspectives on ecological integrity*, Dordrecht: Kluwer Academic Publishers, 60–76.

Okey, B.W. (1996) 'Systems approaches and properties, and agroecosystem health', *Journal of Environmental Management* 48: 187–99.

Olden, J.D. and Rooney, T.P. (2006) 'On defining and quantifying biotic homogenization', *Global Ecology and Biogeography* 15: 113–20.

Olden, J.D., Poff, N.L., Douglas, M.R., Douglas, M.E. and Fausch, K.D. (2004) 'Ecological and evolutionary consequences of biotic homogenization', *Trends in Ecology and Evolution* 19: 18–24.

O'Neill, J., Holland, A. and Light, A. (2008) *Environmental values*, Abingdon: Routledge.

OTA – U.S. Congress and Office of Technology Assessment (1987) *Technologies to maintain biological diversity*, Washington, D.C.

Peretti, J.H. (1998) 'Nativism and nature: rethinking biological invasion', *Environmental Values* 7: 183–92.

Pickett, S.T.A. and White, P.S. (1985) *The ecology of natural disturbance and patch dynamics*, Orlando, FL: Academic Press.

Piechocki, R., Wiersbinski, N., Potthast, T. and Ott, K. (2004) 'Vilmer Thesen zum "Prozeßschutz"', *Natur und Landschaft* 79: 53–6.

Pielke, R.A. (2005) 'Misdefining "climate change": consequences for science and action', *Environmental Science and Policy* 8: 548–61.

Pielke, R.A. (2007) *The honest broker. Making sense of science in policy and politics*, Cambridge: Cambridge University Press.

Potschin, M.B., Klug, H. and Haines-Young, R.H. (2010) 'From vision to action: framing the Leitbild concept in the context of landscape planning', *Futures* 42: 656–67.

Potthast, T. (2000) 'Funktionssicherung und/oder Aufbruch ins Ungewisse? Anmerkungen zum Prozessschutz', in K. Jax (ed.) *Funktionsbegriff und Ungewissheit in der Ökologie*, Frankfurt am Main: Lang, 65–81.

Potthast, T. (2007) 'Biodiversität, Ökologie, Evolution – Epistemisch-moralische Hybride und Biologietheorie', in T. Potthast (ed.) *Biodiversität – Schlüsselbegriff des Naturschutzes im 21. Jahrhundert*, Bonn: Bundesamt für Naturschutz, 57–88.

Prendergast, J.R., Quinn, R.M., Lawton, J.H., Eversham, B.C. and Gibbons, D.W. (1993) 'Rare species, the coincidence of diversity hot spots and conservation strategies', *Nature* 365: 335–7.

Rejmánek, M., Richardson, D.M., Barbour, M.G., Crawley, M.J., Hrusa, G.F., Moyle, P.B., Randall, J.M., Simberloff, D. and Williamson, M. (2002) 'Biological invasions: politics and the discontinuity of ecological terminology', *ESA Bulletin* 83: 131–3.

Richardson, D.M., Pyšek, P., Rejmánek, M., Barbour, M.G., Panetta, F.D. and West, C.J. (2000) 'Naturalization and invasion of alien plants: concepts and definitions', *Diversity and Distributions* 6: 93–107.

Sagoff, M. (1995) 'The value of integrity', in L. Westra and J. Lemons (eds) *Perspectives on ecological integrity*, Dordrecht: Kluwer Academic Publishers, 162–76.

Sagoff, M. (2005) 'Do non-native species threaten the natural environment?', *Journal of Agricultural and Environmental Ethics* 18: 215–36.

Sala, O.E., Chapin III, F.S., Armesto, J.J., Berlow, E., Bloomfield, J., Dirzo, R., Huber-Sanwald, E., Huenneke, L.F., Jackson, R.B., Kinzig, A., Leemans, R., Lodge, D.M., Mooney, H.A., Oesterheld, M., Poff, N.L., Sykes, M.T., Walker, B.H., Walker, M. and Wall, D.H. (2000) 'Global biodiversity scenarios for the year 2100', *Science* 287: 1770–4.

Sarkar, S. and Margules, C. (2002) 'Operationalizing biodiversity for conservation planning', *Journal of Biosciences* 27: 299–308.

Scherzinger, W. (1990) 'Das Dynamik-Konzept im flächenhaften Naturschutz. Zieldiskussion am Beispiel der Nationalpark-Idee', *Natur und Landschaft* 65: 292–8.

Schüttler, E., Rozzi, R. and Jax, K. (2011) 'Towards a societal discourse on invasive species management: a case study of public perceptions of mink and beavers in Cape Horn', *Journal for Nature Conservation* 19: 175–84.

Shrader-Frechette, K.S. and Mccoy, E.D. (1994) 'How the tail wags the dog – how value judgments determine ecological science', *Environmental Values* 3: 107.

Simberloff, D. (2005) 'Non-native species do threaten the natural environment!', *Journal of Agricultural and Environmental Ethics* 18: 595–607.

Simberloff, D., Alexander, J., Allendorf, F. et al. (2011) 'Non-natives: 141 scientists object', *Nature* 475: 36.

Soulé, M.E. (1996) 'Are ecosystem processes enough?', *Wild Earth* 6: 56–9.

Subramaniam, B. (2001) 'The aliens have landed! Reflections on the rhetoric of biological invasions', *Meridians* 2: 26–40.

Takacs, D. (1996) *The idea of biodiversity. Philosophy of paradise*, Baltimore, MD: Johns Hopkins University Press.

UNEP (United Nations Environment Programme) (2000) *Decisions adopted by the Conference of the Parties to the Convention on Biological Diversity at its fifth meeting, Nairobi (UNEP/CBD/COP/5/23), 15–26 May 2000, Decision V/6.*

UNEP (United Nations Environment Programme) (2002a) *Decisions adopted by the Conference of the Parties to the Convention on Biological Diversity at its sixth meeting, The Hague (UNEP/CBD/COP/6/20), 7–19 April 2002, Decision VI/23.*

UNEP (United Nations Environment Programme) (2002b) *Decisions adopted by the Conference of the Parties to the Convention on Biological Diversity at its sixth meeting, The Hague (UNEP/CBD/COP/6/20), 7–19 April 2002, Decision VI/26.*

van Dijk, T. (1980) *Textwissenschaft. Eine interdisziplinäre Einführung*, München: dtv.

van Driesche, J. and van Driesche, R. (2000) *Nature out of place: biological invasions in the global age*, Washington, D.C.: Island Press.

Wallace, K.J. (2007) 'Classification of ecosystem services: problems and solutions', *Biological Conservation* 139: 235–46.

Warren, C.R. (2007) 'Perspectives on the "alien" versus "native" species debate: a critique of concepts, language and practice', *Progress in Human Geography* 31: 427–46.

White, P.S. and Bratton, S.P. (1980) 'After preservation – philosophical and practical problems of change', *Biological Conservation* 18: 241–55.

Wilcove, D.S., Rothstein, D., Dubow, J., Phillips, A. and Losos, E. (1998) 'Quantifying threats to imperiled species in the United States', *Bioscience* 48: 607–15.

Williamson, M. (1999) 'Invasions', *Ecography* 22: 5–12.

Wöbse, H.H. (2002) *Landschaftsästhetik*, Stuttgart: Ulmer.

Xu, W. and Mage, J.A. (2001) 'A review of concepts and criteria for assessing agroecosystem health including a preliminary case study of southern Ontario', *Agriculture, Ecosystems and Environment* 83: 215–33.

5 Community-level biodiversity

An inquiry into the ecological and cultural background and practical consequences of opposing concepts

Thomas Kirchhoff

Introduction

Biodiversity is an ambiguous concept. Its basic ambiguity results from the circumstance that it refers to both facts and values. This ambiguity can be traced back to the given fact that the concept of biodiversity was coined in the context of conservation biology to draw attention to species loss. Secondary ambiguities ground in its reference to different kinds of facts and values. These ambiguities bring about that lifeworldly as well as scientific discussions of biodiversity issues are fairly susceptible to semantic and evaluative confusions (cf. e.g. Takacs 1996; Callicott, Crowder and Mumford 1999; Sarkar 2005).

The basic ambiguity might be absent from technical definitions of biodiversity like that of the Convention on Biological Diversity: '"Biological diversity" means the variability among living organisms from all sources' (CBD: Art. 2). However, as soon as biodiversity is defined operationally to address any empirical biodiversity issue, a criterion has to be applied, be it explicitly or implicitly, that determines *which* variability shall be relevant and which shall be ignored. Such a criterion is inevitably involved because there are no two living beings that would not differ in countless respects. Gottfried Wilhelm Leibniz has famously highlighted this fact with regard to the leaves in a park among which no two were perfectly alike (Leibniz 1996: book II, ch. 27, § 3). Consequently, every description and measure or index of biodiversity is necessarily selective and involving value judgements, which determine whether or not two biological entities are *regarded* as identical or equivalent in a specific respect. Thus, practically *the* biodiversity does not exist but only a plurality of concepts, characterisations, measures and indices of biodiversity (cf. Sarkar 2005: 178–84; Maclaurin and Sterelny 2008: ch. 1). Accordingly, the goal to preserve *the* biodiversity does not make sense at all but has to be qualified (what implicitly is always already the case); an unqualified concept of biodiversity – just as an unqualified concept of nature (Birnbacher 2004: 180) – cannot serve as a rule for human action.

Some of the ambiguities of 'biodiversity' have already been analysed by distinguishing perspectives on biodiversity, e.g. taxonomic versus functional (cf. Callicott, Crowder and Mumford 1999), levels of biodiversity, e.g. genes, species, and ecological communities or ecosystems (cf. CBD: Art. 2), and values of biodiversity, e.g. external instrumental values as a means to an end (renewable resource etc.), subjective or intersubjective intrinsic values for what it is rather than for what it can bring about, however based on human preferences and patterns of interpretation (aesthetic, symbolic, spiritual values etc.), and inherent or objective intrinsic values for what it is independent of anyone's preferences (cf. Elliot 1992; Sandler 2012). Nevertheless, numerous ambiguities have remained. One kind of ambiguity results from the fact that the concepts 'gene', 'species', and 'ecological community' or 'ecosystem', which are used to distinguish levels of biodiversity, are themselves ambiguous. For example, there are longstanding discussions about the species concept, in which morphological, biological, evolutionary, ecological species concepts are competing; and obviously the amount of species diversity measured in an area will depend on the species concept used (cf. Maclaurin and Sterelny 2008: ch. 2), an extreme example being that of the blackberry, where one biological species splits into hundreds of morphological species.

In this chapter, I address semantic and evaluative ambiguities as regards community-level biodiversity. The assessment of this level of biodiversity – which comprises within-community diversity and between-communities diversity at different scales – is subject to even more difficulties than that of species or genetic diversity. This is due to the fact that the objects and units, namely ecological communities or ecosystems, are still less unambiguously defined at this level than at the two other levels. Among the numerous ambiguities regarding community-level biodiversity, I reveal an opposition that I consider to be the root of many fundamentally differing assessments of community-level biodiversity that often remains unrecognised.

First, I characterise four theories about the organisation of ecological communities, beginning with an extreme opposition (organicism versus individualism) and continuing with a more sophisticated but still fundamental one (idiosyncratic versus generic interactionism). *Second*, I propose a distinction of two concepts of community-level biodiversity that correlates to the latter opposition: 'biodiversity as mosaic of coevolved local uniqueness' versus 'biodiversity as structured continuum of assembled local differences'. This simplifying, ideal-typical distinction is intended to serve as a heuristic tool for the analysis of discussions about biodiversity issues. To demonstrate its practical relevance and heuristic capacity, I touch on opposing assessments of neobiota. *Third*, changing from characterisation of ecological concepts to meta-theory of ecology, I propose an explanation for the existence and persistence of those opposing ecological concepts that highlights the influence of culturally shaped patterns of interpretation on the formation of ecological concepts. Thereby I limit myself to idiosyncratic interactionism and to the concept of biodiversity as mosaic of coevolved local uniqueness respectively. *Fourth*, I develop the thesis that the claim to maintain unique local biodiversity for its instrumental values might often be motivated by the interest in its intrinsic values. I conclude with summarising and evaluating theses.

Four concepts of ecological communities

Throughout its history, the science of ecology has experienced fundamental controversies about the causes that determine the distribution of species, about the character of ecological interrelations among coexisting species and, thus, about the nature or principles of the organisation of ecological communities and ecosystems (Whittaker 1975; Trepl 1987; Hubbell 2001; Hengeveld and Walter 1999; Jax 2006; Kirchhoff 2007). The different positions in these discussions lead to different concepts, surveys and valuations of community-level biodiversity.

To portray those ecological controversies and the related concepts of biodiversity, I use the method of ideal types. These are formed by one-sidedly accentuating fundamental ideas of real existing theories while disregarding secondary hypotheses that qualify these ideas. (For details on this method, which goes back to Max Weber, cf. Hekman 1983; for its application to ecological concepts cf. Eisel 2004b; Kirchhoff 2007.)

As a basis I take the typing in Kirchhoff (2007: 77–116) that substitutes the common dichotomic comparison of organicism and individualism by a trichotomic distinction of organicism, interactionism, and aggregationism or individualism. I will refine this trichotomic typing by distinguishing two types of interactionism so that four ideal types emerge (cf. the synopsis in Table 5.1).

Organicism versus individualism

The classical controversy about the nature of ecological communities is that between organicism and aggregationism or individualism. Ecological organicism conceives of ecological communities in analogy to individual organisms: they consist of species that are mutually interdependent as the organs of an individual organism are.[1] Ecological aggregationism or individualism holds that species distribute largely independently from what particular species are present or absent in a particular place. This is not to deny ecological interactions (which are necessary for heterotrophic organisms) but to contradict that they ordinarily have a species selecting effect: in most cases they do lead to neither positive nor negative biotic filtering; thus, species distribution is subject to abiotic filtering and dispersal only.[2]

For quite some time, both of these extreme concepts have been largely dismissed in scientific ecology (but organicism is still present in nature conservation and environmental management). Instead, a wide and complex spectrum of interactionistic concepts has developed, all of which emphasise the influence of biotic filtering on community composition. However, within this spectrum a fundamental opposition has remained. It relies on different hypotheses about the character of the ecological and evolutionary interactions among the species that form an ecological community. To highlight this opposition I built two ideal types of interactionism: idiosyncratic interactionism and generic interactionism.

Idiosyncratic interactionism versus generic interactionism

The core hypothesis of idiosyncratic interactionism is:[3] if ecological communities or ecosystems have realised an undisturbed development, they consist of species that have been coexisting for a long time and that have passed through a shared evolutionary history. In the course of this collective history unilateral and mutual evolutionary adaptations have taken place. These have resulted in niche specialisation and co-speciation, and, thus, in internal differentiation and in a large variety of specialised ecological interactions. In this way, almost everywhere in the world unique ecological communities have emerged that feature a characteristic combination of species (many of which are endemic) and exhibit ecological interactions that tend to be idiosyncratic so that the species are tightly bound to one another. The ecological community as a whole has adapted itself to the specific conditions of its habitat and, at the same time, has changed its habitat such that both ecological community and habitat together form a superordinate unity: an ecosystem that represents a natural unity of the biosphere. These ecosystems are not regarded as static but as dynamic self-organising systems that modify their organisation if slight modifications of superordinate environmental conditions occur.

A possible objection to idiosyncratic interactionism is that many ecological communities have many species in common. This objection is put down by the additional hypothesis that the local populations of these common species differ characteristically from one another in their genotype and phenotype due to local coevolution and adaptation to the specific local conditions.

Generic interactionism[4] coincides with idiosyncratic interactionism in the emphasis on biological filtering. However, it disputes that local coevolutionary adaptations through shared evolutionary history represent a major principle of community organisation. Instead 'ecological fitting' (Janzen 1985: 309) is identified as its basic principle; that is, 'a major part of the earth's surface may be occupied largely by organisms that are rich in ecological interactions and have virtually no detailed evolutionary history with one another' (Janzen 1985: 309). In other words: those species of the regional species pool are building a local ecological community that are fitting together as regards their abiotic and biotic ecological requirements and tolerances – whereby the selected species have evolved the suite of ecological traits that enable the fitting beforehand, elsewhere, independently from one another and under different environmental conditions. Thus, the ecological interactions are not based on adaptations among the coexisting species but represent aptations, which may have originated as adaptations to other species (cf. Mahner and Bunge 2000: 153–61 for the distinction between adaptation and aptation). When species come into a new setting of ecological interactions, they exhibit so-called phylogenetic niche conservatism and preadaptations to the novel conditions; if environmental conditions change, the adaptive traits of the species remain unchanged and the species survive by moving into areas that (now) display conditions that fit their old adaptations.

The concept of ecological fitting presupposes or, seen the other way around, implies the hypothesis that ecological interactions have rather generic than idiosyncratic character. That is to say: in most cases several species are more or less ecologically equivalent and thus interchangeable. Namely, polyphagous generalists outweigh oligophagous and monophageous specialists. Hubbell's (2001, 2005) neutral theory is an extreme version of this view. This hypothesis of generic interactions is qualified, but not suspended, by the assumption of phenotypic plasticity organisms that allows for short-term *onto*genetic adaptation without phylogenetic, evolutionary adaptation.

The hypothesis of generic interactions is confronted with the following objection: evolution of coexisting species always leads to specialised species and idiosyncratic interactions because specialisation enhances competitiveness and divergent evolution reduces interspecific competition. This objection is put down by a theory of convergent or parallel evolution (cf. especially Zamora 2000; Hubbell 2001, 2005). Already in one single community, there are always many other species to which any one species has to adapt itself; further, this set of species is permanently changing grace to environmental fluctuations; further, this variable set of species is different for every local population of a species' metapopulation so that each of its local populations is confronted with a different selection regime, while gene flow within the metapopulation hampers the differential adaptation of the local populations. Consequently, many species or metapopulations adapt themselves to the average biotic conditions of their environment. As these average conditions are similar for many species, natural selection leads to a convergence of the traits of different species so that groups of ecologically similar species emerge.

Table 5.1 Four idealised concepts of ecological communities

	Organicism	*Idiosyncratic interactionism*	*Generic interactionism*	*Aggregationism / individualism*
Characteristics of ecological interactions	Obligate interdependence (very strong biotic filtering)	Idiosyncratic, determinate (strong biotic filtering)	Generic, flexible (moderate biotic filtering)	Generic, very flexible (no/ minor biotic filtering; only abiotic filtering)
Emergence of traits; characteristics of evolution	Collectivistic evolution of superorganisms	Shared evolutionary history of individual species; divergent coevolution of local populations	Phenotypic plasticity; phylogenetic niche conservatism; convergent or parallel (co-) evolution of metapopulations	Independent; individualistic

Opposing concepts of community-level biodiversity and their practical consequences

We have seen that in ecology – despite a wide dismissal of the extreme positions of organicism and aggregationism or individualism – a fundamental opposition has remained between two types of interactionism: idiosyncratic interactionism and generic interactionism. I hold the view that this opposition is, explicitly or implicitly, present and influential in contemporary discussions of biodiversity issues. It forms the ecological basis of an antagonism between two concepts of community-level biodiversity that give rise to opposing surveys and evaluations of stocks of biodiversity and of their direct and indirect changes by human activity. I will first contrast and label these opposing concepts of biodiversity and then give an example of an opposing evaluation that they imply. Please note that all these simplifying dichotomic comparisons are not intended to deny the existence of intermingled views. However, I deem them to capture a main borderline in contemporary discourses on biodiversity (cf. Kirchhoff and Trepl 2001; Kirchhoff and Haider 2009 for a similar distinction).

'Biodiversity as mosaic of coevolved local uniqueness' versus 'biodiversity as structured continuum of assembled local differences'

According to *idiosyncratic interactionism* (and organicism as well), the biosphere naturally consists of a diversity of more or less discrete local ecological communities or ecosystems respectively, each being characterised by a unique combination of coevolved species that are closely interlocked by predominantly idiosyncratic interactions. I propose to label the concept of community-level biodiversity that corresponds to idiosyncratic interactionism *biodiversity as mosaic of coevolved local uniqueness*.

I use the term 'uniqueness' to indicate the assumption of a special form of qualitative individuality that has resulted from a process of historical integration and exhibits an internal principle of individuality (cf. Eisel 1992, 2004b; Kirchhoff 2007: 17–20; Kirchhoff 2012b). I use the term 'mosaic' and not 'system' because not all proponents of idiosyncratic interactionism (and organicism) conceive of the diversity of local ecological communities or ecosystems in a hierarchy-theoretical way as interlinked subsystems of one superordinate system – a view that is nowadays fostered by, for example, the Resilience Alliance's panarchy-framework (Gunderson and Holling 2002).

According to *generic interactionism*, the biosphere naturally consists of assemblages of species that are flexibly interacting as evolutionary 'predefined' but phenotypically plastic 'interactors' according to predominantly generic possibilities of interaction and, thus, it consists of species assemblages that are far from fortuitous but continually changing in space and time. I propose to label the concept of community-level biodiversity that corresponds to generic interactionism *biodiversity as structured continuum of assembled local differences*.

Practical consequences of the opposing concepts: opposing assessments of neobiota

To point out that these opposing concepts of community-level biodiversity imply opposing assessments of biodiversity issues, I address the issue of neobiota. Analogous evaluative differences could be demonstrated as regards, for example, the consequences of climate change or the possibilities and limitations of human construction of novel ecosystems.

Neobiota are organisms that – assisted by human agency or not – have for the first time established in a region. There is an extensive, highly controversial debate on the consequences of anthropogenic introductions of neobiota (cf., e.g. Richardson 2011). In this debate there is one wide consensus: they are leading to biotic homogenisation; that is, a decrease in regional distinctiveness of the Earth's biota. However, there is no consensus about the effects of biotic homogenisation and neobiota on the organisation of ecological communities and on the instrumental services of ecosystems. Simplifying, two opposing views can be identified (cf. Kirchhoff and Haider 2009; cf. the distinction of two images of world society in Beck 2000: 85).

Proponents of the first view argue that the current anthropogenic introductions of neobiota differ not only quantitatively (in frequency, distance and number of species) but also qualitatively, that is in their consequences, from historical natural and historical anthropogenic introductions (e.g. Cassey et al. 2005). While the relatively few ancient neobiota could be integrated into the local ecosystems of native species, the numerous current neobiota cannot. They disrupt their structure and functioning, and thereby damage their provisioning, regulating and supporting services. For Phillip Cassey et al. (2005), introductions of neobiota 'are undesirable for maintaining the function of natural ecological and evolutionary patterns and processes' (cf. Angermeier and Karr 1994; Simberloff 2005). Jason and Roy van Driesche (2004: 2) speak of 'Nature Out of Place' and deplore that 'an unending stream of invasions is changing ... ecosystems from productive, tightly integrated webs of native species to loose assemblages of stressed native species and aggressive invaders.' Some proponents of this view conclude that a 'native' criterion has to be included into the definition of 'biodiversity', or even exclude non-native organisms and 'artificial biodiversity' from the definition of biodiversity (e.g. Angermeier 1994; Sala et al. 2000). Such an assessment of neobiota might mostly be grounded in the concept of biodiversity as a mosaic of coevolved local uniqueness.

Proponents of the second view admit the quantitative differences between historical and current introductions of neobiota but deny that they have qualitatively different consequences (e.g. Brown and Sax 2004, 2005). They emphasise that biotas and ecological communities have always been subject to abiding and omnipresent natural changes and state that, in the great majority of ecological respects, native and non-native species are indistinguishable (e.g. Thompson, Hodgson and Rich 1995; Davis 2003; Davis et al. 2011). Some even question the distinction between native and non-native species, arguing that all species have spread into new territory at some point in their history (Sax et al. 2005). Just like native species, neobiota can be both harmful and useful so that 'management practices regarding the occurrence of "new" species could range from complete eradication to tolerance and even consideration of the "new" species as an enrichment of local biodiversity and key elements to maintain ecosystem services' (Walther et al. 2009: 686). Similarly, Mark Davis et al. (2011: 154) end up asking 'conservationists to focus much more on the functions of species, and much less on where they originated.' Such an assessment of neobiota might mostly be grounded in the concept of biodiversity as a structured continuum of assembled local differences.

On the cultural background of the opposing ecological concepts

What provokes the longstanding debates about the nature of ecological communities and the opposing concepts of community-level biodiversity? One possible explanation is: the scientists who promote opposing concepts have studied different kinds of ecological communities; the controversy results from unjustified generalisations of empirical insights into the character of just one kind of community organisation. Although this may be true in some cases, this cannot explain that the controversy about the nature of ecological communities, despite a wealth of empirical arguments, has persisted (with transformations) for over a century, without exclusive ranges of validity having been assigned to the opposing concepts of ecological communities. This fact, together with two additional findings – namely that analogous, likewise persisting controversies have been identified in other subdisciplines of biology (cf., e.g. Cooper 2001 for population ecology and Skipper 2002 for population genetics) and that all these biological controversies show conspicuous structural analogies with the long-lasting controversies about the nature of or ideals of human individuality and human societies – suggests a different hypothesis. The opposing concepts of ecological communities are inspired by and reflect those controversies about concepts of human individuality and society.

Cultural a priori *as context of discovery of scientific theories*

Before I develop this hypothesis, I sketch the insights of anti-positivistic philosophy of science (e.g. Kuhn 1970; Lakatos 1970; Hacking 2002; Eisel 2004b) and historical epistemology (e.g. Foucault 2002; Rheinberger 2010) that provide its basis: (1) Scientific observation is always motivated and guided by theory (theory-ladenness of observation). (2) Scientific hypotheses are generated not in isolation but within a theoretical framework of ontological and methodological presumptions, which normally remain unquestioned (Kuhn's disciplinary matrix, Lakatos' hard core of a scientific research programme, Hacking's style of scientific reasoning). This framework determines which kind of questions are (not) being posed, how empirical data are (not) generated, and how they are (not) interpreted.

(3) This framework and the hypotheses generated therein are neither derived inductively from empirical data (as representatives of empiricism hold) nor the product of a researcher's subjective creativity (as Popper thought). Rather, they rely on culturally shaped patterns of interpretation, which have been variously labelled, e.g. epistemological field and positive unconscious of knowledge by Michel Foucault (2002: xi, xxiii, 398) or social imaginary (e.g. J.B. Thompson 1984: 6, 16ff.; Taylor 2004). (4) These patterns remain mostly unconscious. As they are constitutive of (the validity of) experience but not Kantian necessary conditions for the possibility of experience, they have paradoxically been denominated historical, or relative, or cultural *a priori* (e.g. Foucault 1966/2002: xxiii, 172, 346; Eisel 2004b: 39). (5) Competing cultural patterns of interpretation engender competing research programmes and theories (cf. Lakatos 1970; Eisel 2004b; Kirchhoff 2007). Or, as Richard Lewontin (1974: 29) has expressed it: 'Schools of thought about unresolved problems ... derive ... from deep ideological biases reflecting social and intellectual world views'. (6) Competing research programmes can coexist for long because one and the same set of observations is explainable by different, even by contradictory, theories (underdetermination of theory by evidence) so that the possibilities of falsification are largely restricted. Further, if empirical data are encountered that contradict the prognoses derived from a theory, neither the theory nor the research programme will normally not be regarded as refuted by its proponents; rather, they will add a 'protective belt' of falsifiable auxiliary hypotheses such that the 'hard core' of the research programme remains unscathed (sophisticated falsificationism) (Lakatos 1970). Please note that these insights of anti-positivistic philosophy of science and historical epistemology do *not* imply an 'anything goes' relativism, inter alia because cultural *a priori* do not emerge ad libitum and scientific theories are subject to criteria of empirical relevance (cf. Laudan 1986; Eisel 2004b).

Herder's theory of culture as cultural a priori *of idiosyncratic interactionism*

Now I can come back to my specific hypothesis. In the case of opposing concepts of ecological communities, the relevant cultural *a priori* and context of discovery are opposing concepts or ideals of human individuality and society (Trepl 1994; Eisel 2004a, 2004b; Kirchhoff 2007; Voigt 2009; Kirchhoff et al. 2010; cf. the extensive controversies in sociology on holism/collectivism/communitarianism/democratism and individualism/atomism/elementarism, universalism and particularism etc.). I confine myself to develop this hypothesis for idiosyncratic interactionism and the concept of biodiversity as a mosaic of coevolved local uniqueness respectively. I will argue: The cultural background of these ecological concepts is a specific concept of cultural development that is paradigmatically exemplified by Johann Gottfried Herder's (Herder SW) theory of culture, which has been – and implicitly still is – highly influential not only in the German-speaking culture. Similar ideas could be drawn, for example, from George Perkins Marsh's (1865) 'Man and nature', Carl Sauer's writings on landscape and cultural geography (republished in Sauer 1976) and Aldo Leopold's (1949) 'Land Ethic', or, to name a more recent example, concepts of bioregionalism. In the form of a methodological reflection, Heinrich Rickert (1986) has characterised the underlying cultural pattern of interpretation under the term 'individualizing concept formation' as against 'generalizing concept formation'.[5]

Herder's theory of cultural development was directed against the enlightenment universalism (cf. for the following Herder SW, especially IV, 204f.; V, 509; VIII, 210; XIII, 253–318, 347–49, 363, 370; XIV, 38, 227; XVII, 122, 287; XVIII, 308; for the interpretation cf. Eisel 1980, 1992; Kirchhoff 2005, 2012b). In the name of individual freedom and authenticity, Herder agitated against despotism and moribund traditions; in this respect he was in line with the Enlightenment. He rejected, however, all the different Enlightenment concepts of society, individuality and freedom – be it that of English liberalism, French democratism or Kantian republicanism – and likewise denied the Enlightenment's claim that the goal of human history is to emancipate from tradition and nature in order to realise one universal form of social organisation all over the world. In contrast, Herder claimed that the goal of human history is to develop unique cultures all over the world.

Cultural uniqueness, according to Herder, is the inherent result of reasonable cultural development. Cultural development is reasonable if and only if it is governed by two interdependent determinants: the people's individual character and the particular natural conditions of its land or place of living. A people has to reveal the naturally determined particular possibilities of land use that its place of living offers and to realise them according to its own character. At the same time, the particular environmental conditions of the place of living shape the people's character: its sensibility, language, habits, way of thinking etc. Likewise at the same time, the people modifies the natural conditions of its place of living without eliminating its special, naturally given character. This process of cultural development requires – and enables – that every individual of the people unfolds his or her individual natural talents. This perfection of individuality represents a service for and integration into the whole community, and at the same time the realisation of individual freedom. Thus, Herder's political philosophy is all at once individualistic and holistic: it represents an individualistic holism or holistic individualism (cf. Spencer 1996), which is directed against despotism and collectivism as well as against elementaristic individualism.

The overall result of this cultural development is a harmonic unity of a people and its place of living that exhibits a unique cultural diversity. This cultural diversity amalgamates a unique land(scape), a people that consists of unique individuals, and a unique cultural regime of land-use techniques, building types, social order, customs etc. If this process of cultural development takes place everywhere in the world, then a manifoldness of unique cultures arises, each of which is aiming at its own perfection: cultures or 'nations modify themselves by place, time, and their inner character; each bears within itself the harmony of its perfection, incommensurable with that of others' (Herder SW XIV, 227; my translation). Thereby, Herder interprets beauty as the sensible expression of this perfection (SW XXII, especially 104), thus held an objectivist-functionalist aesthetic (Kirchhoff and Trepl 2009: 25–7; cf. Norton 1991). Herder focused on language and tradition as expressions of cultural uniqueness. Later, the emphasis shifted to unique cultural landscapes (e.g. Ritter 1817; Riehl 1854; Rudorff 1897).

The description of Herder's theory reveals a structural analogy: idiosyncratic interactionism conceives of the development of ecosystems analogously to Herder's theory of cultural development. In both cases a double-determined development is assumed that leads to local uniqueness. In detail, the ecological community corresponds to the people, the ecological community's habitat to the people's place of living or land; the coevolving species of the ecological community parallel the humans of the people that develop their individual talents in harmony with one another; the mosaic of unique ecosystems matches the manifoldness of unique cultures.

In pointing out this structural analogy, I do not claim a direct inspiration of ecology by Herder's theory of culture. Rather, I assume an indirect influence: Herder's theory formed the intellectual basis of the research programme of classical modern geography (Eisel 1980: 274–92, 568–71; 1992; Hard 2011). This has been professed by its founder, Carl Ritter (1817), and is obvious in the fact that classical geography had cultural landscapes as its fundamental subject and aimed at explaining their unique character and physiognomy as a result of the 'organic' interaction of human and physical factors. This geographical research programme has become highly influential in Germany, in large parts of Europe, and in North America as well (Marsh 1865: 8; Martin 2005: 107–28, 141–3, 305). It dominated geography until the 1970s. Community ecology, the branch of ecology that studies the organisation of ecological communities, emerged by transformation of physiognomic phytogeography. In about 1900, scientists began to reinterpret the theretofore aesthetically defined physiognomic units of the Earth's vegetation (cf., e.g. von Humboldt 1958) as units of interacting and interdependent species that are characteristically shaped by the interactions of the species with one another and with their abiotic environment (Trepl 1987: 103–38, 1997; cf. Blandin 1998; Kirchhoff 2007: 487–97; Jax 2010: 192–4; Hard 2011; for the influence of Herder on Humboldt cf. Graczyk 2004, especially 290f., 302, 410f.).

Why look so far back?

Perhaps it appears to be beside the point to relate current concepts of ecological communities and biodiversity to a more than 200-year-old theory of culture. Should one not, rather, look for relations of idiosyncratic interactionism and the concept of biodiversity as a mosaic of coevolved local uniqueness to current expressions of anti-universalism and critique of globalisation like the demand for an ethics of place, bioregionalism, or biocultural diversity conservation?

While such an analysis might be instructive, it cannot substitute the foregoing historical retrospect. First, idiosyncratic interactionism is simply older than those current movements so that it cannot be 'explained' with reference to those movements. Second, the majority of the current regionalistic movements may have their intellectual roots in ideals of culture and the human–nature relationship like that expressed in Herder's writings. These roots are normally neither conscious nor, if demonstrated, gladly acknowledged within these movements. This 'keeping at a distance' might be motivated by the fact that Herder's theory of culture and even more the consequent movements of homeland and nature protection (e.g. Rudorff 1897) have a politically conservative background which contradicts the mainly progressive self-image of those current movements. Third, looking backward to Herder's theory sheds light on the historical fact that the appreciation of local uniqueness has by no means arisen in the context of 'ecological' problems, environmental management, biodiversity issues etc., but rather as a critique of political and cultural universalism and in the context of an ethics of authenticity of which Herder, striving against an assumed cultural and personal alienation, was the major early articulator (Taylor 1992: 28). Nota bene: Herder's theory of culture might represent a glorifying, social romantic interpretation of pre-industrial societies.

Unique local biodiversity: congruence of instrumental ecosystemic and intrinsic scenic values?

In the first part of this chapter, I have derived the concept of biodiversity as a mosaic of coevolved uniqueness from a scientific concept of ecological communities. So far, this concept of biodiversity is an ecological concept. Accordingly, resting upon idiosyncratic interactionism, ecologically based arguments can be (and are) raised for the maintenance of unique local biodiversity: it is said to be constitutive for the integrity of ecosystems and, thus, for instrumental ecosystems services. (I do not include cultural services because I deem it a category mistake[6] to ascribe such services to ecosystems; cf. below.)

In the previous part, I have characterised a theory of culture that determines cultural uniqueness as the goal of cultural development, and a unique cultural landscape as its key expression. In several European countries, especially in Germany, this or a similar cultural ideal has been a major basis of the nature and homeland protection movements that arose about 1900 (cf., e.g. 1894 *National Trust for Places of Historic Interest or Natural Beauty;* 1904 *Bund Heimatschutz;* 1909 *Naturskyddsföreningen*).[7] These movements basically aimed at the protection of unique cultural (and natural) landscapes and of historic buildings because they were regarded as *cultural* assets that had to be preserved from destruction by urbanisation, industrial agriculture, technical infrastructure etc.; that is, these movements were essentially cultural, not ecological movements (cf. Körner and Eisel 2003; van Koppen and Markham 2007). Species were protected if they were characteristic of landscapes and, thus, contributed to their aesthetic and symbolic qualities. Against this historical background of nature protection, the concept of unique local biodiversity turns out to be a cultural concept: unique local biodiversity is protected as a scenic symbol with intersubjective intrinsic values, not as a constituent of ecosystems with instrumental values.

One interpretation of the findings is that unique local biodiversity has both outstanding instrumental and outstanding intersubjective intrinsic value. This view was and still is widespread in discussions on nature conservation. Usually it goes along with a so-called ecological aesthetic (cf., e.g. Böhme 1989; Thorne and Huang 1991; Parsons and Carlson 2008) which assumes a close link between the ecological integrity, sustainable usefulness and beauty of landscape – whereby the aesthetic pleasure is thought to derive from knowing somehow about this integrity and usefulness. Such an ecological aesthetic is in line with Herder's objectivist-functionalist aesthetics and inherent in Leopold's (1949: 262) famous claim that a 'thing is right when it tends to preserve the integrity, stability, and beauty of the biotic community.'

A different interpretation indeed acknowledges outstanding intersubjective intrinsic values of unique local biodiversity but doubts a congruence with outstanding instrumental values. Historically shaped unique local biodiversity might in many cases neither be a necessary requirement nor optimal as regards sustainable ecosystem services. Indeed, several nowadays cherished cultural landscapes which feature unique biodiversity have originated from degrading land use practices, examples being semi-dry grasslands, heaths and, to some degree, dehesas (cf. Häpke 1990; Plieninger and Bieling 2012). Further, far from all components of local biodiversity contribute significantly to ecosystem services, and research has shown that rare and endemic species often do not play important roles as regards instrumental ecosystem services (Kremen 2005; Ridder 2008; Jax 2010: 190 f.; Ingram, Redford and Watson 2012). Thus, many of the intrinsically valuable species that constitute unique local biodiversity might have insignificant instrumental value – what corresponds to the fact that historically shaped high and unique local biodiversity were nothing more than unintended by-products of traditional forms of land use. If species that belong to the unique local biodiversity have significant instrumental value, they may often be substitutable by species that do not belong to this biodiversity. And useful complementarity and insurance effects of high community-level biodiversity (Loreau 2000) might neither be bound to nor be optimally realised by historically shaped unique local biodiversity.

Given (i) these presumable discrepancies between instrumental and intersubjective intrinsic values of unique local biodiversity, (ii) that nature conservation, starting from the 1960s, has featured a significant shift in argumentation from the original arguments that referred to landscapes as cultural assets towards ecologically based arguments that refer to instrumental services of ecosystems (cf. Körner and Eisel 2003; van Koppen and Markham 2007), and by (iii) that the ecological theory of idiosyncratic interactionism might be inherently connected with the above-mentioned ideal of cultural uniqueness, I conclude: The ecologically based arguments to preserve unique local biodiversity for its instrumental values (and to fight neobiota because they are detrimental to instrumental values) may often be ultimately motivated by the original goal of nature protection to maintain the aesthetic and symbolic qualities of unique landscapes (cf. Trepl 1997; Kirchhoff, Brand and Hoheisel 2012).

Conclusion

To conclude, I establish eight summarising and evaluating theses:

1. In lifeworldly and scientific discourses on biodiversity issues, semantic and evaluative ambiguities arise because 'biodiversity' refers to both facts and values, and, more importantly, to different facts and different values.
2. As regards community-level biodiversity, an important conceptual borderline exists that corresponds to the opposition between two scientific concepts of ecological communities that I call 'idiosyncratic interactionism' and 'generic interactionism'. To clarify this borderline I propose to distinguish between the concept of 'biodiversity as a mosaic of coevolved local uniqueness' and the concept of 'biodiversity as a structured continuum of assembled local differences'.
3. The existence and persistence of these oppositions can be explained by pointing to the existence of opposing cultural patterns of interpretation that influence the formation of ecological concepts.
4. Idiosyncratic interactionism and the corresponding concept of biodiversity as a mosaic of coevolved local uniqueness might both be inherently linked to a specific anti-universalistic theory of culture that regards developing unique human-nature-unities as the goal of human history, which express themselves in unique cultural landscapes.
5. This theory of culture is an essential basis of the intersubjective intrinsic values that unique landscapes and unique local biodiversity have in our culture because of their aesthetic and symbolic qualities (scenic beauty, feelings of cultural identity, sense of belonging etc.).
6. The claim that unique local biodiversity has to be protected because of its instrumental values might often be motivated by the goal to protect the aesthetic and symbolic qualities of unique landscapes because of their intersubjective intrinsic values against the consequences of a global economy, industrial agriculture etc. It might represent an attempt to provide a (seemingly) objective, scientifically founded, universally valid justification for the legitimate, but only particularistic interest in these qualities.
7. However, the assumption that unique local biodiversity features a congruence of outstanding instrumental and outstanding intersubjective intrinsic values is doubtable and prone to category-mistakes, which entail methodological errors. These arise if it is ignored that intrinsic and instrumental values of unique local biodiversity refer to categorically different objects (cf. Kirchhoff 2012a, 2013 as regards so-called cultural ecosystem services; and Kirchhoff, Trepl and Vicenzotti 2013 as regards certain concepts of landscape ecology). At the level of biodiversity above that of single species, the intrinsic values refer essentially to *landscapes* in the original modern sense of the word; that is, to aesthetic units that emerge as culturally shaped pictorial representations of a section of the Earth's surface by a sentient beholder (Simmel 2007; Cosgrove 1984; Kirchhoff and Trepl

2009; Kirchhoff 2012a; Trepl 2012; Kirchhoff, Trepl and Vicenzotti 2013); they do not refer to (complexes of) ecosystems which are causal material systems or functional units of interacting biotic and abiotic components. (Many confusions arise because in some disciplines, e.g. geography and landscape ecology, landscapes are defined as complexes of ecosystems; cf. Kirchhoff, Trepl and Vicenzotti 2013.) In contrast, the instrumental values of biodiversity do refer to (complexes of) *ecosystems*.

8 Thus, the term 'unique local biodiversity' has an ambiguous ontological reference: with respect to intersubjective intrinsic values it refers to pictorial components of aesthetic representations of our environment, namely landscapes; with respect to instrumental values it refers to material components of causal systems, namely ecosystems. If issues of community-level biodiversity are discussed, both objectives are important. However, each objective has to be addressed according to its specific characteristics.

Notes

1 Authors whose theories come close to the *organismic ideal type* are, e.g., August Thienemann (1939), who stated: 'The biocoenosis is not only an aggregate … of *juxta*posed … organisms, but rather a (supra-individual) wholeness, a *with*-each-other and a *for*-each-other of organisms' (Thienemann 1939: 275; my translation), and Eugene Odum (1971) (Kirchhoff 2007: 205–33). Frederic Clements (1916), who is usually mentioned as the classical paradigmatic protagonist of ecological organicism, in fact, was not. Admittedly, he compared ecological communities to individual organisms, even called them superorganisms, and assumed a homeostatically preserved equilibrium state; however, he did not regard the species of a community as interdependent but as forming a hierarchy of dominant and subdominant species (Kirchhoff 2007: 169–86). Clements can be regarded as an advocate of organicism only if one construes 'organicism' as a concept of hierarchical organisation and one-sided dependence (this definition dominated in the nineteenth century) and not as a concept of interdependence (this definition emerged about 1800 and is nowadays dominant).

2 Authors whose theories come close to the *individualistic ideal type* are Henry Allan Gleason who stated that 'every species of plant is a law unto itself' (1926: 26) and regarded vegetation as a mere 'juxtaposition of individuals' (1926: 25), Herbert Andrewartha and Charles Birch (1954), and, more recently, Rob Hengeveld and Gimme Walter (1999).

3 I have built the ideal type of *idiosyncratic interactionism* drawing from theories like that of Hutchinson (1959), Ehrlich and Raven (1964), MacArthur (1972), Angermeier and Karr (1994), J.N. Thompson (1999, 2005), Holling and Gunderson (2002) and Traveset and Richardson (2006).

4 I have built the ideal type of *generic interactionism* with reference to theories like that of Janzen (1985), Zamora (2000), Hubbell (2001, 2005), Brown and Sax (2004, 2005), Wilkinson (2004), Agosta (2006), Agosta and Klemens (2008), Donoghue 2008, Agosta et al. (2010) and Wiens et al. (2010).

5 A reconstruction of the cultural background of generic interactionism is less straightforward. It could start from the Enlightenment universalism (criticised by Herder), which comes, however, in at least two forms: the nominalistic English liberalism and the rationalistic French democratism (Eisel 2004a; Kirchhoff 2005; Voigt 2009). Moreover, the reconstruction would have to lean on more recent characterisations of modern societies and economies that reflect phenomena like

flexibilisation of work and flexible specialisation on the one hand and standardisation, institutional/technological isomorphism and product substitutability on the other.

6 A category-mistake (Ryle 1949: 16) is a linguistic or ontological mistake in which an expression is used in a way that does not conform to the logical type of the expression or to the ontological characteristics of the so-termed object, so that the object is attributed with a property that it could not possibly have.

7 Admittedly, some differentiations would be in order here, for example that in some European countries, e.g. Norway and Sweden, the interest in unique natural landscapes outweighed that in unique cultural landscapes, as it does in the USA (cf. van Koppen and Markham 2007).

References

Agosta, S.J. (2006) 'On ecological fitting, plant-insect associations, herbivore host shifts, and host plant selection', *Oikos* 114: 556–65.

Agosta, S.J. and Klemens, J.A. (2008) 'Ecological fitting by phenotypically flexible genotypes: implications for species associations, community assembly and evolution', *Ecology Letters* 11: 1123–34.

Agosta, S.J., Janz, N. and Brooks, D.R. (2010) 'How specialists can be generalists: resolving the "Parasite Paradox" and implications for emerging infectious disease', *Zoologia* 27: 151–62.

Andrewartha, H.G. and Birch, L.C. (1954) *The distribution and abundance of animals*, Chicago, IL: University of Chicago Press.

Angermeier, P.L. (1994) 'Does biodiversity include artificial diversity?', *Conservation Biology* 8: 600–2.

Angermeier, P.L. and Karr, J.R. (1994) 'Biological integrity versus biological diversity as policy directives: protecting biotic resources', *BioScience* 44: 690–7.

Beck, U. (2000) 'The cosmopolitan perspective: sociology of the second age of modernity', *British Journal of Sociology* 51: 79–105.

Birnbacher, D. (2004) 'Limits to substitutability in nature conservation', in M. Oksanen and J. Pietarinen (eds) *Philosophy and biodiversity*, Cambridge: Cambridge University Press, 180–95.

Blandin, P. (ed.) (1998) *The European origins of scientific ecology (1800–1901)*, Amsterdam: Éditions des archives contemporaines.

Böhme, G. (1989) *Für eine ökologische Naturästhetik*, Frankfurt am Main: Suhrkamp.

Brown, J.H. and Sax, D.F. (2004) 'An essay on some topics concerning invasive species', *Austral Ecology* 29: 530–6.

Brown, J.H. and Sax, D.F. (2005) 'Biological invasions and scientific objectivity: reply to Cassey et al. (2005)', *Austral Ecology* 30: 481–3.

Callicott, J.B., Crowder, L.B. and Mumford, K. (1999) 'Current normative concepts in conservation', *Conservation Biology* 13: 22–35.

Cassey, P., Blackburn, T.M., Duncan, R.P. and Chown, S.L. (2005) 'Concerning invasive species: reply to Brown and Sax', *Austral Ecology* 30: 475–80.

CBD – United Nations (1992) *Convention on biological diversity*. Online. Available HTTP: <http://www.cbd.int/doc/legal/cbd-en.pdf> (accessed 12 June 2013).

Clements, F.E. (1916) *Plant succession. An analysis of the development of vegetation*, Washington, D.C.: Carnegie Institution of Washington.

Cooper, G. (2001) 'Must there be a balance of nature?', *Biology and Philosophy* 16: 481–506.

Cosgrove, D.E. (1984) *Social formation and symbolic landscape*, London: Croom Helm.

Davis, M.A. (2003) 'Biotic globalization: does competition from introduced species threaten biodiversity?', *BioScience* 53: 481–9.

Davis, M.A., Wallace, D., Chew, M.K., Hobbs, R.J., Lugo, A.E., Ewel, J.J., Vermeij, G.J., Brown, J.H., Rosenzweig, M.L., Gardener, M.R., Carroll, S.P., Thompson, K., Pickett, S.T.A., Stromberg, J.C., Tredici, P.D., Suding, K.N., Ehrenfeld, J.G., Grime, J.P., Mascaro, J. and Briggs, J.C. (2011) 'Don't judge species on their origins', *Nature* 474: 153f.

Donoghue, M.J. (2008) 'A phylogenetic perspective on the distribution of plant diversity', *Proceedings of the National Academy of Sciences of the United States of America* 105: 11549–55. Online. Available HTTP: <http://www.pnas.org/content/105/Supplement_1/11549.full.pdf+html> (accessed 13 June 2013).

Ehrlich, P.R. and Raven, P.H. (1964) 'Butterflies and plants: a study in coevolution', *Evolution* 18: 586–608.

Eisel, U. (1980) *Die Entwicklung der Anthropogeographie von einer 'Raumwissenschaft' zur Gesellschaftswissenschaft*, Kassel: Gesamthochschulbibliothek.

Eisel, U. (1992) 'Individualität als Einheit der konkreten Natur: Das Kulturkonzept der Geographie', in B. Glaeser and P. Teherani-Krönner (eds) *Humanökologie und Kulturökologie*, Opladen: Westdeutscher Verlag, 107–51.

Eisel, U. (2004a) 'Naturbilder sind keine Bilder aus der Natur – Orientierungsfragen an der Nahtstelle zwischen subjektivem und objektivem Sinn', *Gaia* 13: 92–8.

Eisel, U. (2004b) 'Politische Schubladen als theoretische Heuristik. Methodische Aspekte politischer Bedeutungsverschiebungen in Naturbildern', in L. Fischer (ed.) *Projektionsfläche Natur. Zum Zusammenhang von Naturbildern und gesellschaftlichen Verhältnissen*, Hamburg: Hamburg University Press, 29–43.

Elliot, R. (1992) 'Intrinsic value, environmental obligation, and naturalness', *The Monist* 75: 138–60.

Foucault, M. (2002) *The order of things. An archaeology of the human sciences*, London: Routledge.

Gleason, H.A. (1926) 'The individualistic concept of the plant association', *Bulletin of the Torrey Botanical Club* 53: 7–26.

Graczyk, A. (2004) *Das literarische Tableau zwischen Kunst und Wissenschaft*, München: Fink.

Gunderson, L.H. and Holling, C.S. (eds) (2002) *Panarchy: understanding transformations in human and natural systems*, Washington, D.C.: Island Press.

Hacking, I. (2002) *Historical ontology*, Cambridge: Harvard University Press.

Häpke, U. (1990) 'Die Unwirtlichkeit des Naturschutzes. Böse Thesen', *Kommune* 2: 48–53.

Hard, G. (2011) 'Geography as ecology', in A. Schwarz and K. Jax (eds) *Ecology revisited. Reflecting on concepts, advancing science*, Dordrecht: Springer, 351–68.

Hekman, S.J. (1983) *Weber, the ideal type, and contemporary social theory*, Notre Dame, IN: University of Notre Dame Press.

Hengeveld, R. and Walter, G.H. (1999) 'The two coexisting ecological paradigms', *Acta Biotheoretica* 47: 141–70.

Herder, J.G. (1877–1913) *Sämtliche Werke*, B. Suphan (ed.), Berlin: Weidmann, SW.

Holling, C.S. and Gunderson, L.H. (2002) 'Resilience and adaptive cycles', in L.H. Gunderson and C.S. Holling (eds) *Panarchy: Understanding transformations in human and natural systems*, Washington, D.C.: Island Press, 25–62.

Hubbell, S.P. (2001) *The unified neutral theory of biodiversity and biogeography*, Princeton, NJ: Princeton University Press.

Hubbell, S.P. (2005) 'Neutral theory in community ecology and the hypothesis of functional equivalence', *Functional Ecology* 19: 166–72.

Hutchinson, G.E. (1959) 'Homage to Santa Rosalia; or, why are there so many kinds of animals?', *American Naturalist* 93: 145–59.

Ingram, J.C., Redford, K.H. and Watson, J.E.M. (2012) 'Applying ecosystem services approaches for biodiversity conservation: benefits and challenges', *S.A.P.I.EN.S (Online)* 5. Online. Available HTTP: <http://sapiens.revues.org/1459> (accessed 13 June 2013).

Janzen, D.H. (1985) 'On ecological fitting', *Oikos* 45: 308–10.

Jax, K. (2006) 'Ecological units: definitions and application', *The Quarterly Review of Biology* 81: 237–58.

Jax, K. (2010) *Ecosystem functioning*, Cambridge: Cambridge University Press.

Kirchhoff, T. (2005) 'Kultur als individuelles Mensch-Natur-Verhältnis. Herders Theorie kultureller Eigenart und Vielfalt', in M. Weingarten (ed.) *Strukturierung von Raum und Landschaft. Konzepte in Ökologie und der Theorie gesellschaftlicher Naturverhältnisse*, Münster: Westfälisches Dampfboot, 63–106.

Kirchhoff, T. (2007) *Systemauffassungen und biologische Theorien. Zur Herkunft von Individualitätskonzeptionen und ihrer Bedeutung für die Theorie ökologischer Einheiten*, Freising: Technische Universität München. Online. Available HTTP: <http://mediatum2.ub.tum.de/node?id=685961> (accessed 13 June 2013).

Kirchhoff, T. (2012a) 'Pivotal cultural values of nature cannot be integrated into the ecosystem services framework', *Proceedings of the National Academy of Sciences of the United States of America* 109: E3146. Online. Available HTTP: <http://www.pnas.org/content/109/46/E3146.full.pdf+html> (accessed 13 June 2013).

Kirchhoff, T. (2012b) 'Räumliche Eigenart. Sinn und Herkunft einer zentralen Denkfigur im Naturschutz', *Schriftenreihe der TLUG* 103: 11–22.

Kirchhoff, T. (2013) '"Biodiversität erhalten" – "Ökosystemdienstleistungen sichern". Begriffliche Klarstellungen zu zwei aktuellen Zielen des Naturschutzes', in Forschungsstätte der Evangelischen Studiengemeinschaft (ed.) *Jahresbericht 2012*, Heidelberg: FEST, 43–50.

Kirchhoff, T. and Haider, S. (2009) 'Globale Vielzahl oder lokale Vielfalt: zur kulturellen Ambivalenz von "Biodiversität"', in T. Kirchhoff and L. Trepl (eds) *Vieldeutige Natur. Landschaft, Wildnis und Ökosystem als kulturgeschichtliche Phänomene*, Bielefeld: transcript, 315–30.

Kirchhoff, T. and Trepl, L. (2001) 'Vom Wert der Biodiversität. Über konkurrierende politische Theorien in der Diskussion um Biodiversität', *Zeitschrift für angewandte Umweltforschung* S13: 27–44.

Kirchhoff, T. and Trepl, L. (2009) 'Landschaft, Wildnis, Ökosystem: zur kulturbedingten Vieldeutigkeit ästhetischer, moralischer und theoretischer Naturauffassungen. Einleitender Überblick', in T. Kirchhoff and L. Trepl (eds) *Vieldeutige Natur. Landschaft, Wildnis und Ökosystem als kulturgeschichtliche Phänomene*, Bielefeld: transcript, 13–66.

Kirchhoff, T., Brand, F. and Hoheisel, D. (2012) 'From cultural landscapes to resilient social-ecological systems: transformation of a classical paradigm or a novel approach?', in T. Plieninger and C. Bieling (eds) *Resilience and the cultural landscape: Understanding and managing change in human-shaped environments*, Cambridge: Cambridge University Press, 49–64.

Kirchhoff, T., Trepl, L. and Vicenzotti, V. (2013) 'What is landscape ecology? An analysis and evaluation of six different conceptions', *Landscape Research* 38 (1): 33–51.

Kirchhoff, T., Brand, F., Hoheisel, D. and Grimm, V. (2010) 'The one-sidedness and cultural bias of the resilience approach', *Gaia* 19: 25–32.

Körner, S. and Eisel, U. (2003) 'Naturschutz als kulturelle Aufgabe – theoretische Rekonstruktion und Anregungen für eine inhaltliche Erweiterung', in S. Körner, A. Nagel and U. Eisel (eds) *Naturschutzbegründungen*, Bonn: Bundesamt für Naturschutz, 5–49.
Kremen, C. (2005) 'Managing ecosystem services: what do we need to know about their ecology?', *Ecology Letters* 8: 468–79.
Kuhn, T.S. (1970) *The structure of scientific revolutions*, 2nd edn, Chicago, IL: University of Chicago Press.
Lakatos, I. (1970) 'Falsificationism and the methodology of scientific research programmes', in I. Lakatos and A. Musgrave (eds) *Criticism and the growth of knowledge*, Cambridge: Cambridge University Press, 91–196.
Laudan, L. (1986) *Science and values: The aims of science and their role in scientific debate*, Berkeley, CA: University of California Press.
Leibniz, G.W. (1996) *New essays on human understanding*, ed. P. Remnat and J. Bennett, Cambridge: Cambridge University Press.
Leopold, A. (1949) *A sand county almanac, and sketches here and there*, New York: Oxford University Press.
Lewontin, R.C. (1974) *The genetic basis of evolutionary change*, New York: Columbia University Press.
Loreau, M. (2000) 'Biodiversity and ecosystem functioning: recent theoretical advances', *Oikos* 91: 3–17.
MacArthur, R.H. (1972) *Geographical ecology. Patterns in the distribution of species*, New York: Harper & Row.
Maclaurin, J. and Sterelny, K. (2008) *What is biodiversity?*, Chicago, IL: University of Chicago Press.
Mahner, M. and Bunge, M. (2000) *Philosophische Grundlagen der Biologie*, Berlin: Springer.
Marsh, G.P. (1865) *Man and nature: Or, physical geography as modified by human action*, New York: Scribner's sons.
Martin, G.J. (2005) *All possible worlds: A history of geographical ideas*, New York: Oxford University Press.
Norton, R.E. (1991) *Herder's aesthetics and the European Enlightenment*, Ithaca, NY: Cornell University Press.
Odum, E.P. (1971) *Fundamentals of ecology*, Philadelphia, PA: Saunders.
Parsons, G. and Carlson, A. (2008) *Functional beauty*, New York: Oxford University Press.
Plieninger, T. and Bieling, C. (2012) 'Connecting cultural landscapes to resilience', in T. Plieninger and C. Bieling (eds) *Resilience and the cultural landscape: Understanding and managing change in human-shaped environments*, Cambridge: Cambridge University Press, 3–26.
Rheinberger, H.J. (2010) *On historicizing epistemology: An essay*, Stanford, CT: Stanford University Press.
Richardson, D.M. (ed.) (2011) *Fifty years of invasion ecology: The legacy of Charles Elton*, Oxford: Wiley-Blackwell.
Rickert, H. (1986) *The limits of concept formation in natural science. A logical introduction to the historical sciences*, Cambridge: Cambridge University Press.
Ridder, B. (2008) 'Questioning the ecosystem services argument for biodiversity conservation', *Biodiversity and Conservation* 17: 781–90.
Riehl, W.H. (1854) *Die Naturgeschichte des Volkes als Grundlage einer deutschen Social-Politik. Erster Band: Land und Leute*, Stuttgart: Gotta'scher Verlag.
Ritter, C. (1817) *Die Erdkunde im Verhältniß zur Natur und zur Geschichte des Mensche. Erster Theil*, Berlin: Reimer.

Rudorff, E. (1897) *Heimatschutz*, München: Müller.
Ryle, G. (1949) *The concept of mind*, London: Hutchinson's University Library.
Sala, O.E., Chapin III, F.S., Armesto, J.J., Berlow, E., Bloomfield, J., Dirzo, R., Huber-Sanwald, E., Huenneke, L.F., Jackson, R.B., Kinzig, A., Leemans, R., Lodge, D.M., Mooney, H.A., Oesterheld, M., Poff, N.L., Sykes, M.T., Walker, B.H., Walker, M. and Wall, D.H. (2000) 'Global biodiversity scenarios for the year 2100', *Science* 287: 1770–4.
Sandler, R. (2012) 'Intrinsic value, ecology, and conservation', *Nature Education Knowledge* 3: 4.
Sarkar, S. (2005) *Biodiversity and environmental philosophy: An introduction*, Cambridge: Cambridge University Press.
Sauer, C.O. (1976) *Land and life. A selection from the writings of Carl Ortwin Sauer*, ed. J. Leighly, Berkeley, CA: University of California Press.
Sax, D.F., Gaines, S.D. and Stachowicz, J.J. (2005) 'Introduction', in D.F. Sax, J.J. Stachowicz and S.D. Gaines (eds) *Species invasions: Insights into ecology, evolution, and biogeography*, Sunderland: Sinauer Associates, 1–7.
Simberloff, D.S. (2005) 'Non-native species *do* threaten the natural environment!', *Journal of Agricultural and Environmental Ethics* 18: 595–607.
Simmel, G. (2007) 'The philosophy of landscape', *Theory, Culture & Society* 24: 20–9.
Skipper, R.A. (2002) 'The persistence of the R.A. Fisher-Sewall Wright controversy', *Biology and Philosophy* 17: 341–67.
Spencer, V. (1996) 'Towards an ontology of holistic individualism: Herder's theory of identity, culture and community', *History of European Ideas* 22: 245–60.
Takacs, D. (1996) *The idea of biodiversity: Philosophies of paradise*, Baltimore, MD: Johns Hopkins University Press.
Taylor, C. (1992) *The ethics of authenticity*, Cambridge, MA: Harvard University Press.
Taylor, C. (2004) *Modern social imaginaries*, Durham: Duke University Press.
Thienemann, A.F. (1939) 'Grundzüge einer allgemeinen Ökologie', *Archiv für Hydrobiologie* 35: 267–85.
Thompson, J.B. (1984) *Studies in the theory of ideology*, Berkeley, CA: University of California Press.
Thompson, J.N. (1999) 'Specific hypotheses on the geographic mosaic of coevolution', *The American Naturalist* 153: S1–S14.
Thompson, J.N. (2005) *The geographic mosaic of coevolution*, Chicago, IL: University of Chicago Press.
Thompson, K., Hodgson, J.G. and Rich, T.C.G. (1995) 'Native and alien invasive plants: more of the same?', *Ecography* 18: 390–402.
Thorne, J.F. and Huang, C.-S. (1991) 'Toward a landscape ecological aesthetic: methodologies for designers and planners', *Landscape and Urban Planning* 21: 61–79.
Traveset, A. and Richardson, D.M. (2006) 'Biological invasions as disruptors of plant reproductive mutualisms', *Trends in Ecology & Evolution* 21: 208–16.
Trepl, L. (1987) *Geschichte der Ökologie. Vom 17. Jahrhundert bis zur Gegenwart*, Frankfurt am Main: Athenäum.
Trepl, L. (1994) 'Competition and coexistence: on the historical background in ecology and the influence of economy and social sciences', *Ecological Modelling* 75–76: 99–110.
Trepl, L. (1997) 'Ökologie als konservative Naturwissenschaft. Von der schönen Landschaft zum funktionierenden Ökosystem', in U. Eisel and H.-D. Schultz (eds) *Geographisches Denken*, Kassel: Gesamthochschulbibliothek, 467–92.
Trepl, L. (2012) *Die Idee der Landschaft. Eine Kulturgeschichte von der Aufklärung bis zur Ökologiebewegung*, Bielefeld: transcript.

van Driesche, J. and van Driesche, R. (2004) *Nature out of place: Biological invasions in the global age*, Washington, D.C.: Island Press.

van Koppen, C.S.A. and Markham, W.T. (eds) (2007) *Protecting nature. Organizations and networks in Europe and the USA*, Cheltenham: Edward Elgar Publishing.

Voigt, A. (2009) *Die Konstruktion der Natur. Ökologische Theorien und politische Philosophien der Vergesellschaftung*, Stuttgart: Steiner.

von Humboldt, A. (1958) 'Ideen zur Physiognomik der Gewächse', in R. Zaunick (ed.) *Alexander von Humboldt. Kosmische Naturbetrachtung. Sein Werk im Grundriß*, Stuttgart: Kröner, 196–217.

Walther, G.-R., Roques, A., Hulme, P.E., Sykes, M.T., Pyšek, P., Kühn, I., Zobel, M., Bacher, S., Botta-Dukát, Z., Bugmann, H., Czúcz, B., Dauber, J., Hickler, T., Jarošík, V., Kenis, M., Klotz, S., Minchin, D., Moora, M., Nentwig, W., Ott, J., Panov, V.E., Reineking, B., Robinet, C., Semenchenko, V., Solarz, W., Thuiller, W., Vilà, M., Vohland K. and Settele, J. (2009) 'Alien species in a warmer world: risks and opportunities', *Trends in Ecology & Evolution* 24: 686–93.

Whittaker, R.H. (1975) *Communities and ecosystems*, 2nd rev. edn, New York: Macmillan.

Wiens, J.J., Ackerly, D.D., Allen, A.P., Anacker, B.L., Buckley, L.B., Cornell, H.V., Damschen, E.I., Davies, T.J., Grytnes, J.-A., Harrison, S.P., Hawkins, B.A., Holt, R.D., McCain, C.M. and Stephens, P.R. (2010) 'Niche conservatism as an emerging principle in ecology and conservation biology', *Ecology Letters* 13: 1310–24.

Wilkinson, D.M. (2004) 'The parable of Green Mountain: Ascension Island, ecosystem construction and ecological fitting', *Journal of Biogeography* 31: 1–4.

Zamora, R. (2000) 'Functional equivalence in plant–animal interactions: ecological and evolutionary consequences', *Oikos* 88: 442–7.

6 Ethics of nature and biodiversity

Dieter Sturma

Introduction

In recent years, the loss of biodiversity has become the focus of increasing interest in the bioethical debate on the relationship between man and nature. Human interventions pose a threat to biodiversity in a number of ways. So-called modern civilisations are associated with a continuously growing consumption of raw materials, energy, land and water. In addition to the general environmental pollution caused by modern civilisations, this association applies especially to large-scale land exploitation and to the expansion of industrial production in the areas of agriculture, forestry and fishery.

Irrespective of the ways in which the consequences are evaluated, there is no doubt that human beings are having a large-scale effect on nature. It is apparent that the capacity to interfere with the environment and to manipulate nature is not always accompanied by an equivalent degree of willingness to consider the consequences thoroughly, i.e. to apply ethical standards, to revise attitudes and to correct actions accordingly. Often enough, people behave as conquistadors toward their environment, overthrowing and exploiting nature without knowing or appreciating it. With regard to the reasons and motivations for this kind of behaviour, it should be noted that a considerable part of environmental damage can be attributed directly to social constraints. Social and natural conditions are more closely interconnected than is generally assumed. Too little attention has been paid to this connection in the relevant debates in environmental ethics.

Environmental damage is, to a large extent, the result of behaviour and actions that are generally oriented towards short-term benefits. A diverse range of species not only represents a necessary condition for life as a whole, but also has a decisive influence on the quality of life of *human* beings. These facts tend to be overlooked, or even deliberately ignored. Although the short-sightedness of this practice is apparent from the economic, ecological and social standpoints, establishing ethically sound justifications for the preservation of biodiversity nevertheless proves to be difficult. Since the phenomenon of biodiversity does not itself provide a direct starting point for normative assessments, access must be gained indirectly. The project is thus to identify the normative consequences arising from the place of the human life-form in the system of nature,[1] which can then be used for an ethical justification and protection of biodiversity.

Apart from the political and pragmatic dominance of considerations of short-term benefits, the problem of justification is further complicated by epistemic and semantic problems in identifying biodiversity and by exceedingly asymmetric forms of recognition. Recognition takes on a particular asymmetric form if those addressed in normative deliberations are not persons.

Ethics of nature

Morality is a distinctive feature of persons, who in ethical deliberations and reasonings – in a reflective manner – inquire into their attitudes and actions and subject them to normative assessments. Within the ethical community of persons – to which on earth, as far as we know, only human individuals belong – normative orders have been developed and defined. For epistemological and moral reasons, persons have a distinguished position: they have moral consciousness, and they are the addressees of high ethical protection as expressed in human rights as well as in the formula of persons as ends in themselves and in the principle of non-instrumentalisation (cf. Kant 1999: IV 429; cf. Sturma 2004: 174ff.). This makes for a difficult starting point when it comes to extending the scope of ethical recognition beyond that of human individuals.

The ethical community of persons is not entirely dominated by the epistemic and moral status of its members. The scope of ethical acknowledgement is certainly not restricted to *face-to-face* relationships. People commonly accept moral obligations with respect to past and future persons or to persons without the capacity for autonomy in the same way as moral obligations with respect to non-natural persons and non-human forms of life (cf. Sturma 2008c: 305ff.).

When considering the addressees of moral obligation, difficult questions arise as to how much the scope *has to be* extended and how much it *can be* extended. In the case of natural conditions, it can be assumed that a categorical normative obligation would not be widely accepted. It could nevertheless be admitted that the so-called rest of the natural world can be a factor in the determination of obligations. In doing so, decisions have to be made on issues of relevance and precedence: *What counts?* and *What counts more than other things?*

Extending the field of addressees of moral recognition and obligations beyond the ethical community of persons involves not only difficult normative considerations, but also a large number of semantic and epistemic challenges. These difficulties are already apparent in the basic concept of the ethics of nature.

No consensus exists to date with respect to the normatively relevant meaning of the term 'nature'. It can be used in different ways with different extensions. For instance, it can designate the entirety of identifiable physical objects and processes, the entirety of the non-human world, the entirety of living beings, or the entirety of the organic world that is not bound up with humanity. There is a philosophical tradition dating from antiquity that assumes a more comprehensive understanding of nature. It takes the human life-form as an integral part of nature, without denying the special status of its subjects in terms of an epistemic, ethical or aesthetic point of view. This position of an integrative naturalism (cf. Sturma 2008b) can be attributed to a wide variety of positions, such as those of Aristotle and McDowell on the one hand and Spinoza and Schelling on the other, but it is not especially prominent as a position in the current debate.[2]

Investigations in environmental ethics are concerned with the descriptive identification and the normative evaluation of the relationship of persons as subjects of the human life-form to nature. The ethics of nature does not enquire into what may be good or bad *in* nature or *about* nature. Rather, it analyses ethically justified limitations to action and thus opens up a methodological access – irrespective of the unclear semantic starting position.

In seeking ethical reasons for imposing limitations to action, the question of the value of nature takes on a decisive role. To date, this question has been approached from two fundamentally different directions: the anthropocentric approach on the one hand and the non-anthropocentric or physiocentric approach on the other hand. The anthropocentric approach may take the form of an epistemological or ethical theory. Whereas ethical anthropocentrism includes, in one way or another, epistemological anthropocentrism, it is possible to argue for an epistemological anthropocentrism without necessarily postulating that the human life-form should have an overriding ethical status. The non-anthropocentric or physiocentric position can be further differentiated into pathocentrism, biocentrism and ecocentrism or radical physiocentrism.[3] In current discussions, the non-anthropocentric approaches have to battle with the difficulty that, so far, there is no well-founded basis for justification, and only little motivational potential available. Therefore, physiocentric theories frequently engage in radical criticism of anthropocentric positions, as well as evocative arguments, without putting much emphasis on semantic clarification and justification. Under the conditions of current ethical standards, references to a transcendental ecosphere have, however, little capacity to motivate or to convince.

Ethics deals with descriptive identifications and normative evaluations of phenomena, institutions and behaviour in the social realm. In the case of the ethics of nature, asymmetric forms of recognition have to be established, which then allow for justifiable connections between descriptive identifications and normative evaluations. In this task, the problem of the naturalistic fallacy – which the ethics of nature must come to terms with in all of its fields – must be avoided.

David Hume[4] has already pointed out the dangers of uncritically inferring normative sentences from descriptive sentences. Contrary to what is often assumed, he does not assert that descriptive and normative sentences can never be connected to each other. If such a relation is to be established then this requires a particular kind of explanation and specific methodological moves.

With respect to argument control for ethical argumentations and justifications, a critical attitude to naturalistic fallacies is of considerable methodological importance. But it does not imply an unbridgeable gap between *is*-sentences and *ought*-sentences or between descriptive and prescriptive propositions. Indirectly, the naturalistic fallacy draws attention to the theoretically and practically precarious place of the human life-form in nature. Human beings share many features with descriptively accessible non-human life-forms and at the same time have epistemically, practically and normatively a special status in nature as we know it.

Criticism of the naturalistic fallacy is an important instrument of argumentative control in dealing with descriptive and normative propositions. Such criticism makes evident that ethical evaluations cannot be derived from descriptive identifications. In this regard, objections to arguments that take the form of the naturalistic fallacy are justified. Nevertheless, there is always the possibility of taking an evaluative attitude toward facts or descriptions. The way in which we react normatively to situations and circumstances is an issue of justification at all times. There are normative or ethical reasonings that can be identified as better or worse reactions to descriptions. In this respect, descriptions can provide grounds for applying normative evaluations.

Taking seriously the problem of the naturalistic fallacy can prevent both conceptual dogmatism as well as ontological reifications. This does not mean that descriptive and normative propositions cannot be used in the same argumentative context under any circumstances. In the case of the ethics of nature, the criticism suggests an extended framework, which would indirectly dispense with narrow semantic and methodological restrictions. This theoretical framework includes, among other aspects, the critique of exploitative behaviour, and the critique of narrow or one-sided conceptions of human benefits. Not least, it will gain ground for justifiable self-interest under the condition of limited resources.

Biodiversity

The normative emphasis on biodiversity was originally a project of non-anthropocentric approaches. In recent years, it has advanced to the status of a basic bioethical problem. The semantic profile of the expression 'biodiversity' is not always clearly defined in politics or in discussions of environmental ethics. In particular, its descriptive and normative components are often not sharply delineated. The expression is used both for capturing the wide variety of genetic material, species and ecosystems, and also for defining normative guidelines to protect species diversity. Furthermore, demands for species protection always include value judgements and prioritisations, which are often not specifically justified. It is generally presupposed, for example, that microorganisms dangerous to humans are to be excluded from any considerations about protecting biodiversity. Plausible though it may appear, no argument is generally developed that could provide a reason for such exclusion apart from human interests. Thus demands for species preservation are accompanied, at least tacitly, by projections based on sympathy or expectations of benefit.

The idea of variety contained within the term 'biodiversity' can refer equally to species, genetic blueprints, biotopes or whole ecosystems. 'Biodiversity' is a relational concept encompassing the quantity, quality or distribution of individuals and species in a local environment. To this extent, it is a measure rather than a clearly identifiable object of attributions. A statement about biodiversity is something to which one then has to respond in a certain way. In practical terms, this means that not simply *something* has to be protected but something in a *certain state* has to be protected.

From the epistemological point of view, biodiversity turns out to be elusive. In view of the large number of relational determinations and dependencies that are always implicitly involved whenever statements are made in connection with cases of biodiversity, exhaustive knowledge of all of the interactions and consequences would be needed – which obviously goes beyond the epistemic capabilities and the technical options employed by persons.[5]

Irrespective of the question of the part played by humans in changing the environment, epistemic and epistemological issues represent a considerable obstacle to dealing with biodiversity both in a theoretical and practical context. The interactions between human beings and the environment are both far-reaching and complex. The effects of human interventions on the environment cannot be fully identified with respect to their impact on the totality of the natural interrelations of life, especially since it is not even known within what variations natural processes actually take place 'naturally'. In order to make a reliable determination of the impact of human activities on nature or the environment, comprehensive insights would be required into the systematic relations of natural processes – which is too great an epistemic challenge to current scientific approaches. Descriptive and normative analyses of diversity must therefore always be carried out under the condition of epistemic uncertainty.

The opposition between anthropocentric and non-anthropocentric approaches becomes apparent within the framework of motivational considerations and considerations of the theoretical justification for the preservation of biodiversity. This results in inverse perspectives. The anthropocentric approach, which directly links the preservation of biodiversity to human interests, has a promising motivational starting point at its disposal. However, this means that an ethical stance is taken which is regarded as controversial in environmental ethics due to its narrow concepts of human interests and benefits. The non-anthropocentric approach does not unquestioningly accord ostensible human benefits priority in matters of ethical assessment, and can therefore lay claim to a greater degree of consideration towards the ethical addressees, since it treats nature as an independent source of normative obligations. Though due to the subordination of human interests to what is regarded as intrinsically valuable in nature, it can be objected that this approach does not have a justifiable epistemological basis and that it, moreover, leads to implausible and excessive demands.

If biodiversity is to become the subject of normative issues, then, for methodological reasons, use must be made of asymmetric recognitions. Ethical consideration can only be formulated from the perspective of the ethical community of persons capable of taking part in the normative discourse. The normatively challenging task consists in establishing a justifiable relationship within this community between the interests of its members and the normative status of non-persons.

The most advanced expressions of intrinsic value can be found in the formula of the end in itself and the prohibition of instrumentalisation. According to these principles, persons as *rational nature* (cf. Kant 1999: IV 429) exist as ends in themselves and may never be treated merely as means. In ethics of nature the attribution of intrinsic values cannot be restricted to persons, otherwise the argumentation would not get off the ground. The ethical implementation of the term 'intrinsic value' can take on a strong or a weak meaning: either as applying to the essence of an object, or as treating an object as a convergence point for the application of the principle of non-instrumentalisation. But the idea of intrinsic value is encumbered with difficulties: it cannot be grounded epistemologically and attributions of intrinsic value often take the form of the naturalistic fallacy. Thus only in the case of the weak meaning of the concept of intrinsic value is there a favourable perspective for ethical justification.

Asymmetric acknowledgement

Naturalistic extensions are not groundless projections. From an ethical point of view, they share, at least to a certain extent, the way into determining the ontological contexts of the human life-form. For a long time during their lives its members are under normal conditions *subjects as persons* (cf. Sturma 2007). They inhabit the *one* world in which the entirety of the system of nature unfolds. But the fact of co-presence does not, as such, supply any ethical argument – this is what must be learned from the problem of the naturalistic fallacy.

The naturalistic extensions are normative interpretations of the fact that the human life-form remains an integral part of the system of nature irrespective of its far-reaching potential for reflection, knowledge and intervention. The methodological problem inherent in these extensions arises from the successive decline of the reference points for normative recognitions. If the boundaries of the ethical community of persons are crossed, then the contours of identifiable addressees of moral obligations gradually lose contours – as if moving along the great chain of being (cf. Lovejoy 2001). They simply fade away one by one. The indicators for ethical recognition are then no longer the characteristics and capabilities of living organisms but rather the possibilities for the development of natural processes, scarcity of resources or criticism of exploitative behaviour.

Within the framework of ethical extensions into the realm of nature, the methodological point to be borne in mind is that in normative evaluation the burden of proof should be handled in such a way as to avoid a one-sided speciesism. If the possibility of a significant scope for development and flourishing cannot be excluded, the implementation of ethical protection has to be an option. As a consequence of this, a re-evaluation of human self-interest has to be seriously considered. In the case of environmental damage, it has to be shown that non-human developments are not curbed without good reasons, and actions are responding to exceptionally challenging situations not taken without due consideration.

The ontological commitments that the ethics of nature needs to address find their expression in the form of a non-eliminative naturalism in which the human life-form is taken as an integral part of the system of nature. Such an integrative naturalism (cf. Sturma 2008b) is based on the unity of reality and the unity of science (cf. Rescher 1984). It rejects supranaturalism just as much as it rejects a simple compatibilism which allows for quasi-scientific assertions contradicting methodologically well-founded assumptions in other disciplines.

From the normative point of view, the conceptual approach of integrative naturalism has far-reaching consequences. It excludes both ruthless anthropocentrism and an indifferentialism that does not admit normative differentiation in the treatment of highly developed ecosystems and forms of life. Integrative naturalism takes account both of the natural contexts of human life as well as the special significance of consciousness and self-awareness. Just like everything else that is the case, the human life-form is part of the natural world. However, due to its intellectual and scientific achievements, it takes a reflective perspective in nature. Leaving the individual standpoint aside, one could say that in the human life-form, nature is referring to itself, or, in other words, the human life-form is the self-reference of nature (cf. Schelling 1857).

The capabilities of the human life-form do not come with a licence for arbitrary intervention in nature. The special epistemic status is by no means restricted to knowledge of what is the case, but rather also includes knowledge of what ought to be done. Epistemic abilities and normative obligations cannot be separated from each other.

For reasons of motivation and justification, naturalistic extensions and asymmetric recognitions must begin with the human life-form. The paradigm for such a justification is given in the normative treatment of issues relating to the problem of other minds in general and to other forms of life in particular. In accordance with this, animal ethics can be seen as providing a normative bridge to the rest of nature. From an evolutionary point of view, the great apes are closely associated with humanity. For this reason, they share a number of properties and capabilities with human beings, including social skills, but they lack the capacity to express moral or ethical standards and to establish ethical institutions. They remain in the narrow field of their needs and interests. From the perspective of the human life-form, this fact evokes asymmetric recognitions. Beyond the boundaries of the ethical community of persons, there are no basic normative concepts such as justice, equality or reciprocity. They must be compensated for by provisions such as protection or care. These provisions display the special feature of being applicable not only in the asymmetric recognition of persons with limited capabilities but also in the asymmetric recognition of non-persons.

The ethical protection of biodiversity

In the case of ethical protection of biodiversity, asymmetric recognition must do without any clearly identifiable addressees as well as without the overlap of at least some properties or capabilities – apart from basic biological functions. This is why justifications are much more difficult to develop than, for example, the application of asymmetric recognitions in animal ethics. Ethical protection of biodiversity is not a matter of establishing norms for the treatment of living beings, but for securing the conditions of local situations within some degree of variation. It is not the local situation as such that is to be protected, but rather the situation in a specific state, which can only be grasped in approximation.

In the case of biodiversity, the epistemic boundaries of environmental ethics become very obvious, for the relevant conditions and dependencies cannot be limited in their extension to a closed realm or a local area, nor can their long-term impact be clearly defined. The ethical preservation of biodiversity must therefore take as its starting point the critical analysis of human behaviour with regard to nature. Even if no obligations towards nature can be derived directly from nature itself, there are good reasons for rejecting ruthless and exploitative behaviour (cf. Sturma 2008a: 51ff.).

Exploitative behaviour is under any circumstances an unjustified practice. The fact that resources are easily or cheaply accessible is not a justification for them to be thoughtlessly consumed – especially as low extraction and processing costs are often associated with socially unacceptable working conditions. In calculating the economic cost of a practice, the social and ecological burdens must also be taken into account. So far, exploitative behaviour is too often encouraged by the fact that responsibility for the economic and ecological violations associated with it is often not borne by the persons who caused it.

The systematic core of criticism of exploitative behaviour is to be found in the clarification as to what constitutes reasonable self-interest and social justice. Self-interest has far-reaching epistemic and normative implications. It is not simply a matter of immediate preferences; evaluations of self-interest have to consider extended time frames. These temporal perspectives in epistemic and normative evaluations make it necessary to establish relations to the different stages of one's own life and the perspectives of past, present and future persons as well as to nature, in which the human life-form is embedded. Only if these relations and dependencies are considered can self-interest claim to be well founded (cf. Sen 1977; cf. Sturma 2008c: 316ff.) – this applies equally to the standpoint of the individual person and the social or cultural standpoint.

Persons have the ability to behave in accordance with reasons. They are receptive to moral considerations, which places them in a position to critically evaluate the presumed benefit of individuals or groups of persons. They are thus – at least in principle – epistemically and practically open to the fact that there are always good reasons for protecting the natural environment or at least not subjecting it to thoughtless behaviour. However, this form of self-limitation does not mean that resources are to be passed on to later generations largely untouched. Access is always possible in cases of emergency or if well-founded self-interest in using certain resources actually exists.

From the point of view of the ethics of nature, the criticism of exploitative behaviour is mainly concerned with the fact that it misappropriates the physical conditions and the quality of the resources it uses. This disregard also has repercussions on the human life-form. In response, bioethical enlargements can be oriented formally on three provisions – on the usage of the term 'good', on the comparability with other well-founded applications, and on identifying a perspective or a point of view.

According to the criteria of introducing norms into the ethics of nature, recognition can be established if it is possible to express better or worse conditions for addressees and thus to assign a good to them. This does not imply that the addressee should have explicit consciousness or feelings. In the case of the preservation of biodiversity, it would be sufficient to demonstrate that it would be better if an ecosystem were to be in a very specific state – and not in another state.

The better or worse conditions must, furthermore, be comparable with goods that are ascribed to other addressees in a normatively appropriate manner. Such comparability is not necessarily restricted to persons. As a condition for comparison, it is sufficient to take life-forms or ecosystems which, for good reasons, are already regarded as a starting point for normative obligations.

Finally, with respect to the addressee, it must be possible to identify a perspective or a standpoint in the sense of an intentional correlate for attributing better or worse states. In the case of the protection of biodiversity, it is sufficient if an organisational unity – as it were, a quasi-standpoint – could be identified, which could also function as an identifiable framework for the corresponding descriptive and normative provisions.

If the ethics of nature is to include asymmetric recognition, the semantics of the expression 'good' are not allowed to be reduced to the predicate 'good for us'. The various usages for the term 'good' mean that, in each case, the semantics depend on the characteristics and capacities to which the attribution refers (cf. Foot 2001: 25ff.; Thompson 2008: 63ff.). Otherwise the intended extension would merely amount to calculated self-interest. Incidentally, it does not follow from this semantic restriction that extensions in the form of asymmetric recognition could not also be due to an understandable interest on the part of the human life-form. But for the ethics of nature, it is of decisive importance that ethical norms are not bound to anthropocentric interests.

It is indisputable that extensions in accordance with the ethics of nature fundamentally involve epistemic difficulties. This situation is, however, not a sufficient justification for retaining the current practice of dealing with nature in general and biodiversity in particular. Simple self-interest and sheer ruthlessness are sure signs of an unjustified practice, which must be replaced by revising our attitude toward nature such that we would be able to establish asymmetric recognition. The burden of proof for ethical obligations is to be designed in such a way that ethical protection comes in, irrespective of epistemic uncertainties, whenever the applicability of the predicate 'good', comparability with other applications, and a quasi-standpoint are present.

Normative statements concerning biodiversity will always be made from a certain point of view that is characterised on the one hand by narrow epistemic boundaries and by specific interests on the other. The decision as to whether the development of a certain ecosystem is better or worse must unavoidably be made from the outside or from a distance. In this context, the practical difficulty is that criteria such as quality, quantity or distribution cannot be implemented in individual cases in such a way that a reliable decision could be made on what is good for the human life-form, good for life-forms overall, and good for individual persons or individual things affected by biodiversity.

The epistemological difficulties do not represent a reason for treating biodiversity as if it were without any normative content. The need for normative protection is the outcome of the criticism of exploitative behaviour as well as of asymmetrical recognitions. Accordingly, it is not good for the human life-form to pursue aims that lead to exploitative behaviour or short-term benefits.

Due to the large number of epistemic difficulties in generating a reliable decision on the preservation of biodiversity, recourse should initially be taken to indirect applications. Thorough criticism of exploitative behaviour for obvious economic reasons would identify limitations which would have a positive effect on the preservation of biodiversity. Exploitative behaviour does not exist wherever humans treat the environment in a sustainable manner. Conversely, it can also be said that considerate behaviour towards nature corresponds to a lack of exploitative behaviour. Sustainability and exploitative behaviour are apparently mutually exclusive. It seems that what is good for the *justifiable* development of the human life-form also proves to be good for nature as a whole in the long run. The options for behaviour also include the possibility of specifically maintaining ecosystems and further developing them with caution. This ability distinguishes the human life-form from all other life-forms.

The human life-form is an integral part of natural processes, to which it can and should make its specific contribution in accordance with its distinguished capabilities. In particular, it disposes of epistemic and moral capacities of a sort that we have not found in other forms of life. If persons behave in a manner corresponding to their epistemic and moral potentials then no unbridgeable gap should emerge between persons and non-persons or culture and nature. Irrespective of answers to the difficult questions of cause and effect, it is apparent that disturbed natural conditions and violations of social justice go hand in hand. That is why an attempt to secure the preservation of biodiversity on simply economic grounds falls short of the mark. These approaches are part of the problem syndrome, not its solution.

The consequences of investigating the justifiability of human behaviours may be expected to consist of normative limitations to and restrictions of the scope of human activity. In the case of biodiversity, the criterion of the justification of human behaviour towards nature involves both fundamental and concrete questions of why damage to ecosystems could not be avoided. In normative terms, this question leads inevitably to the demand for giving justifiable reasons for particular actions. Thoughtlessness, ruthlessness or short-term benefits always stand in contradiction to justifiable practice. Critical examinations and revisions which meet the standards of ethical justification will lead to normative restrictions on the options of human activities. These restrictions will, at least to some extent, suggest a withdrawal from certain habitats.

Notes

1 The term 'system of nature' refers to the interdependencies of forms of existence and the coherence of the various scientific nomologies; for the ethical use of the term cf. Taylor 1986: 116ff.
2 Schelling, in particular, took great pains in developing his early natural philosophy to join up nature and subjectivity on a developmental level. In his natural philosophy of the mind, he assigns both the ideality and materiality of the world to be properties of nature. According to this view, nature is visible mind and mind is invisible nature. This means, in his view, that nature is the logical and historical precondition for subjectivity, or human life, that has developed from originally non-living processes into life and consciousness, cf. Schelling 1857: 38ff.; Sturma 2008b.
3 The differences between these positions are sometimes hard to discern, cf. Taylor 1986: 117: 'When one accepts the biocentric outlook, the whole realm of life is understood to exemplify a vast complex of relationships of interdependence similar to that found in each ecosystem. All the different ecosystems that make up the Earth's biosphere fit together in such a way that if one is radically changed or totally destroyed, an adjustment takes place in others and the whole structure undergoes a certain shift. The biological order of the planet does not remain exactly what it was.'
4 See Hume 1978: 469 (III. 1. 1): 'In every system of morality which I have hitherto met with I have always remark'd, that the author proceeds for some time in the ordinary way of reasoning, and establishes the being of a God, or makes observations concerning human affairs; when of a sudden I am surpriz'd to find, that instead of the usual copulations of propositions, is, and is not, I meet with no proposition that is not connected with an ought, or an ought not. This change is imperceptible; but is, however, of the last consequence. For as this ought, or ought not, expresses some new relation or affirmation, 'tis necessary that it shou'd be observ'd and explain'd; and at the same time that a reason should be given, for what seems altogether inconceivable, how this new relation can be a deduction from others, which are entirely different from it.'
5 This was pointed out by Rousseau in his criticism of Voltaire's poem on the Lisbon earthquake, cf. Rousseau 1969; cf. Sturma 2011: 155ff.

References

Foot, P. (2001) *Natural goodness*, New York: Oxford University Press.
Hume, D. (1978) *A treatise of human nature*, ed. L.A. Selby-Bigge, 2nd edn, New York: Oxford University Press.
Kant, I. (1999) *Grundlegung zur Metaphysik der Sitten*, ed. B. Kraft and D. Schönecker, Hamburg: Meiner.
Lovejoy, A.O. (2001) *The great chain of being. A study of the history of an idea*, Cambridge, MA: Harvard University Press.
Rescher, N. (1984) *The limits of science*, Berkeley, CA: University of California Press.

Rousseau, J.-J. (1969) 'Lettre à Monsieur de Voltaire', in B. Gagnebin and M. Raymond (eds) *Jean-Jacques Rousseau, Œuvres complètes*, vol. 4, Paris: Gallimard.

Schelling, F.W.J. (1857) 'Ideen zu einer Philosophie der Natur', in K.F.A. Schelling (ed.) *Friedrich Wilhelm Joseph von Schellings sämmtliche Werke*, vol. 2, Stuttgart: Cotta.

Sen, A. (1977) 'Rational Fools: A Critique of the Behavioral Foundations of Economic Theory', *Philosophy and Public Affairs* 6: 317–44.

Sturma, D. (2004) 'Kants Ethik der Autonomie', in K. Ameriks and D. Sturma (eds) *Kants Ethik*, Paderborn: Mentis, 7–16.

Sturma, D. (2007) 'Person as Subject', *Journal of Consciousness Studies* 14: 77–100.

Sturma, D. (2008a) 'Die Gegenwart der Langzeitverantwortung', in C.F. Gethmann and J. Mittelstraß (eds) *Langzeitverantwortung. Ethik – Technik – Ökologie*, Darmstadt: Wissenschaftliche Buchgesellschaft, 40–57.

Sturma, D. (2008b) 'Die Natur der Freiheit. Integrativer Naturalismus in der theoretischen und praktischen Philosophie', *Philosophisches Jahrbuch* 115: 385–96.

Sturma, D. (2008c) *Philosophie der Person. Die Selbstverhältnisse von Subjektivität und Moralität*, Paderborn: Mentis.

Sturma, D. (2011) 'Rousseau', in T.-L. Eissa and S.L. Sorgner (eds) *Geschichte der Bioethik*, Paderborn: Mentis, 151–63.

Taylor, P.W. (1986) *Respect for nature. A theory of environmental ethics*, Princeton, NJ: Princeton University Press.

Thompson, M. (2008) *Life and action: elementary structures of practice and practical thought*, Cambridge, MA: Harvard University Press.

7 The values of biodiversity

Philosophical considerations connecting theory and practice

Thomas Potthast

Introduction

The appreciation and ensuing call for protection of the life forms beyond – or besides – humans has a complex cultural history, which is deeply embedded in religious and philosophical as well as scientific and mundane contexts (e.g. Meyer-Abich 1997; Worster 1995). From the beginning of the 1990s, the term "biodiversity" has become the umbrella concept for most of the issues dealt with before under the rubrics of, among others, "nature protection" and "nature conservation" (Myers 1979), but with decisive new implications (Potthast 1996; Takacs 1996; see also below). With the United Nations' "Convention on Biological Diversity" (CBD; UN 1992), biodiversity became an accepted good to be protected in terms of international law and its national implementations, as well. But, philosophically speaking, what kinds of values are being attached to the multiplicity of life forms? From the perspective of moral philosophy but also of epistemology, this chapter will explore the diverse notions of values related to biodiversity. Firstly, conceptual clarifications will be made regarding the twin terms ethics/morals and value/price. Then the invention of the term "biodiversity" as a science-policy concept in the context of the Convention on Biological Diversity is sketched and its status will be characterised as an *epistemic-moral hybrid*, which is to be explained in some detail. The diversity of values attached to biodiversity will be analysed with regard to three important general issues of moral philosophy: a) the current situation and groundwork for a future possible axiology of values, b) the contested physiocentric extension of the moral community, and c) the limits and problems of economic valorisation depending on different understandings of "value". It shall be shown that philosophical issues have major repercussions on practical questions. An open concept of values, although advantageous in many ways, appears to have two major shortcomings: i) The reduction of values to individual preferences only makes sense for a restricted neoclassic economic and utilitarian ethics approach to moral issues; it neglects all aspects of biodiversity beyond preferences and might foster an alienation in human-nature relationships by commodification and monetarisation. ii) The opaqueness of the notion of values does not give criteria for weighing up in order to guide developing moral rules and specific decisions. However, the multiple values of biodiversity can at the same time be regarded as providing common moral ground – despite differences in detail – for joint environmental decisions and governance. Furthermore, they can serve as a starting point to once again address questions of axiology beyond preference aggregation, which have been neglected by moral philosophy ever since the metaphysics-based attempts had been discarded. Investigating the values of biodiversity elucidates the necessary link between practical activities to protect the diminishing multiplicity of life forms and philosophical clarification as to how and why we can and should give good reasons for such an endeavour.

Philosophical clarifications on morals and ethics, and on values and prices

Morals and ethics

In science, business and politics, a growing need for – or at least appeal to – "ethics" is noted. But more often than not, it is rather unclear what exactly is meant by this term. In addition, morals and ethics are used almost interchangeably in English, and to an increasing degree in German as well. Facing this situation, I will suggest the following usage of terms in order to differentiate various levels of discourse.

By "morals", I shall refer to individual and/or collective ideas about the Good (and the good life, *eudaimonia*) as well as the Right (norms and principles for action). In this vein, the morally good or right means that which is to be done for its own sake, and not (solely) in an instrumental perspective for other purposes. For example, being a just person or acting altruistically in favour of the good life of others or respecting human dignity would be good in itself, although it might be good for the functioning of social groups, too. Even if the latter purely functional perspective of moral sociology or moral psychology is a valid description, this does not explain away the other, the normative, dimension: we act in accord with moral rules or virtues because we judge them as being good in themselves and/or leading to morally favourable consequences. This is rather complementary to functional explanations.

Many, but by far not all, scholars would understand "ethics", then, as the philosophical theory of morals. Ethics consists of different domains. 1) The part of descriptive ethics analyses existing morals and moral systems and their historical and current contexts; this domain has strong links to moral sociology, moral psychology and the like. 2) Metaethics comprises the philosophical, among other things the logical, analysis of moral language and moral theories. It focuses mainly on the structure, on coherence and consistence of moral theories. 3) Normative ethics (moral philosophy s.str.), then, is the domain for establishing and justifying certain moral systems, principles, norms, virtues or values. Single judgements on the moral acceptability of a specific case are part of normative ethics, as well (for an overview on the field cf. Düwell et al. 2011). In the mid twentieth century, normative ethics had almost disappeared, but due to many contested issues in medicine and modern technologies, application-oriented normative ethics returned to academic philosophy (cf. Toulmin 1982).

Understanding ethics as the systematic philosophical reflection of morals implies that morals and ethics are not the same – despite the current usual usage of terms. However, the separation is not absolute (not binary) but gradual: everyone also reflects – more or less – while enacting morals! And there is another consequence: moral philosophers are experts for ethics, i.e. moral theory, not for morals per se. Hence, moral questions cannot be delegated to philosophical experts alone but remain to be reflected by every person.

Values and prices

One of the major terms used in current application-oriented ethics as well as in politics, is "value". In the philosophical tradition, especially since Immanuel Kant, there is a major distinction between two kinds of value. On the one hand, a value can be understood as an attribute of (positive) significance of something for someone. On the other hand, in moral philosophy there is the concept of an absolute value. For Kant only human beings have an absolute value, which designates what is also understood as the dignity of persons (Kant 1991: BA78). Departing from this distinction between value in general and absolute value, another notion, that of "price", comes into play. A price designates an exchange value, which ultimately – but not necessarily – could be expressed in monetary terms, since money would appear as the universal exchange unit. It should be noted that there are fundamental differences between value, absolute value and price with regard to their calculability, quantifiability, exchangeability and substitutability. In short: comparing – but not equating – the values of apples and pears with regard to their quantitative and qualitative traits is fine. But only if they have a price, they can be exchanged and substituted on the basis of a common denominator. The notion of an absolute value of fruits would make no sense to us – maybe with the exception of cases like the biblical apple of Adam and Eve.

Very often, the following scene from Oscar Wilde's play "Lady Windermere's Fan" (Wilde 1892/2011) is used to illustrate the distinction between value and price:

> CECIL GRAHAM:. What is a cynic? [Sitting on the back of the sofa.]
>
> LORD DARLINGTON: A man who knows the price of everything and the value of nothing.

The critical punch-line of this quote is rather clear. However, there is an immediate continuation of the dialogue, which is hardly ever reported:

> CECIL GRAHAM: And a sentimentalist, my dear Darlington, is a man who sees an absurd value in everything, and doesn't know the market price of any single thing.

In a nutshell, this conversation depicts what is at stake also in the political and ethical debates on values of biodiversity, as will be explicated in the following sections.

In philosophy, a further complex and often contested relation is that between values, norms and actions. The various (philosophical) traditions and perspectives regarding this issue are far from unanimous. According to the view adopted here, a value is *not a norm* nor does it immediately lead to the prescription of specific actions. While a norm entails deontic operators like "should (not)", "must (not)", "ought (not)", etc., a value does no such thing. Nonetheless, values do *influence* wishes, interests, preferences and even norms in a complex way. Therefore, values ultimately *enable* actions without directly *demanding* them.

Values have a binding force for people. According to Hans Joas (1999), this entails a seeming paradox: people feel bound to values and by values; but at the same time, they are more or less free to act at odds with their values, or to transform their values. It has to be noted in this context that, of course, not all values are *moral* values in a strict sense. Some values have their primary relevance in e.g. a technical or an aesthetic sense. Moral values s. str. relate to the morally good and/or the highest moral good (see above).

From a metaethical point of view, it has to be discussed what values are, precisely, and how they are generated. Three major positions can be distinguished. 1) Value idealism regards values as ideal validities in a non-empirical space (Plato's *kosmos noetos*) which have to be viewed and contemplated intuitively. 2) Value realism assumes that values are empirical traits of objects and systems which have to be discovered and recognised (cf. Jonas 1979). 3) Value individualism considers values as individual (not collective) valuations which have to be generated and/or attached by individuals *only*, shared values in collectives of individuals notwithstanding. Value idealism and value realism come at some costs in terms of ontological premises; value individualism seems to avoid these problems and appears to be an apt perspective for morally pluralist liberal theories and societies. However, value individualism does not leave much space for acknowledgement of values beyond an individual's immediate – and maybe ideologically modulated – valuations. In this vein, there is a tendency to equate values with (mere) preferences for the individual.

In an ethical taxonomy, there are three types of values: use value, existence value and moral value s. str. *Use values* (= instrumental values = functional values) apply to entities which can be used as a resource for human economic or social purposes. To give a mundane example: for persons drinking tea, a teapot has an instrumental value in order to prepare and keep the hot fluid. If the teapot cracks the use value disappears. *Existence values* (= intrinsic values = eudaimonistic values) constitute a value in itself beyond immediate use. Such values are formed in the specific relation between humans and the valued entity in question. In such cases, no immediate functional substitution is possible. For example, the teapot which my grandmother used is of intrinsic value to me because it reminds me of my childhood visits to her. I will keep and value the teapot despite a crack, due to which I can no longer use it for preparing tea.

However, the separation between use values and existence values is contested, because especially utilitarians and neoclassic economists would claim that the existence value is also of use for the valuing person, because the old cracked teapot functions as an instrument for generating good feelings. Despite this critique, the case example mentioned shows an important difference in the process of how values come about. Finally, *moral values s. str.* (= inherent values) are values of direct moral significance and obligation, completely independent from relations to humans. This moral value s. str. is an absolute value as mentioned above. It cannot be withdrawn by a valuing person but has to be acknowledged, because moral significance does not come from the valuing person but lies in the object itself. It is very important to note that only moral values s. str. can by no means be included in economic valuations, because the latter rely on human valuations. If someone or something has a moral value s. str., s/he or it is part of the moral community and hence there is a duty for moral actors to *directly* morally consider them. In contrast, moral duties with regard to use values and existence values exist only to the valuing persons (and not to the objects). This debate on moral values s. str. and on who – or what – is to be included into the moral community is an unresolved contested issue in environmental ethics (cf. Eser & Potthast 1999; Ott 2008).

Notwithstanding different positions, even the use of terms applied to the different types of values is often rather messy. Therefore, it is advisable to clearly point out the meaning and content of the term one uses, as e.g. intrinsic values may sometimes be referred to as inherent values and vice versa by different authors:

- *Value beyond (immediate) human use* = Eigenwert = existence value = intrinsic value (but sometimes also = inherent or extrinsic value).
- *Value in and for itself; value of direct moral obligation* = Selbstwert (but sometimes also Eigenwert) = moral value s.str. = inherent value (but sometimes also = extrinsic or even = intrinsic value).

I shall return to these different notions of values with regard to biodiversity in the following sections.

The invention of biodiversity as an epistemic-moral hybrid term

The 1992 United Nations Conference on Environment and Development in Rio de Janeiro/Brazil was a landmark event towards international acknowledgement of the need to maintain globally threatened biodiversity. A multiplicity of different groups and institutions claimed their parts in worldwide preservation efforts: biologists from taxonomy and systematics, ecology, evolutionary biology, plant genetics and also molecular biology; non-governmental environmental organisations, multinational trusts of industry, crop-breeders, developmental organisations and business managers. Both countries from the global South and industrialised countries negotiated under the flag of the United Nations. One of the ensuing documents was the "Convention on Biological Diversity" (CBD; UN 1992), which was signed by almost all countries of the world and – some years later – even by the United States of America. The term "biological diversity" and even more the short version of "biodiversity" came to be the key concept for international and, as a consequence of legal implementation, for national protection of biota and ecosystems. This career of a concept can be traced back to an initiative of international, mainly American biologists and conservation scientists organising a "National Forum on BioDiversity" in Washington, D.C. in 1986. Sponsored by the National Academy of Sciences and the Smithsonian Institution, the loss of species and habitats, particularly in the tropical rainforests of Central and South America, was discussed and the explicit aim was mobilising both public and political interest and support. Takacs (1996) has reconstructed the history of this conference and the coining of the term in detail. Biologist Dan Janzen noted: "The Washington Conference? That was an explicit political event", and Paul Ehrlich, a leading ecologist, said: "I'll tell you that it's not a scientific argument. One of the silly things is the idea that science is somehow separate from society. There is no value-neutral science". And according to Walter Rosen, main organiser, it was a decisive measure to take the 'logical' out of biological diversity for the publication of the widely received conference contribution (cf. Takacs 1996).

Bio-scientifically speaking, the term biodiversity refers to variability on all levels of biological hierarchies, often highlighted as gene – species – community/ecosystem. However, in a socio-political context, "biodiversity" has come to epitomise *endangered* life on earth. Thus, the term biodiversity quickly became a powerful metaphor in the conservation discourse between science and the public (Väliverronen 1998). It was introduced as a governance concept of nature, invented by conservation biologists and taken up for politics. In that sense, "biodiversity" implies a conservation and safe-guard element in its core meaning. Talking about biodiversity means discussing *endangered* diversity and the need to protect it. This link between a scientific and a political dimension could be dubbed an *epistemic-moral hybrid* (Potthast 1996; 2007).

Epistemic-moral hybrids are historically contingent specific blends of scientific practices and normative agendas. They comprise a merging of practical actions, concepts, implicit and explicit agendas of protagonists and the epistemological and moral substructure which emerges. Note, however, that this kind of account stands in sharp contrast to some philosophical groundwork. Any strict logical inference of normative propositions from purely factual ones is forbidden because of the "Is-Ought Fallacy", the "Naturalistic Fallacy", and the "Genetic (= historical) Fallacy". However, beyond those rather simple explicit fallacies and thus – hopefully – beyond "science wars" and "Werturteilsstreit", epistemic-moral hybrids reveal intricate connections of epistemology and morals within a complex field of biology, politics and ethics, as in the case of biodiversity. Systematic juxtapositions of science and morals notwithstanding, this approach shall allow us to investigate the link between contested fields of epistemological reflections, philosophical boundary building, practical as well as theoretical blurs, and different ontological agendas shaping any ethics of nature in close connection with knowledge from ecology and biology.

Values of biodiversity as an exemplary case of general issues in moral philosophy

In its Preamble, the Convention on Biological Diversity gives a number of reasons for the conservation of biodiversity. Firstly, biodiversity's "intrinsic value" is mentioned. Furthermore, other relevant dimensions of biodiversity for human well-being are listed: ecological, genetic, social, economic, scientific, educational, cultural, recreational and aesthetic. The Preamble mentions the importance of biodiversity for evolution and for maintaining life-sustaining systems of the biosphere, as well. Finally, biodiversity conservation is classed as a common concern for humankind.

In Article 1, the CBD names three goals driven by the above-mentioned values: a) conservation of biological diversity, b) sustainable use of its components, and c) fair and equitable sharing of the benefits arising from the utilisation of genetic resources, e.g. by adequate access to genetic resources and by appropriate transfer of relevant technologies. Thus, with regard to the text and context of the CBD, biodiversity is inextricably linked to and hence considered as a matter of values.

It should be noted at this point that in the CBD, there is no general evaluative difference between natural and anthropogenic diversity, or more broadly speaking, between nature and culture. One could interpret the CBD's explicit mentioning of natural as well as cultured biodiversity in the following way: it is essential to avoid the wrongly constructed opposition of "humans vs. nature" for science, ethics and politics. If this cautionary remark is taken seriously, it has important consequences for the sciences and for conservation: "*Nature does not always do or know best.*", and "*Wild(er)ness is not always best.*" (cf. Potthast 1999; Sarkar 1999).

It is necessary to remember that biodiversity is only a *part* of nature, albeit a major one. Biodiversity includes alleles/genes, species and communities as levels of *biological* hierarchy; however, landscapes and ecosystems belong to a different category. Furthermore, it can be argued that biodiversity is more a perspective than a material object or even a material object cluster. It is not even unequivocal whether biological diversity is mainly a process or mainly a trait. Finally, biodiversity is an epistemic-moral hybrid – nature is not, or at least not necessarily and not always.

Certain ethical principles seem to match the values attached to biodiversity and its conservation. The first refers to *anthropocentric prudence*: it is instrumentally prudent and hence a justified advice (for states and other institutions as well as individuals) to sustain biodiversity and ecosystem functions for human survival. This reasoning implies that both biodiversity itself and its role in providing ecosystem services have essential functions for guaranteeing human survival. A second ethical principle comprises *anthropocentric moral obligations* for justice and basic prerequisites for a good life within and between generations of humans. Biodiversity conservation thus becomes part of a general moral obligation towards the present and future generations. So far, use values and existence values come into play, but only with regard to the moral obligation towards human life on earth. Lastly, there might also be a norm deriving from the *non-anthropocentric (?) "intrinsic value" of biodiversity*. This intrinsic value is neither specified in the CBD nor e.g. in German conservation law (BNatSchG 2009). It seems to imply some moral obligation to protect biodiversity, but the extent of that obligation is far from clear. If we understand "intrinsic value" as an anthropocentric existence value, there are no further conceptual problems. However, the explicit mentioning has also been understood as a moral value s. str. of biodiversity (see below).

Towards an axiology of the values of biodiversity and its deliberation

Leaving the contested issue of "inherent/intrinsic" value aside for the moment, the CBD and most of the biodiversity discourse are linked to anthropocentric arguments. Biodiversity can be understood as a so-called critical (scarce and unrecoverable) good, hence a natural resource to be preserved, which is in the self-interest of humans. There are various instrumental, aesthetic and existence value arguments for the conservation of distinct segments of biodiversity. These arguments are: a) dependence on biodiversity (cf. Tilman 2000), b) an innate positive valuing of biodiversity (biophilia), c) health and well-being of humans, d) aesthetic dimensions (for an extended interpretation of the aesthetic value of species cf. Carter 2010), e) issues of an ethics of place or "Heimat" connected with specific biodiversity elements, f) a transformative value of biodiversity (transforming people in a positive way by experience of biodiversity; could be interpreted in a virtue ethics perspective), g) the value of experiencing various and different environments, often in contrast to urban life ("Kontrasterfahrung"), h) the possibility to have religious or spiritual experiences in/with biodiversity.

In addition to these values, a human right to nature and duties to present and future generations are mentioned. In sum, a collective duty to conserve more or less the entirety of biodiversity can be expressed, unless there are existential reasons not to do so. Hence we have a prima facie duty to protect biodiversity, to be derived from the space of reasons of environmental ethics (cf. Ott 2010; Krebs 1999).

Beyond immediate life-sustaining elements of biodiversity, however, it is far from clear which values count in which situation. The first question is how anyone can have the "right" values if values are solely generated by individuals and their experiences. This would call for an axiology that goes beyond calculating preferences. One way to treat the problem is to show that not only one individual but many share the same values. This would lead to the identification of shared values, but again, will not provide criteria for decision. Deliberative procedures will have to be established in order to explain and critically discuss different ideas about what kind of biodiversity is needed by whom to live a good life.

For arising inner- and inter-value conflicts, a rational solution is required which allows to identify and trade off competing values. Several tasks are relevant: (1) The first task would be the identification of accepted values in the relevant context. It might be possible that different groups of people hold the same value but fill it with different meanings. It seems therefore reasonable not to deal with value conflicts in an abstract way but to consider them in context. Thus, we have to open the black box of "values" and clarify the multiple meanings of its content. In this early stage a rational dialogue about value contents might already be helpful. If sustainability, for instance, is interpreted as a value set, some participants might emphasise subordinated ecological values, whereas others stress economical values. Hence, with regard to the criticised three-pillar model of sustainability, a dialogue might help to prioritise conflicting aims. Here, reference to the basic moral principle of intra- and intergenerational justice appears as the preferred option. (2) As in certain contexts relevant values might clash, the identification of value conflicts would be the next step. (3) Their solution must be provided in continuation of the pragmatic value approach. Therefore, a rational discourse about the trade-off between conflicting values should be launched aiming at a huge consensus among the participants of the value discourse. From consensual (or fair compromise) compliance patterns follows the possibility to contrast and relate values within a value system. For instance, we should point out that individuals need to participate actively in democratic institutions to enable them to achieve a rational ordering of their preferences for collective choices. Certain aspects of a non-violent discourse have to be considered in negotiation processes. Although the results of value identifications and trade-offs may differ between dialogue groups, the participation of value acceptors guarantees the realisation of value-based decisions. In that way, the suggested approach is pragmatically justified by successful application options (cf. Beck et al. 2012).

The contested physiocentric extension of the moral community

One of the most contested issues in environmental ethics is the question who or what is part of the moral community, in other words: to whom, i.e. to which moral objects (patients) do "we" as moral subjects (agents) have a direct moral obligation? The qualifier "direct" is important, because all environmental ethics approaches agree that protection of biodiversity is a duty at least to some extent. But do we have to protect it for the sake of and as a moral obligation only to our fellow humans (anthropocentric ethics) or are we directly morally committed to, at least parts of, non-human nature (several forms of physiocentric ethics) as well? Sentientism or pathocentric ethics means that all sentient beings are part of the moral community, not only humans. Biocentric positions state a direct moral obligation to all living beings, ecocentrism includes units like biological communities or ecosystems, and holism includes all (natural) entities. Acknowledging an inherent value (*moral value s. str.*) of a being or a thing means including it in the moral community.

The major criticism regarding biocentric, ecocentric and holistic positions concerning the respective inclusions of non-sentient beings relates to metaethical and political problems. For example, Wolters (1995) chided the United Nations for setting up a completely new ethics which he sees as absolutely incompatible with all existing anthropocentric ethics. He interprets the "intrinsic value" of the Preamble of the CBD as an inherent value. In Wolters' view, we have no possibility to "prove" such an inherent value because this would require strong metaphysical claims which are not substantiated. This argument is ultimately based on value individualism. Apart from these metaethical problems, the assumption of an inherent value of biodiversity is also seen as politically counterproductive and de-motivating, because humans are completely left out of this moral obligation.

However, the metaethical opposition might simply be going the wrong way. Usually, it is said that either values can be derived only from human valuation or – in all physiocentric approaches – from (non-human) nature or biodiversity itself. *Tertium non datur.* But it should be differentiated here between "anthropogenic/physiogenic" and "anthropocentric/physiocentric": The latter addresses the extent of the moral community as explained above. The former asks where moral significance/values come from. Authors like Wolters state that we could only conceive of values as being anthropogenic. One might put biodiversity into the category of existence values – nevertheless this would still remain in the anthropocentric realm. Yet a second position would be to maintain the anthropogenic origins of all morals but opt for some physiocentric extent of the moral community, because humans may acknowledge an inherent value in, e.g., biodiversity. Such an approach would be anthroporelational but not anthropocentric. Still, this would mean that a bio-diverse world was completely value-free before humans evolved – and will be value-free again once humans will have disappeared. In some sense, I would hold that this is not plausible, or even counter-intuitive.

An alternative – dialectical – approach would start from the idea that the genesis of values could be understood as a co-process between humans and non-human nature. This, however, would mean to acknowledge some moral significance of *natura naturans* (the creative process of nature) in itself, even beyond humans. By open-mindedly looking for the moral dimensions of biodiversity, humans enter into a co-process and encounter aspects of moral significance which are not solely generated by the human side but co-emerging from the non-human living world. In that sense, the diversity of life would have been without moral value (this would be the difference to Hans Jonas' (1979) biocentric account); but ever since humans started interacting with nature, the moral significance has co-evolved in a broad sense – not restricted to biological evolutionary theory but also including cultural histories. Thinking along these lines, the sharp and complete separation between humans and biodiversity, which has been criticised on both empirical and moral grounds, will have to be replaced by a more inclusive approach.

Such a dialectic theory of nature and of environmental ethics cannot be expanded here, but it would transcend the approaches of "extending" the moral community from inside (the Kantian absolute value of rational beings) or outside (holism), as Ott (2008) has suggested. The next step in a dialectical ethics of nature would be to identify moral significance by asking for the forms of interaction, be it material and/or symbolic. Considering biodiversity in that sense would mean taking into account moral significance with regard to natural as well as cultural processes that connect humans and biodiversity – even beyond moral obligations to fellow humans alone.

Economic valorisation of the values of biodiversity

In recent years, a comprehensive economic valuation of biodiversity has been attempted (TEEB 2010; Brock & Xepapadeas 2001; cf. Christie et al. 2006 on difficulties attached to such attempts). Ultimately, these approaches try to calculate what is often called the Total Economic Value (TEV) of biodiversity. This TEV combines a) (all) *direct use values* (food, fuel, fibre,...), b) (all) *indirect use values* (erosion prevention, CO_2–sink,...), c) (all) *option/insurance values* (future pharmaceuticals, genes,...), d) (all) *bequest values* (values for future generations) and e) (all) *existence values* (values of being there). As noted above, economic valuation is always related to human valuations and hence for methodological reasons cannot include moral values in the strict sense. Furthermore, it does not deal with absolute but always with marginal values: state x´ is preferred over state x; there is no context-free (absolute) value of something. Being linked to individual human valuations, TEV is always partial and related to certain interests.

With a polemical intention, Frederic Vester (1983) calculated the economic monetary value of a Bluethroat (*Luscinia svecica*). He gave the bird's cost of biological material at 1.5 Euro cents and its aesthetic value at 5 cents/day (this is the monetary value of a valium pill). Furthermore, the Bluethroat's role in seed dispersal, insect predation, as a partner for symbiosis and an ecological/biological indicator was monetarised. In conclusion, Vester stated the total value of an individual Bluethroat at 154.09€ per year. In a different calculation, but with serious intentions for political communication, Costanza et al. (1997) gave the complete pricing of the world's ecosystem services at 33×10^{12} US$ per year. These authors have been heavily criticised for methodological reasons, but the underlying logic seems to be much less contested. If one could show the economic and monetary value of biodiversity, we would have objective data, even prices to compare (Hampicke et al. 2009). But is this attempt to price biodiversity morally acceptable in the first place?

Part of the problem of economic valuation of biodiversity is based on the fact that there are different concepts and resulting implications of "valuation". The German term "Inwertsetzung" can be identified with three meanings: valuing, valorisation and commodification. Valuing refers to the identification of a temporarily/contingently given value/price. This concept presumably entails no further moral problems. Valorisation is about generating a value, i.e. enabling the valuation of a hitherto non-"value-able" object or process. It is not quite clear whether this concept contains moral problems or not, but the issue is at least debated in conservation ethics. The major reason for critique is that this valuation might show the importance of biodiversity in comparison to other goods. But what if biodiversity is only second in importance? Commodification is the term for the private "enclosure" of formerly common goods. Therefore, commodification is not about generating value but a private capitalisation/monetarisation of hitherto non-monetary values. Hence, this concept of valuation entails the considerable moral problem of shifting one value type to a very different one.

In this context, it becomes evident that there are some moral limits with respect to the values of biodiversity and their economic valuation. Firstly, there are some difficulties regarding the issue of value as price vs. absolute moral value (dignity or moral value s. str.), e.g. with reference to the idea of an absolute value of (each) species. Secondly, it is problematic to deal with un-substitutable biodiversity in terms of (mainly substitutable) functions, as this conceptualisation means losing the perspective of irreplaceability. Thirdly, dealing with values of the invaluable causes further complications. An example for such an invaluable entity is "home landscape" or "Heimat", which gets its meaning precisely from the fact that it has no solely economic or monetary value. Therefore, attaching a price tag results in the loss of values or at least value dimensions.

Economic instruments for biodiversity valuation, especially cost-benefit analyses comparing different states and options, can be excellent political tools, but only if the goals of protection are more or less clear and if there is a given budget which is to be optimised. Nunes & van den Bergh (2001) suggest that available economic valuation estimates should be considered as providing only a very incomplete perspective on the value of biodiversity.

As far as a political economy framing is concerned, it has to be asked whether it can be truly desirable to allow property or patenting of genes/alleles or species/cultivars and property and access and benefit rights to ecosystems. Furthermore, it needs to be considered to what extent biodiversity and ecosystem services *have* to be transformed from commons to private and/or state property – i.e. commodified – in order to facilitate their valuation/valorisation. There is currently an ethical and political debate on this issue.

To take a similar example, no one calculates the monetary costs of the solar energy supply to the earth as a *solar system service*. However, this would start to change as soon as people could actually place shades in the stratosphere (geoengineering) or take other drastic measures. The "being at (human) hand" ("Verfügbarkeit") of nature is a prerequisite for valorisation and monetarisation. Unfortunately, this means the loss of the dimension of "Unverfügbarkeit" (untouchability/unavailability).

As with the above-mentioned geoengineering, there might be a slippery slope in economic valuations of biodiversity: if one starts with valuing, one may slip to valorisation and eventually to commodification. Hence, there is a danger of ending up with economic or monetary one-dimensionality towards biodiversity, i.e. a reduction of the values of biodiversity to ecological utility functions. Therefore, there is a strong need for ethical framings. "Limiting rules" and/or self-restriction against totalising monetarisation should be enforced, e.g. explicit ethical and political limits or frames for cost-benefit analysis. Finally, a new political economy of nature beyond neoclassic economics should be employed. This political economy should incorporate theories of human-nature interactions and valuation beyond the total economic value (even if that is calculated as a "marginal" value in the economic sense) and preferences alone. The heterogeneity of values of biodiversity beyond the market has to be taken into account.

Values of biodiversity – an inclusive approach

In forming an inclusive approach towards the valuation of biodiversity, two metaethical issues seem to be important. 1) The anthropocentric-physiocentric divide of the inclusion debate should be overcome to comprise more integrative models. But this would mean establishing a dialectical *moral* (meta)theory of human-nature interaction as well as being willing to enter ontological debates of philosophy of nature. 2) It should be attempted to get beyond the conceptualisation of values as either realistic or solely individualistically constituted. Ultimately, the interaction of humans and nature creates value dimensions beyond both the moral community and eudaimonistic individual values and virtues.

Policy targets of present (sometimes naïve) conservationism have to be revised not least with regard to the concept of biodiversity framed by the CBD (cf. Potthast 2007). This includes the role of "naturalness" as the main or only focal point for the derivation of values. Naturalness has to be reassessed as an important but not all-encompassing goal and criterion. Species and habitat changes should not per se be viewed as negative in relation to earlier "historical" benchmarking. The evaluation of biological invasions and "alien" taxa has to be revised. At the same time, existing tendencies of uncritically welcoming all change will have to be questioned. Most notably, the targets need to be expanded with regard to human-nature interaction for sustainable development. On the other hand, processes as goals need at least some indication of the pathways and trajectories to be taken, notwithstanding that no fixed goals might be targeted. Nature conservation and sustainability do not overlap completely: the *differentia specifica* of the former lies in some sense of eudaimonistic (good life) and/or intrinsic value of biodiversity not to be covered completely by sustainability. "Good change" (cf. Potthast 2012) shall thus provide input for more encompassing notions of sustainable development with perspectives on biodiversity reaching beyond ecosystem services and other functional approaches.

Investigating the values of biodiversity elucidates the necessary link between practical activities to protect the diminishing multiplicity of life forms and philosophical clarification as to how and why we give good reasons to such an endeavour. Although the variety of meanings sometimes appears to be a problem, the diversity of values of biodiversity poses productive opportunities for both – political action and philosophical reflection.

Acknowledgement

I would like to thank my colleague Margarita Berg for productive discussions on earlier drafts of this paper and for linguistic improvement.

References

Beck, R., Meisch, S. and Potthast, T. (2012) 'The value(s) of sustainability within a pragmatically justified theory of values: Considerations in the context of climate change', in T. Potthast and S. Meisch (eds) *Climate change and sustainable development: Ethical perspectives on land use and food production*, Wageningen: Wageningen Academic Publishers, 49–54.

BNatSchG – *Gesetz über Naturschutz und Landschaftspflege (Bundesnaturschutzgesetz – BGBI. I S. 148)*, Berlin: Deutscher Bundestag, last revised 29 July 2009, in force since 1 October 2010.

Brock, W. and Xepapadeas, A. (2003) 'Valuing biodiversity from an economic perspective: a unified economic, ecological and genetic approach', *American Economic Review* 93: 1597–614.

Carter, A. (2010) 'Biodiversity and all that jazz', *Philosophy and Phenomenological Research* 80: 58–75.

CBD – United Nations (1992) *Convention on biological diversity*. Online. Available HTTP: <http://www.cbd.int/doc/legal/cbd-en.pdf> (accessed 12 June 2013).

Christie, M., Hanley, N., Warren, J., Murphy, K., Wright, R. and Hyde, T. (2006) 'Valuing the diversity of biodiversity', *Ecological Economics* 58: 304–17.

Costanza, R., d'Arge, R., de Groot, R., Farberk, S., Grasso, M., Hannon, B., Limburg, K., Naeem, S., O'Neill, R.V., Paruelo, J., Raskin, R.G., Suttonkk, P. and van den Belt, M. (1997) 'The value of the world's ecosystem services and natural capital', *Nature* 387: 253–60.

Düwell, M., Hübenthal, C. and Werner, M.H. (2011) *Handbuch Ethik*, 3rd edn, Stuttgart: Metzler.

Eser, U. and Potthast, T. (1999) *Naturschutzethik – eine Einführung für die Praxis*, Baden-Baden: Nomos.

Hampicke, U. et al. (2009) *Memorandum Ökonomie für den Naturschutz – Wirtschaften im Einklang mit Schutz und Erhalt der biologischen Vielfalt*. Online. Available HTTP: <www.bfn.de/fileadmin/MDB/documents/themen/oekonomie/MemoOekNaturschutz.pdf> (last accessed 25 March 2013).

Joas, H. (1999) *Die Entstehung der Werte*, Frankfurt am Main: Suhrkamp; English edn: *The genesis of values*, Cambridge: Polity Press, 2000.

Jonas, H. (1979) *Das Prinzip Verantwortung*, Frankfurt am Main: Suhrkamp; English edn: *The imperative of responsibility: In search of an ethics for the technological age*, Chicago, IL: University of Chicago Press, 1984.

Kant, I. (1991) *Grundlegung zur Metaphysik der Sitten*, ed. B. Kraft and D. Schönecker, Hamburg: Meiner (Orig. 1785).

Krebs, A. (1999) *Ethics of nature – A map*, Berlin: de Gruyter.

Meyer-Abich, K.M. (1997) *Praktische Naturphilosophie – Erinnerungen an einen vergessenen Traum*, München: Beck.

Myers, N. (1979) *The sinking ark – a new look at the problem of disappearing species*, Oxford: Pergamon Press.

Nunes, P.A.L.D. and van den Bergh, J.C.J.M. (2001) 'Economic valuation of biodiversity: Sense or nonsense?', *Ecological Economics* 39: 203–22.

Ott, K. (2008) 'A modest proposal of how to proceed in order to solve the problem of inherent moral value in nature', in L. Westra, K. Bosselmann and R. Westra (eds) *Reconciling human existence with ecological integrity*, London: Earthscan, 39–60.

Ott, K. (2010) *Umweltethik zur Einführung*, Hamburg: Junius.

Potthast, T. (1996) 'Inventing biodiversity: Genetics, evolution, and environmental ethics', *Theory in Biosciences* (formerly: *Biologisches Zentralblatt*) 115: 177–85.

Potthast, T. (1999) *Die Evolution und der Naturschutz – zum Verhältnis von Evolutionsbiologie, Ökologie und Naturethik*, Frankfurt am Main: Campus.

Potthast, T. (ed.) (2007) *Biodiversität – Schlüsselbegriff des Naturschutzes im 21. Jahrhundert?*, Naturschutz und Biologische Vielfalt 48, Bonn: Bundesamt für Naturschutz.

Potthast, T. (2012) '"Good change" in the woods: Conceptual and ethical perspectives on integrating sustainable land-use and biodiversity protection', in T. Potthast and S. Meisch (eds) *Climate change and sustainable development: Ethical perspectives on land use and food production*, Wageningen: Wageningen Academic Publishers, 142–50.

Sarkar, S. (1999) 'Wilderness preservation and biodiversity conservation – keeping divergent goals distinct', *BioScience* 49: 405–12.

Takacs, D. (1996) *The idea of biodiversity – philosophies of paradise*, Baltimore, MD: Johns Hopkins University Press.

TEEB (2010) *The economics of ecosystems and biodiversity: Mainstreaming the economics of nature: A synthesis of the approach, conclusions and recommendations of TEEB*. Online. Available HTTP: <http://www.teebweb.org/wp-content/uploads/Study%20and%20Reports/Reports/Synthesis%20report/TEEB%20Synthesis%20Report%202010.pdf> (accessed 2 July 2013).

Tilman, D. (2000) 'Causes, consequences and ethics of biodiversity', *Nature* 405: 208–11.

Toulmin, S. (1982) 'How medicine saved the life of ethics', *Perspectives in Biology and Medicine* 25: 736–50.

Väliverronen, E. (1998) 'Biodiversity and the power of metaphor in environmental discourse', *Science Studies* 11: 19–34.

Vester, F. (1983) *Der Wert eines Vogels*, München: Kösel.

Wilde, O. (2011) *Lady Windermere's fan*, London: Penguin Books (Orig. 1892).

Wolters, G. (1995) '"Rio" oder die moralische Verpflichtung zum Erhalt der natürlichen Vielfalt – zur Kritik einer UN-Ethik', *Gaia* 4: 244–9.

Worster, D. (1995) *Nature's economy: A history of ecological ideas*, 2nd edn, Cambridge: Cambridge University Press.

Part II
Policies and justice

This second part of the book will concentrate on questions of policies, legislation and justice associated with the various efforts of the protection of biodiversity and how to regulate transmissions of benefits that emerge from the usage of natural resources. A major aspect of the debate on allowing a suitable benefit and profit of natural resources while ensuring at the same time the protection of species and their habitats can be found in the discussion on access and benefit sharing, and the 'biopiracy' debate. Biological, legal, political and philosophical perspectives are necessary to understand these elaborated and complex debates.

Wolfgang Erdelen provides an overview on the IUCN Red Lists and from the Global Biodiversity Assessment to COP 10 in Nagoya and from 'Classical' to 'Modern' approaches of sustainable development. A sustainable development in actions to protect environment and its biological diversity needs a different view on the interactions within all system levels involved: from a regional to a global and from biotope to ecosystems and the Earth as a whole. Living in a more globalised world provides new challenges but offers also new chances and opportunities. On a local level Manfred O. Hinz and Oliver C. Ruppel provide a synoptic overview of the legal biodiversity protection efforts in Namibia where internationally recognised biodiversity hotspots exist. The article provides an interesting example of the application of international agreements on the protection of biodiversity and provides detailed information on the most important current regulations on the local as well as on the national level, which have been developed within the framework of international agreements. Based on this systematic compilation a presentation of the current political efforts in respect to further developing and implementing environmental laws in Namibia is given.

Many of the chapters of this book are in one way or another related to the UN Convention on Biological Diversity (CBD) of 1992. The Convention is based on three basic principles: the maintenance of biodiversity, the sustainable use of its components and the equitable sharing of the benefits deriving from the utilisation of genetic resources. Many further regulations followed. However, as Tade Spranger points out in his chapter, irrespective of the numerous and extensive regulative activities that have been engaged in particularly by the WIPO and by other members of the UN family, there is still a lack of consistent and internationally binding regulations that are able to assure an appropriate balance of all conflicting interests. While companies in industrial countries are trying to ensure their profit potentials by making use of the incredible richness of biological materials that are to be found in many developing countries, the latter express concerns over the sellout of their genetic heritage and traditional knowledge. Spranger's article provides a number of prominent examples of areas that caused concern regarding 'biopiracy'. 'Biopiracy', 'bioprospecting' or 'biocolonialism' have severe practical consequences.

Without adequate legislative tools, traditional knowledge of genetic resources has often been exploited without any benefits for indigenous holders of this knowledge. In recent years international agencies (particularly under the CBD) tried to overcome this problem of access and benefit sharing, generally through the patent system. In this context we must not forget that indigenous ecological knowledge may be a valuable tool in biodiversity protection. Karen Bubna-Litic points to these aspects as she explores the unique relationship between biodiversity, indigenous communities and indigenous ecological knowledge in Australia. Based on an overview of the legal framework for the protection of biodiversity relevant for Australia, the article discusses if patents are suitable tools to protect indigenous ecological knowledge at all.

Klara H. Stumpf expresses similar reservations, namely that the access and benefit sharing requirements of international instruments such as the CBD and the Nagoya Protocol might in practice turn out to be too narrow to understand all claims of (in)justice in the 'biopiracy' debate. Her philosophical analysis of selected claims in the biodiversity debate focuses on arguments which either explicitly use the term 'biopiracy' or refer explicitly to claims of (in)justice connected to patenting, genetic resources and traditional knowledge. Claims of moral desert or human rights and arguments from reciprocity, to mention just some examples of the debate's typical argumentative figures, are equally challenged. Based upon this analysis of typical claims and argumentative figures and foundations, Stumpf puts a finger on currently yet unsolved justificatory questions and points out the structures and open questions that might deserve more attention.

8 The future of biodiversity and sustainable development

Challenges and opportunities

Walter R. Erdelen

Introduction

The Convention on Biological Diversity (CBD) defines biodiversity as '*the variability among living organisms from all sources including terrestrial, marine and other aquatic ecosystems and the ecological complexes of which they are part; this includes diversity within species, between species and of ecosystems*'. This had been the definition already used at the 1992 UN Earth Summit, held in Rio de Janeiro.

Biodiversity on planet Earth is the product of 3.5 billion years of evolution having resulted both in the creation of new species but also in the extinction of many species. In other words, species turnover has been a permanent feature throughout the history of our planet. It is estimated that the extant species may represent less than 1% of all species which have existed on Earth. Several mass extinction events have shaped the biota of our planet. Current extinction rates of species, essentially due to human impacts, are estimated at 100 to 1000 times the 'background' or average rates during the Earth's past (Lawton and May 1995; SCBD 2010).

Despite the enormous increase in global conservation efforts, pressure on biodiversity continues to erode biodiversity of virtually all our ecosystems. The '2010 Target', agreed by Governments at the 2002 World Summit on Sustainable Development in Johannesburg, South Africa, to 'achieve a significant reduction of the current rate of biodiversity loss', has not been met.

The assessment of extinction risks carried out by the International Union for Conservation of Nature (IUCN) draws a bleak picture of the status of the species assessed to date. At the same time we still do not even know the order of magnitude of the number of species inhabiting our planet. Estimates of the total number of eukaryotic species, for instance, range from two million to 100 million. A more recent estimate is about 8.7 million species of eukaryotes of which 86% of species on Earth and 91% of marine species have not yet been described (Mora et al. 2011). Moreover, bacteria and archaea would possibly comprise further tens of millions of species (Stuart et al. 2010).

Efforts to address biodiversity issues and problems have most recently been significantly reinforced. This is underscored by the United Nations having declared 2010 the International Year of Biodiversity, a year which mobilised tremendous efforts at national, regional and global levels for biodiversity conservation.

In 2010, Heads of State and Government for the first time in the history of the United Nations debated biodiversity issues at the UN General Assembly and at the COP 10 meeting of the Convention on Biological Diversity, held in October 2010 in Nagoya, Japan. In Nagoya decisive steps were taken for the future of the planet's biodiversity. The meeting culminated in the adoption of the 'Nagoya Protocol on Access to Genetic Resources and the Fair and Equitable Sharing of Benefits Arising from their Utilization (ABS)', a new strategic plan for 2011–2020, and a resource mobilisation strategy.

This paper highlights the interconnectedness of different approaches to biodiversity conservation, outlines activities envisaged to reduce biodiversity loss, and suggests systemic approaches to address biodiversity and sustainable development issues.

Species – major elements of biodiversity

The classical definitions of biodiversity refer to different levels within a hierarchical system. These are:

- *Genetic diversity*: the variety of genetic information contained in all of the individual microorganisms, plants and animals.
- *Species diversity*: the variety of living species, both extant and extinct.
- *Ecosystem diversity*: the variety of habitats, biotic communities, and ecological processes as well as the tremendous diversity present within ecosystems in terms of habitat differences and the variety of ecological processes.

Even a fourth level, i.e. the level of *molecular biodiversity*, the variety of molecules found in life, has been proposed (e.g. Campbell 2003).

Despite our ignorance of species numbers, species are still considered the cornerstone of biodiversity. Moreover, though different species concepts are in use (for an overview cf. e.g. Mayr 2001), we have had a long tradition of studying species and have meanwhile compiled an enormous knowledge base about them. They are more easy 'to handle and analyse' than ecosystems and are easier to identify than genes. As a consequence, meaningful strategies for biodiversity conservation for pragmatic and operational reasons usually target the species level of biological diversity. For these reasons most of the discussions of biodiversity centre around biodiversity at the species level.

Ecosystem services

The latest biodiversity discussions focused on the economic aspects of biodiversity, the most famous being the Millennium Ecosystem Assessment (MEA 2005a), in particular its biodiversity synthesis report (MEA 2005b) and, more recently, the so-called TEEB studies on the economics of ecosystems and biodiversity (e.g. TEEB 2010).

Discussions of the relationship between economy and biodiversity, however, date back over twenty years (e.g. McNeely 1988; Pearce and Moran 1994). What is in modern jargon referred to as *ecosystem services* was touched upon over thirty years ago. For instance, the World Conservation Strategy (IUCN 1980) already identified the maintenance of 'ecological processes and life-support systems' as one of its priority areas.

The Millennium Ecosystem Assessment (MEA 2005a) distinguishes four categories of ecosystem services, viz. (1) provisioning services, essentially the material outputs from ecosystems, such as food, raw materials, freshwater, medicinal resources, (2) regulating services such as local climate and air quality regulation, carbon sequestration and storage, moderation of extreme events, waste-water treatment, prevention of erosion and maintenance of soil fertility, pollination, biological control, (3) habitat or supporting services such as habitats for species, maintenance of genetic diversity, and (4) cultural services including those for recreation, mental and physical health, tourism, aesthetic appreciation and inspiration for culture, art and design, the spiritual experience and sense of place (from TEEB 2010; cf. also figures provided for the economics of ecosystem services in TEEB 2010). Increasingly biodiversity is also seen in the context of providing models for innovation and technological development as illustrated by terms such as 'biomimicry' or 'bionics'.

From this naturally follows that if species, with their genetic makeup and being the basic unit of ecosystems, are the central element of biodiversity, their loss diminishes the quality of our lives and our basic economic security.

Species – our ignorance

Despite all our efforts to catalogue the species on planet Earth, to date less than two million species have been scientifically described. The different taxonomic groups amongst these are not known equally well. Almost completely known are the mammals, birds and coniferous tree species. Though many species of plants and invertebrates, insects being the most species-rich amongst them, are still to be described, our ignorance is greatest as regards groups like the lichens, mushrooms, bacteria and archaea. Most recent research, based on analyses of metagenomic data, even indicates that there may be a fourth domain of life, in addition to eukaryotes, bacteria and archaea (Wu et al. 2011).

As a consequence of our limited knowledge, estimates of the total number of eukaryotic species in existence on Earth today vary greatly, ranging from two million to 100 million. A generally accepted working estimate seems to be between eight and nine million species (cf. references in Hilton-Taylor et al. 2009).

A big challenge results from the fact that while we continue to describe species and at the same time extinction continues at a massive scale, we lose expertise in scientists with special knowledge of particular groups of organisms, i.e. the taxonomists and systematists. Initiatives like the Global Taxonomy Initiative, the European Distributed Institute of Taxonomy, the Encyclopaedia of Life, the Census of Marine Life and others have tried to address aspects of these trends. The demands, however, are overwhelming as underscored by the fact that, as just outlined, we do not even know the order of magnitude of extant species on Earth and at the same time species disappear at an ever increasing rate. As James Mallet of the University College London put it: *'We are wiping out species at a terrible rate while we are finding thousands of new ones. But although it may be encouraging to find new animals, we often only come across a new beetle or spider because its habitat is being destroyed for a new farm. Its prospects aren't good'* (*The Observer*, Sunday 25 September 2005).

The IUCN Red Lists

Though certainly incomplete, the IUCN Red Lists have provided useful indicator function for trends in species development and for triggering conservation action. The number of species which have been assessed has grown considerably over the last decade: from 16,500 in 2000 to 44,800 in 2008 and 47,677 in 2009. Of the latter already 17,291 species (36%) were considered threatened with extinction. The 2012 IUCN Red List of Threatened Species gives a figure of 65,518 species evaluated with an estimate of 20,219 species (31%) threatened with extinction.[1]

Insufficient coverage of many taxonomic groups, however, poses severe difficulties in interpreting such information. According to Hilton-Taylor et al. (2009), *'the conservation status for most of the world's species remains poorly known, and there is a strong bias in those that have been assessed so far towards terrestrial vertebrates and plants and in particular those species found in biologically well-studied parts of the world'* (Hilton-Taylor et al. 2009: 15).

Some of the major findings of the 2008 IUCN Red List are (after Hilton-Taylor et al. 2009):

- A complete reassessment of the world's mammals showed that nearly one-quarter (22%) of the world's mammal species are globally threatened or extinct.
- One of eight species of birds (12.5%) is considered threatened or extinct; birds are one of the best known groups.
- Amphibians face a global extinction crisis: nearly one-third (31%) are threatened or extinct.
- 24% of the 359 freshwater fish species endemic to Europe are listed as threatened.
- Of 845 species of warm water reef-building corals, more than 25% are listed as threatened.

These are just a few examples which, despite possible shortcomings of the Red List, clearly show that numbers of species under threat of extinction continue to grow. This is mainly a direct or indirect result of human activities having led in particular to habitat loss and/or habitat degradation.

These figures should be seen as indicative. They will change continuously as more information and more detailed analyses become available. For instance, a comprehensive study of the conservation status of the world's vertebrates (Hoffmann et al. 2010) has shown that 20% of all vertebrate species are threatened, percentages which are increasing. Percentages range from 13% for birds to 41% for amphibians, and species of threatened vertebrates are mainly from tropical regions. The same study also highlights that climate change may relate to the deteriorating status of some of the species and may have an accelerating effect acting in concert with other factors contributing to species extinction.

Species – future biodiversity assessments

The current biodiversity assessments naturally have severe limitations (Collen et al. 2009). Amongst these we find that (1) they have been restricted to the better known taxonomic groups, (2) the conservation status of less than 3% of the world's described biodiversity is known and (3) of the higher plant species, some 85% have been described but for only 4% of them conservation status has been assessed.

An epistemological issue may be whether we really need complete knowledge for the assessment of conservation status and trends or whether sample approaches would allow for generalisations across a broad spectrum of taxa. Recently new approaches have been developed, based on random sampling and inclusion of large species-rich taxonomic groups (Baillie et al. 2008). Similar efforts towards better schemes for global monitoring of biodiversity (e.g. Collen et al. 2009) were developed in view of the 2010 target to significantly reduce the rate of biodiversity loss by the year 2010. We have not met this goal and, especially following the COP 10 meeting of the CBD in October 2010 in Nagoya, Japan, a set of new targets and sub-targets, the 'Aichi Targets', has been developed.

In addition to loss and degradation of habitats there is growing evidence that climate change will become a major driver of extinction in the twenty-first century (e.g. Foden et al. 2009; Lovejoy 2010; SCBD 2010; Thomas et al. 2004). This may reinforce the quest for new approaches which give us a better picture of the overall situation of biodiversity across all groups of organisms.

From global biodiversity assessment to COP 10 in Nagoya

The fact that 2010 was probably the most successful year as regards global efforts towards biodiversity conservation has been the result of a longer process, in a sense a gestation process which prepared the ground for a 'quantum leap' forward.

A comprehensive review of global biodiversity was for the first time compiled by the World Conservation Monitoring Centre (WCMC) and published in 1992. This report, entitled '*Global Biodiversity: Status of the Earth's living resources*' (WCMC 1992), was jointly published with the Natural History Museum, London, the

International Union for Conservation of Nature (IUCN), the United Nations Environment Programme (UNEP), the World Wide Fund for Nature/World Wildlife Fund (WWF), and the World Resources Institute (WRI).

Though concerns about the status of global biodiversity had been raised earlier (for overviews cf. e.g. Reaka-Kudla, Wilson and Wilson 1997; Wilson 1988), a major breakthrough came with the publication of the results of the *Global Biodiversity Assessment* (GBA; UNEP 1995). The GBA is the result of a two-and-a-half-year effort (1993–1995) of over a thousand experts worldwide to scientifically analyse issues, theories and views related to biodiversity. The assessment was commissioned by UNEP.

A next step towards reinforcing efforts of biodiversity conservation was the Rio +10 meeting in 2002, the World Summit on Sustainable Development (WSSD), held in Johannesburg, South Africa. The Summit endorsed the so-called '2010 Target' which was actually an element of a decision on a strategic plan at the COP 6 of the CBD in April 2002 with the commitment to 'achieve by 2010 a significant reduction of the current rate of biodiversity loss at the global, regional and national level as a contribution to poverty alleviation and to the benefit of all life on earth' (CBD decision VI/26).

WSSD was followed by two major assessments, the Millennium Ecosystem Assessment (MEA 2005a), in particular the *Biodiversity Synthesis Report* (MEA 2005b) and the *International Assessment of Agricultural Knowledge, Science and Technology for Development* (IAASTD 2009). The IAASTD has been, in fact, the last of a series of global assessments carried out with contributions from an enormous number of experts in the relevant fields.

Meanwhile, as illustrated by the *Bonn International Conference on Biodiversity Research: Safeguarding the Future* (2008) and the German Presidency of COP 9 of the CBD, a process of scaling up towards the UN International Year of Biodiversity, the UN General Assembly in the same year, and the famous COP 10 meeting in Nagoya, Japan, was set in motion.

Increasingly, the economic aspects of biodiversity received attention from the stakeholder communities. This culminated in the series of reports on the economics of ecosystems and biodiversity, the so-called TEEB Reports (e.g. TEEB 2010). This was a 'natural continuation' of the focus on ecosystem services already emphasised in the MEA and its reports.

A decadal effort of research into the diversity of marine life, the census of marine life (2000–2010),[2] added to making our limited understanding of marine ecosystems public, in particular the least known of them, the deep sea systems.

Finally, the year 2010, the UN International Year of Biodiversity, saw an enormous number of biodiversity-related events around the globe, the release of the *Global Biodiversity Outlook 3* (SCBD 2010), and the famous COP 10 meeting. The UN General Assembly, for the first time in the history of the UN, discussed biodiversity issues at the Heads of State and Government level. The end of the year saw the adoption of an ambitious set of recommendations for the future coming out of the CBD COP 10 meeting which now needs urgent implementation.

COP 10 in Nagoya and Rio+20

The Convention on Biological Diversity (CBD) held its tenth Conference of the Parties (COP 10), from 18–29 October 2010 in Nagoya, Japan. COP 10 adopted an impressive package and has been considered one of the most successful meetings in the history of CBD and the most important meeting on biodiversity in UN history.

In Nagoya ambitious decisions were taken in three areas of major importance: (1) on new global biodiversity targets, the Aichi Targets, and a strategy on global biodiversity conservation for 2011–2020, (2) on binding financing targets for its implementation, and (3) on a protocol for access to genetic resources and the fair and equitable sharing of the benefits arising from their utilisation.

Examples of major outcomes of COP 10 include, amongst others:

- The Nagoya Protocol on Access and Benefit Sharing (ABS)
- The CBD Strategic Plan 2011–2020
- A Multi-Year Programme of Work (MYPOW) for the CBD
- A Strategy for Resource Mobilization
- The continuation of the process of establishing a Science-Policy Platform on Biodiversity and Ecosystem Services (IPBES)
- The preparations to declare 2011–2020 the UN Decade of Biodiversity.

The outcome document of the 2012 Rio+20 Summit, entitled *The Future We Want*,[3] reaffirmed the values of biodiversity as 'critical foundations for sustainable development and human well-being', recognised the severity of global biodiversity loss and stressed the importance of conserving biodiversity. The importance of these outcomes was reiterated and 'urgent actions that effectively reduce the rate of, halt and reverse the loss of biodiversity' were called for. Finally, the UN Decade of Biodiversity was welcomed, and most interestingly linked to a 'vision of living in harmony with nature'.

Therefore, for implementing the recommendations of COP 10, action needs to be taken at all levels, i.e. the national, regional and global levels. For instance, the national biodiversity strategies and action plans (NBSAPs) are requested to be updated. Regional cooperation will need to be better coordinated and reinforced, and global implementation of the new goals and sub-goals which have replaced the 2010 Target and its associated goals is now of the essence, especially after our failure to meet the 2010 Target.

The most recently established important tool to meet these challenges is the Intergovernmental Science-Policy Platform on Biodiversity and Ecosystem Services (IPBES). The role of IPBES with regard to the Convention on Biological Diversity is comparable to the role of the Intergovernmental Panel on Climate Change (IPCC) for the UN Framework Convention on Climate Change (UNFCCC). IPBES, product of a many-year process, was officially established in 2012 and had its first plenary meeting in January 2013 in Bonn, Germany.

Towards a systemic approach to biodiversity and sustainable development

The current global challenges include poverty alleviation or eradication, global climate change, the mastering of multiple financial and economic crises, global environmental problems, an energy crisis and many more. These are intrinsically systems problems (Meadows 2008). To address these problems we need new approaches or we may need increasingly to use approaches already in place but not yet used to their fullest.

The last century saw approaches with which we discovered and characterised the fundamental particles and learned to use them to synthesise materials, catalysts and pharmaceuticals, for instance. We had developed tools to examine and describe 'simple' phenomena and structures. The demands on science have changed since. The new millennium is the millennium of a world of complexity. As Gallopin et al. (2001) put it already over ten years ago: 'The complexity of the systems to be dealt with in the domain of science as, for instance, science for sustainable development, is one of the most critical arguments for the need for changes in the production and utilization of science'.

The Club of Rome has been at the forefront of analysing these issues. The Club was involved in the production of a series of outstanding reports, which were ahead of their time, i.e. the famous 'Limits to Growth' Report (Meadows et al. 1972), followed by the twenty-year update, entitled 'Beyond the Limits' (Meadows, Meadows and Randers 1992), the thirty-year update of the 1972 report (Meadows, Randers and Meadows 2004), and a global forecast for the next forty years (Randers 2012).

Meanwhile the challenge to reach global sustainability remains and, even worse, our global footprint, becomes larger by the day. Our ecological footprint has overshot the Earth's capacity to provide us with the resources we need. By 2030 humanity will need the capacity of two Earths to cope with the production of waste and keep up with natural resource consumption (WWF 2010). Or, as already said in the thirty-year update (Meadows et al. 2004: 3): 'We believe that if a profound correction is not made soon, a crash of some sort is certain'.

Martin Lees, former Secretary General of the Club of Rome, clearly highlighted the issues and problems we are facing (M. Lees, pers. comm.):

- We cannot manage the scale, complexity and dynamics of the issues and problems of the twenty-first century with the tools of the twentieth century.
- It is not possible to understand the root causes and the dynamics of climate change and the real risks of destabilizing the climate system, the degradation of ecosystems, and instability of our financial and economic systems, implications of a growing world population or the causes of growing food insecurity without adopting a systems approach.
- A holistic, systems-oriented approach is needed which can recognize the connectivity between issues and which can establish a conceptual framework within which the contributions of different disciplines, sectors and actors can combine in the formulation, implementation and evaluation of policy.

- Educational institutions face the challenge to adapt their curricula and methodologies to produce a new generation capable of holistic reasoning and of understanding and acting on integrated, systemic issues.
- The contributions and aspirations of different cultures must be fully respected and taken into account in developing the new approaches and policies needed.

These considerations may have profound implications on how we should address the issue of sustainable development in the future.

Sustainable development – from 'classical' to 'modern'

The classical notion of sustainable development, as repeatedly emphasised during the preparatory process for the Johannesburg Summit, was that of three pillars: the social, the economic and the environmental pillar. These pillars form a new whole (holistic approach), as a cup of tea consists of tea, water and sugar but each of them is not separately distinguishable, a notion which to my knowledge was created by Emil Salim, the Chairman of the World Summit on Sustainable Development (WSSD) 2002.

Naturally, the environmental pillar was essentially perceived as dealing with biodiversity. We now know that biodiversity conservation and its sustainable management would indeed comprise these three aspects, if not a fourth: the cultural dimension. UNESCO, the UN Organization with the mandate for culture, has for long emphasised that the cultural dimension should be considered the fourth pillar of sustainable development. This was particularly highlighted by UNESCO during WSSD in Johannesburg.

I would like to discuss a modified notion of sustainable development, in a sense a more modern and timely notion which could facilitate future concrete action as (still) required to have sustainable development truly globally implemented.

First, the cultural dimension should be part and parcel of the debates on sustainable development. Culture should be seen as even more than a fourth pillar of sustainable development. It should underpin any of the new conceptual approaches which might eventually be developed for sustainable development. The UNESCO World Report *Investing in Cultural Diversity and Intercultural Dialogue* (2009) has devoted a separate chapter to cultural diversity as a key dimension of sustainable development. The report proposes indeed to see culture as a cross-cutting dimension rather than a separate, fourth pillar.

Cultural diversity is seen as a key lever for ensuring sustainable, holistic development strategies. This is reflected in the Nagoya Protocol and its close link to traditional knowledge systems as an essential element of cultural diversity.

Secondly, biodiversity including its biological but also the economic aspects, in particular ecosystem services, should also be seen as an essential and increasingly important component of sustainable development.

In view of global change processes (for overviews cf. Steffen et al. 2004a, 2004b), in particular global climate change, we may need to look at sustainable development differently compared to the past. There continue to be national dimensions, but there are also global responsibilities which require global responses. 'Divides', as they have been seen in past discussions on sustainable development such as the North/South divide, may have new meanings and increased importance in view of these patterns and processes of global change. These processes themselves include all the dimensions which we commonly include in our thinking about sustainable development.

We may also need to view sustainable development differently from a systems point of view. Commonly when we speak of an Earth system, we refer to systems of systems, complex systems which interact, tipping points (e.g. Lenton et al. 2008, SCBD 2010), planetary boundaries (e.g. Galaz et al. 2012) and the like. In short, our rather simplistic approaches may need to be replaced by systemic approaches and views, if we wish to address sustainable development in the future in a meaningful way.

This is much easier said than done. Understanding how complex systems such as our climate system function is not something that is commonly taught. Indeed, there is an urgent need for both the layperson as well as the specialist and in particular the decision makers to have a sound understanding of the systems we are part of: in most cases these are complex systems with their own specific elements and ways these interact, including through internal feedback loops, time-delayed reaction to impacts and so forth. Creating literacy in these domains poses new challenges for our systems of education.

Finally, the latest thinking about sustainable development centres on ideas commonly referred to as 'greening', 'green economy', or 'green development'. Especially within the UN system the movement towards a green economy has become quite strong in the recent past (UNEP 2011).

It may be timely to re-think what sustainable development should mean in the future. Essentially the longitudinal responsibility towards our future generations remains. The means at our disposal to achieve sustainable development may be different now compared to the past.

The fact that we live in an ever more globalised world may provide new challenges but also new opportunities. These may relate to key areas such as renewable sources of energy, new building and other technologies, and paradigm changes in economy. Future population development, questions of mobility, and political instability in certain regions may however pose severe threats to securing quality of life for our children and grand-children. Will the notion of sustainable development itself be replaced by new terminology which takes into account the latest developments in the sciences, technology and innovation? Will *sustainable development* even be replaced by terms related to green development?

Acknowledgements

For comments on earlier drafts of the chapter I am grateful to Al Gramstedt, Diane Klaimi and Jacques Richardson.

Notes

1 IUCN Red List version 2012.2. Retrieved 15 May 2013 from www.iucnredlist.org.
2 For details cf. <www.coml.org> (accessed 13 June 2013).
3 Cf. <http://www.uncsd2012.org/thefuturewewant.html> (accessed 13 June 2013).

References

Baillie, J.E.M., Collen, B., Amin, R., Akcakaya, H.R., Butchart, S.H.M., Brummitt, N., Meagher, T.R., Ram, M., Hilton-Taylor, C. and Mace, G.M. (2008) 'Toward monitoring global biodiversity', *Conservation Letters* 1: 18–26.

Campbell, A.K. (2003) 'Save those molecules! Molecular biodiversity and life', *J. Appl. Ecology* 40: 193–203.

CBD – United Nations (1992) Convention on biological diversity. Online. Available HTTP: <http://www.cbd.int/doc/legal/cbd-en.pdf> (accessed 12 June 2013).

Collen, B., Ram, M., Dewhurst, N., Clausnitzer, V., Kalkman, V.J., Cumberlidge, N. and Baillie, J.E.M. (2009) 'Broadening the coverage of biodiversity assessments', in J.-C. Vie, C. Hilton-Taylor and S.N. Stuart (eds) *Wildlife in a Changing World – An Analysis of the 2008 IUCN Red List of Threatened Species*, Gland, Switzerland: IUCN, 67–75.

Foden, W. B., Mace, G.M., Vie, J.-C., Angulo, A., Butchart, S.H.M., DeVantier, L., Dublin, H.T., Gutsche, A., Stuart, S.N. and Turak, E. (2009) 'Species susceptibility to climate change impacts', in J.-C. Vie, C. Hilton-Taylor and S.N. Stuart (eds) *Wildlife in a Changing World – An Analysis of the 2008 IUCN Red List of Threatened Species*, Gland, Switzerland: IUCN, 77–87.

Galaz, V., Biermann, F., Crona, B., Loorbach, D., Folke, C., Olsson, P., Nilsson, M., Allouche, J., Persson, A. and Reischl G. (2012) '"Planetary boundaries" – exploring the challenges for global environmental governance', *Current Opinion in Environmental Sustainability* 4: 80–7.

Gallopin, G.C., Funtowicz, S., O'Connor, M. and Ravetz, J. (2001) 'Science for the twenty-first century: from social contract to the scientific core', *Intern. Soc. Science J.* 168: 219–29.

Hilton-Taylor, C., Pollock, C.M., Chanson, J.S., Butchart, S.H.M. and Oldfield, T.E.E. (2009) 'State of the world's species', in J.-C. Vie, C. Hilton-Taylor and S.N. Stuart (eds) *Wildlife in a Changing World – An Analysis of the 2008 IUCN Red List of Threatened Species*, Gland, Switzerland: IUCN, 15–41.

Hoffmann, M. et al. (2010) 'The impact of conservation on the status of the world's vertebrates', *Science* 330: 1503–9.

IAASTD (International Assessment of Agricultural Knowledge, Science and Technology for Development) (2009) *Synthesis Report*, Washington, D.C.: Island Press.

IUCN (International Union for Conservation of Nature) (1980) *The World Conservation Strategy*, Gland, Switzerland: IUCN.

Lawton, J.H. and May, R.M. (1995) *Extinction Rates*, Oxford: Oxford University Press.

Lenton, T.M., Held, H., Kriegler, E., Hall, J.W., Lucht, W., Rahmstorf, S. and Schellnhuber, H.J. (2008) 'Tipping elements in the Earth's climate system', *PNAS* 105: 1786–93.

Lovejoy, T. (2010) 'Climate change', in N.S. Sodhi and P.A. Ehrlich (eds) *Conservation Biology for All*, Oxford: Oxford University Press, 153–62.

Mayr, E. (2001) *What Evolution Is*, New York: Basic Books.

McNeely, J.A. (1988) *Economics and Biological Diversity: Developing and Using Economic Incentives to Conserve Biological Resources*, Gland, Switzerland: IUCN.

MEA (Millennium Ecosystem Assessment) (2005a) *Ecosystems and Human Well-being: Synthesis*, Washington, D.C.: Island Press.

MEA (Millennium Ecosystem Assessment) (2005b) *Ecosystems and Human Well-being: Biodiversity Synthesis*, Washington, D.C.: World Resources Institute.

Meadows, D.H. (2008) *Thinking in Systems*, White River Junction, VT: Chelsea Green Publishing.

Meadows, D.H., Meadows, D.L. and Randers, J. (1992) *Beyond the Limits*, White River Junction, VT: Chelsea Green Publishing.

Meadows, D.H., Randers, J. and Meadows, D.L. (2004) *Limits to Growth. The 30-Year Update*, White River Junction, VT: Chelsea Green Publishing.

Meadows, D.H., Meadows, D.L., Randers, J. and Behrens III, W.W. (1972) *The Limits to Growth*, White River Junction, VT: Chelsea Green Publishing.

Mora, C., Tittensor, D.P., Adl, S., Simpson, A.G.B. and Worm, B. (2011) 'How many species are there on Earth and in the ocean?', *PLoS Biol.* 9(8). Online. Available HTTP: <http://www.plosbiology.org/article/info:doi/10.1371/journal.pbio.1001127> (accessed 13 June 2013).

Nagoya Protocol – Secretariat of the Convention on Biological Diversity (2010) *Nagoya Protocol on Access to Genetic Resources and the Fair and Equitable Sharing of Benefits arising from their Utilization to the Convention on Biological Diversity: text and annex, COP 10 Decision X/1*. Online. Available HTTP: <http://www.cbd.int/abs/doc/protocol/nagoya-protocol-en.pdf> (accessed 13 June 2013).

Pearce, D. and Moran, D. (1994) *The Economic Value of Biodiversity*, Gland, Switzerland: IUCN.

Randers, J. (2012) *2052: A Global Forecast for the Next Forty Years*, White River Junction, VT: Chelsea Green Publishing.

Reaka-Kudla, M.L., Wilson, D.E. and Wilson, E.O. (eds) (1997) *Biodiversity II. Understanding and Protecting our Biological Resources*, Washington, D.C.: Joseph Henry Press.

SCBD (Secretariat of the Convention on Biological Diversity) (2010) *Global Biodiversity Outlook 3*, Montreal, Canada.

Steffen, W., Andreae, M.O., Bolin, B., Cox, P.M., Crutzen, P.J., Cubasch, U., Held, H., Nakicenovic, N., Scholes, R.J., Talaue-McManus, L. and Turner II, B.L. (2004a) 'Abrupt changes: the Achilles' heels of the Earth system', *Environment* 46, 3: 9–20.

Steffen, W., Sanderson, A., Tyson, P.D., Jaeger, J., Matson, P.A., Moore III, B., Oldfield, F., Richardson, K., Schellnhuber, H.J., Turner II, B.L. and Wasson, R.J. (2004b) *Global Change and the Earth System: A Planet Under Pressure*, Berlin: Springer.

Stuart, S.N., Wilson, E.O., McNeely, J.A., Mittermeier, R.A. and Rodriguez, J.P. (2010) 'The barometer of life', *Science* 328: 177.

TEEB (The Economics of Ecosystems and Biodiversity) (2010) *The Economics of Ecosystems and Biodiversity: Mainstreaming the Economics of Nature: A Synthesis of the Approach, Conclusions and Recommendations of TEEB*. Online. Available HTTP: <http://www.teebweb.org/wp-content/uploads/Study%20and%20Reports/Reports/Synthesis%20report/TEEB%20Synthesis%20Report%202010.pdf> (accessed 13 June 2013).

Thomas, C.D., Cameron, A., Green, R.E., Bakkenes, M., Beaumont, L.J., Collingham, Y.C., Erasmus, B.F.N., de Siqueira, M.F., Grainger, A., Hannah, L., Hughes, L., Huntley, B., van Jaarsveld, A.S., Midgley, G.F., Miles, L., Ortega-Huerta, M.A., Peterson, A.T., Phillips, O.L. and Williams, S.E. (2004) 'Extinction risk from climate change', *Nature* 427: 145–8.

UNEP (United Nations Environment Programme) (1995) *Global Biodiversity Assessment*, Cambridge: Cambridge University Press.

UNEP (United Nations Environment Programme) (2011) *Towards a Green Economy: Pathways to Sustainable Development and Poverty Eradication*. Online. Available HTTP: <http://www.unep.org/greeneconomy/Portals/88/documents/ger/GER_synthesis_en.pdf> (accessed 13 June 2013).

UNESCO (United Nations Educational, Scientific and Cultural Organization) (2009) *Investing in Cultural Diversity and Intercultural Dialogue. UNESCO World Report*, Paris: UNESCO.

WCMC (World Conservation Monitoring Centre) (1992) *Global Biodiversity: Status of the Earth's Living Resources*, London: Chapman and Hall.

Wilson, E.O. (ed.) (1988) *Biodiversity*, Washington, D.C.: National Academy Press.

WWF (World Wide Fund For Nature/World Wildlife Fund) (2010) *Living Planet Report 2010*, Gland, Switzerland: WWF.

Wu, D., Wu, M., Halpern, A., Rusch, D.B., Yooseph, S., Frazier, M., Venter, J.C. and Eisen, J.A. (2011) 'Stalking the fourth domain in metagenomic data: searching for, discovering, and interpreting novel, deep branches in marker gene phylogenetic trees', *PLoS ONE* 6(3). Online. Available HTTP: <http://www.plosone.org/article/info%3Adoi%2F10.1371%2Fjournal.pone.001801> (accessed 13 June 2013).

9 Access and benefit sharing as a challenge for international law

Tade M. Spranger

Introduction

The access to genetic resources which is sometimes also discussed under the terms of 'biopiracy' and 'biocolonialism' has finally gained public awareness. The high conflict potential of this topic attracted little attention only a few years ago, but now it is realised globally. The developing countries express concerns over the sellout of their genetic heritage, while companies in industrial countries are trying to ensure their profit potentials. Actually, there is still a lack of consistent and internationally binding regulations that are able to ensure an appropriate balance of all conflicting interests. This applies irrespective of the numerous and extensive activities that have been engaged in particularly by the World Intellectual Property Organization (WIPO) and by other members of the UN family.

It stands to reason that a viable solution approach must become legally binding. The parties can only be forced to observe necessary rules of conduct, and the interests of indigenous groups as well as intellectual property rights can only be guaranteed by the imposition of legally binding and enforceable obligations. As far as compensation mechanisms have been developed, they mostly represent detailed regulations that ignore the controversial issues.

Outlining the problem

Access to genetic resources is gaining importance. For industries, there is an increasing need for new genetic materials that can be tested for effective properties or ingredients. With regard to the distribution of the coveted row materials, the North–South divide is clearly visible. While the industrialised north has already exploited its rather barren genetic resources, some of the economically weakest so-called developing countries have an incredible richness of biologic materials – up to 80% (Verma 1997: 205; Straus 1998: 90) to 90% (Cottier 1998: 564) of genetic resources are estimated to be found in developing countries that represent the majority of the so-called megadiversity countries. However, there still is a lack of technical know-how for industrial use of these resources.

Access and access rules

The researchers' focus lies on genetic resources of plant or animal origin (Henne 1998). Around 25% of the prescribed medicine in the US (Blakeney 1997: 94; Stoll and Schillhorn 1998: 630) has been isolated from plant-based resources (Blakeney 1997: 94; Pearce and Puroshothaman 1995: 128; Kawai 1995: 437–8). The total annual turnover of pharmaceutical products, based on traditional medicine, (Mishra 2000: 220–1) is estimated to amount to 32 billion US dollars (Mugabe et al. 1997: 6; Chaves 2004: 224).The global profit, which is achieved by companies operating in the pharmaceutical and agricultural sector by use of substances isolated from plants, amounts to up to 100 billion US dollars (Stevenson 2000: 1119 no. 2). If a pharmaceutical company seeking unknown substances takes advantage of traditional knowledge of indigenous people and thereupon isolates and patents a new substance, the bearers of the traditional knowledge generally do not benefit from the corresponding revenues. Rather, the enforceable right to commercial utilisation belongs to the holder of the patent. These operations are summarised under the term of 'biopiracy', 'bioprospecting' or 'biocolonialism'.

Among the international organisations, solely the *Food and Agricultural Organization of the United Nations* (FAO) has anticipated, i.e. in 1983, associated potential problems at least to some extent that might occur in future and published the 'International Undertaking on Plant Genetic Resources' (FAO 1983), declaring the plant genetic resources as a 'common heritage of humanity' (Straus 1998: IX 101). That undertaking was initially not binding under international law, until it was transferred into the binding 'International Treaty on Plant Genetic Resources for Food and Agriculture' in 2004 (FAO 2004).

A more institutionalised interest occurred only in the 1990s. In the first place, the Convention on Biological Diversity (CBD) of 1992 has to be named (see in more detail in the relevant section of this chapter). This binding instrument attributes the biological diversity to the concept of international law of 'common concern of mankind' and emphasises the crucial need for the development of humankind; highlighting on the other hand the responsibility of humankind for the preservation of that diversity. The CBD is based on three basic principles: the maintenance of biodiversity, the sustainable use of its components, and the equitable sharing of the benefits deriving from the utilisation of genetic resources.

In 1996, the OECD published an overview of the current practices and policies in the access area of genetic resources (Intellectual Property, Technology Transfer and Genetic Resources: An OECD Survey of Current Practices and Policies) (OECD 1996). The WIPO within the UN family (Ipsen 2004: § 33 no. 64) organised a 'Roundtable on Intellectual Property and Indigenous Peoples' in 1998. Moreover, supranational activities within the scope of the Cartagena Agreement (Andean Community 1996) of the so-called Andean Community could be recognised. According to a decision of the Andean Community of 1994, the States have permanent sovereignty over genetic resources and represent the only authorised entities for the question of access regulation. Finally, the Biosafety-Protocol deals with the specific interests of indigenous and local communities in biodiversity under the generic term of 'socio-economic considerations' (Cartagena Protocol: Art. 26).

In October 2000, the WIPO established the *Intergovernmental Committee on Intellectual Property and Genetic Resources, Traditional Knowledge and Folklore* that comprehends currently around 150 organisations acting as ad-hoc consultants. The duties of the institutions are described by the WIPO as follows:

> The WIPO Intergovernmental Committee on Intellectual Property and Genetic Resources, Traditional Knowledge and Folklore (IGC) was established ... as an international forum for debate and dialogue concerning the interplay between intellectual property (IP), and traditional knowledge (TK), genetic resources, and traditional cultural expressions (TCEs)/(folklore).
>
> (WIPO 2000)

The 'Bonn Guidelines' that were adopted at the 6th session of the Conference of the Parties in 2002 are aimed at concretising the Convention on Biological Diversity. The non-binding (Normand 2004: 133–5) guidelines require inter alia the obtainment of a new prior informed consent, and the conclusion of a new usage agreement in case probes have been used for purposes other than those for which they were acquired (Bonn Guidelines: no. 16 b), iv) and v)). Thus, they concretise the informed consent (i.e. according to the fundamental principles of legal certainty and clarity Bonn Guidelines: nos 24ff.) and enumerate different forms of monetary and non-monetary benefits, i.e. up-front payments, salaries, license fees or collaboration, cooperation and contribution in scientific research, education or training (Bonn Guidelines: Appendix II). Nevertheless, the Bonn Guidelines remain vague in relevant fields; this applies particularly to the question of the concrete function of benefit sharing systems (Bonn Guidelines: nos 45ff.).

On October 29th 2010, the 'Nagoya Protocol' to the Convention on Biological Diversity was adopted at the 10th meeting by the Conference of the Parties. It is a legally binding international agreement that has not yet entered into force. Actually, it has been signed by 92 and ratified by eight parties.[1] According to Art. 1, it is aimed at the 'fair and equitable sharing of the benefits arising from the utilization of genetic resources, including by appropriate access to genetic resources and by appropriate transfer of relevant technologies, taking into account all rights over those resources and to technologies, and by appropriate funding, thereby contributing to the conservation of biological diversity and the sustainable use of its components.' Art. 5 provides further details regarding fair and equitable benefit sharing. Access to genetic resources regulated by Art. 6 refers to the necessity to obtain prior informed consent for utilisation of the party providing such resources and concretising the legislative, administrative or policy measures that have to be taken in this regard. Art. 10 addresses the handling of so-called transboundary situations in that prior informed consent cannot be obtained. According to a 'global multilateral benefit sharing mechanism', the conservation of biological diversity and the sustainable use of its components shall be supported globally by the shared benefits. Thus, by setting out core obligations for its contracting parties, the Nagoya Protocol creates greater legal certainty and clarity for both users and providers, in particular for so-called developing countries.

Prominent examples of 'biopiracy'

In the following, two well-known constellations shall clarify the complexity of access and benefit sharing issues.

The neem tree

The Indian neem tree (azadirachta indica) can be cited as a widely known example for the challenge of biopiracy. All components of the neem tree contain pharmacologically effective substances. The plant exercises fungicidal, antibacterial, immune strengthening, antiviral and analgesic antipyretic functions (Spranger 2008: 20). In India, the tree has been known for its pharmacy for thousands of years. Neem substances have not only been given a key role in Ayurvedic medicine since ancient times; neem trees are also planted ring-shaped around villages for the purpose of insect protection. When the first patents on neem substances were applied at the beginning of the 1990s, this led to widespread protests in India. In October 1993, half a million people took part in a protest demonstration in Bangalore (Blakeney 1999: 147).

The criticism relating to the patenting of neem substances is based inter alia on the allegation that Indians are traditionally not conferred the right of patenting neem substances, because they are part of the 'general knowledge'. However, in the US alone, nearly 100 corresponding patents have been granted up until now. As a consequence of that alleged prohibition on patenting for Indians, no written proof about already practised procedures exists that would possibly support any objection against US patents (Primlane 1999: no. 131 23–6). Without anticipating the separate comments on patent law (cf. the section 'Outlining the problem' above), it can already be maintained that those statements give a partially correct impression of the patent situation. The exclusion of traditional medicine from patenting – declared as 'Indian specific feature' – can be found in every relevant patent law in the world. In this case, there is a lack of 'novelty of invention'. The US Patent Office does not offer patent protection either. The United States Patent and Trademark Office (USPTO) has only granted patents to isolated substances that are in some way technically modified and then used for certain applications. In observance of the generally usual requirements, Indian companies also obtained patents for neem inventions: from 1991 to 1994, neem patents were granted in nine cases (Neemfoundation 2008).

This background also explains the frequently quoted revocation of a 'neem patent' by the European Patent Office. According to the press release of May 10th 2000, the patent office reported that the European patent 0436 257 dealing with fungicide properties of the neem tree oil had been revoked. The patent in question had been applied for by the *US Department of Agriculture* and the company *WR Grace*. The revocation was carried out because the applied invention had been used in India for some time not only traditionally, but also industrially. During the two-day consultation in the European Patent Office, *Abhay Phadke*, the founder of the Indian company *Ajay Bio-Tech Ltd.*, was able to prove that he, the director of *Rhone Poulenc Agro-chemical (India) Ltd.*, advised the French corporate management in the 1980s of the corresponding application possibilities, whereupon field tests were carried out in 1985 and 1986 in Pune and Sangli (Nair 2000). Thus, according to patent law, the applied invention lacked the necessity of novelty.

The fact that neem products were already *traditionally* used in India did not play a role in the negative decision of the European Patent Office. Therefore, the revocation cannot be seen as a measure of the authority against extreme expressions of so-called biopiracy (Raj 2000).

Basmati rice

Major attention was also paid to the patent of a variant of basmati rice granted to the Texan company *Ricetec*. On September 2nd 1997, the USPTO patented that variant of basmati rice. Irrespective of the difficult distinction in cases of patent protection for plant variety rights, such a patent needs some clarification with regard to the requirements of patentability of the inventive step and the novation. The Indian Government initially did not take any measures against the patent. This reserved attitude is surprising, as India exports 500,000 tons of that extremely valuable rice each year.

On March 4th 1998, the Research Foundation for Science Technology and Ecology (RFSTE) took action before the Indian Supreme Court, in order to encourage the Indian Government to act quickly against the patent. The Indian Government demanded the USPTO thereupon to review the corresponding patent in June 2000. It is quite obvious that this measure was not based on a specific patent-related argumentation, but on the general reference that the applied rice variant did not satisfy the recognised quality characteristics of basmati and affected only three from twenty patent claims.

On March 27th 2001, the United States Patent and Trademark Office (USPTO) evoked the patent to a large degree. Only three patent claims were maintained. The decision was based on the assumption that in case of the revoked patent claims the invention lacked the necessity of novelty.

Convention on biological diversity

The conflict between access and benefit interests and intellectual property rights is indeed broadly discussed but scarcely regulated. The only legally binding examination of these conflicting interests can be found in the aforesaid convention on biological diversity (CBD).

The convention's history of origins

The CBD developed from out-of-date structures of international environmental law. Following the United Nations Conference on the Human Environment in Stockholm in June 1972, the United Nations Environment Program (UNEP) was established as an organ of the UN general assembly. An ad-hoc task force, formed in 1987, eventually developed the convention on biological diversity.

Early in 1990 the task force came to the conclusion that an effective global regulation of this matter in the form of an 'umbrella convention', meaning an instrument that merely combines existing regulations and thereby tries to close present protection-gaps, was not possible. Thus the elaboration of an extensive international framework was necessary. After five rounds of negotiation in the ad-hoc task force, which was renamed 'Intergovernmental Negotiating Committee (INC) for a convention on biological diversity' in 1991, the convention was accepted on May 22nd 1992 in Nairobi.

In the context of the United Nations Conference on environment and development (UNCED) that took place from June 3rd to 14th 1992 in Rio de Janeiro, the convention was signed by 153 member states and the EU. After 30 ratifications it came into effect on December 29th 1993. Next to the non-legal texts of the Rio Declaration (UNEP 1992), the Forestry Principles Declaration (United Nations 1992a), the Agenda 21 (United Nations 1992b) and the United Nations Framework Convention on Climate Change (UNFCCC) (United Nations 1992c), the CBD is part of the so-called Rio Documents. Counted among the most controversial elements of the convention is the therein established distribution regime.

Access to technology as an essential element

Art. 16 Para. 3 of the convention is of vital importance to the collision of patent rights and rights of access. According to this regulation, each party to the contract has to take jurisdictional, administrational or political measures, if reasonable, in order to grant that the contract parties which provide genetic resources can access and pass on technology which uses these resources in line with agreed conditions. This possibility of access is established particularly with regard to developing countries and comprises even technology protected by patent rights and other rights of intellectual property. It is obvious that intellectual property rights are gravely impaired by this regulation. For this reason the convention further states in Art. 16 Para. 5: 'The Contracting Parties, recognising that patents and other intellectual property rights may have an influence on the implementation of this Convention, shall cooperate in this regard subject to national legislation and international law in order to ensure that such rights are supportive of and do not run counter to its objectives.' This creates a perfect contradiction. On one hand, access to patented technology should be enforced, while on the other hand the Convention and intellectual property rights have to be conciliated.

Conflict with the TRIPS Agreement

It is questionable how this contradiction can be brought to unison with standards of international trade law. Prevalent patent law theories see the reward of the innovator for the gain of technical knowledge of general public as the main reason why patent protection is granted (Grubb 2004: 14). This is based on the idea that one can conclude from the reward of the innovation to technical progress and from technical progress to a stimulating effect on economy. Conversely, in the case of no protection, economic stagnation is imminent. The legal protection of knowledge, gathered by access to genetic resources and revised by protection instruments of intellectual property, ensures amortisation of the – often vast – investments of research facilities. Yet a patent claim does not, as is often affirmed in the current debate, grant a right of property over the invention. Further a patent does not include the right to use the invention. It rather gives the patentee the right to exclusively use the invention and forbids third parties the use for industrial and commercial purposes (Spranger 2012: 1206).

Furthermore scientific studies enjoy the so-called research privilege and are not affected when patents are granted. To clarify the outcome of a patent grant for the matters here dealt with, the following case can be taken as an example: Plant P, which contains substance X, grows in country A. The plant is used for traditional medicine of the indigenous group I. Research company B analyses plant P and isolates X. B synthesises the substance and uses it for the production of a drug. B has the procedures of isolating and synthesising protected by a process patent. For substance X, B gets a substance patent; the novelty of the innovation is, according to the prevailing opinion, the allocation of the isolated substance (Mes 2005: § 3 no. 30).

There are mainly difficulties which arise from the extent of the patent's protection. Substance patents cannot be granted for a substance in its natural surroundings (*in situ*) but only for nature identical – isolated and synthesised – substances. But even if the nature identical substance is patented, conflicts can arise using the substance *in situ*. The extent of protection of an absolute substance patent includes the substance described in the patent's claim as well as every type of application of the substance, whereupon no difference is made between natural and nature identical substance anymore. It is of no concern whether the patentee detected the relevant function by himself, nor if he displayed it in the patent specification. This absolute effect of patents has been increasingly criticised over the past years in the field of innovations based on human genetic substances (Spranger 2002: 399–403). Partly the overcome 'substance paradigm' is considered to be in need of an amendment, whereas others support a broad substance protection. The latter view is based upon the idea that all patents on chemical substance should be treated equally and upon the necessity of an as extensive as possible protection of innovation (Hansen 2001: 477–93). Biopiracy shows similar conflict situations. Substance patents can impair the traditional use of the relevant substance in its country of origin. But, above all, bearers of knowledge are denied the possibility of their own commercial utilisation of the genetic resource. The unapproved use of the substance, meaning the use that is not compensated by payment of license fees to the patentee, would violate the patent and cause an infringement lawsuit.

The machinery of international patent law is often named as the cause of the described situations. Thus the law of intellectual property, which was essentially developed in the legal culture of today's industrial countries and hence shows a clear Western imprint, shall be amended, whereupon the 'legitimate interests of the south' would have to find adequate recognition (Stevenson 2000: 1120; Ramani 2001: 1147). Yet more moderate views are held as well, which – like the Convention on Biological Diversity – want to achieve a just balance of interests without intervening in the existing legal structure of patent law.

Attempts to modify the acknowledged regulations of patent law have to be met restrainedly for a number of reasons. For one: applicable international law has to be complied with. Existing commitments in the field of international patent law should not be erased, because they are perceived to be adverse in the field of genetic resources. Secondly, standards of international law notably reflect the well-balanced result of negotiations that often went on for decades and usually cannot be revived in rapid succession. Finally the view should not be narrowed on biopiracy issues only. Solution statements that give bearable results concerning the access to genetic resources, but lead to unbearable consequences in other fields of intellectual property, lack the required consistency. Hence during the course of the analysis the basic principles of international patent law have to be depicted to begin with. The offered approaches to the problem can then be discussed and analysed on this basis.

Central guidelines of international patent law

The set of rules and regulations in international patent law in today's form is mainly determined by the so-called TRIPS Agreement (TRIPS 1994), which – institutionally established in the WTO – is partly based on the principles of the agreements that were in force until now but partly also creates new law (Senti 2000: no. 1301). For the establishment of the TRIPS Agreement, it was not least of all decisive that many so-called developing countries had at best minimal protection of intellectual property rights in their national legal systems (Levy 2000: 789). At the same time, existing intellectual property rights were infringed more frequently and uninhibitedly by certain nations (Senti 2000: no. 1294), so that the establishment of specific guidelines could not be postponed.

Taking advantage of the possibility for extending the fixed time for coming into force in Art. 65 Para. 2, the TRIPS Agreement is now in force in most developing countries since January 1st 2000 (Levy 2000: 789–95) as well. The establishment of the TRIPS Agreement, which is attributable to the pressure of the industrial nations, was aimed at creating adequate worldwide protection of legitimate interests, especially by stopping widely spreading copyright piracy. However, it does not aim at overcharging indigenous groups in questions of using biological resources. Further, the Agreement merely tries to create a certain minimum standard, which needs to be observed worldwide. A complete standardisation of intellectual property positions, in the variety that they exist in various legal cultures, is neither intended nor practicable (Chiappetta 2000: 361).

The regulation that is vital for the issue of a patent is Art. 27 TRIPS. Disregarding whether it concerns products or a process after this regulation, patents have to be guaranteed for innovations in all fields of technology (Subramanian and Watal 2000: 403). In order for a patent to be granted it is required that the innovation is new, based on inventive work and commercially practicable (Art. 27 Para. 1). Hereby the TRIPS Agreement reflects common international standards of patent grants. Innovations can be excluded from patentability, if prevention from commercial use is necessary for the protection of public order or good manners, including the protection of life or health of humans, animals or plants or for avoidance of severe harm of the environment in the sovereign territory in question. A legal ban on utilisation does not, however, justify an exclusion of patentability. Furthermore, plants and animals, excepting micro-organisms, can be excluded from patentability as well as biological procedures of cultivation of plants and breeding of animals, excepting non-biological and microbiological procedures (Art. 27 Para. 3).

On first view, the previously displayed requirements open up various possibilities for a restrictive practice of patent granting. However, the regulations are meant to be exceptions and therefore have to be interpreted restrictedly. Hence they are not suited to be general guidelines in a patent law system.

The possibility of patent exclusion, as called for in Art. 27 Para. 2 TRIPS, especially for those procedures whose utilisation would infringe on public order or morality or would lead to a severe damage of the environment, is of no significance to the problem of biopiracy: not the prevention of the *commercial utilisation*, but the prevention of the access to the genetic resource, which takes place far before the patent grant, is of importance to environmental protection (Kunczik 2007: 103). Beyond that, it fits international standards to apply the described ban on patentability only if practically every utilisation that is intended or can reasonably come to consideration has to be rated as an infringement. In this context, weighing up risks and expected benefits is of importance (Mes 1997: § 2 no. 6). Also without importance to biopiracy is the ban on patentability of plants and animals. In case of such a restriction the members of the TRIPS agreement have to provide for the protection of plant species either through patents, an effective system sui generis or by combination of both approaches. Even though the members are free to exclude certain plant species from patentability, they are obliged to provide a comparable protection in the end (Footer 1999: 74). Hence Art. 27 Para. 3 TRIPS does not include a substantial limitation of intellectual property rights and cannot be used for questions of access to genetic resources. Regardless of the latter, the regulation only finds application on granting of patents on plants *in toto* anyhow, so that the utilisation of single genes or gene sequences does not meet the field of application of this regulation from the beginning.

The demand to exclude the countries of origin of genetic material from license requirement for patented products, which were developed from genetic resources, was just as unaccepted internationally (Götting 2004: 734) as the attempt to make the indication of origin obligatory for issuing of patents. After all, and this is crucial, the patents concerned are mostly not applied for in the countries of origin, but in those industrial nations that are considered output markets. Hence the restrictive issuing of patents in the countries of origins cannot influence or stop the issuing of patents in other countries.

Critique of a TRR-concept

To take account of the specific interests of the countries of origin in a sufficient way, but especially of the indigenous populations living there, one turns away from the basic principles of the TRIPS Agreement and towards the so-called concept of Traditional Resource Rights (TRR) (Posey and Dutfield 1998: 127). This concept, however, does not present a unified system but is on the contrary characterised by numerous facets and varieties (Federle 2005: 167). However, in the end, the proposals are similar insofar as the generation of a sui generis-patent system in the interest of the bearer of traditional knowledge is aimed at.

One of these variants, for example, contains the proposal to hand out every declaration of the respective village elders on the use of the concerned invention for the purpose of proof within the realm of patent issuance (Singh 1997: 233). Concerning the question of whether a certain procedure is new or has already been applied commercially for a long time, the questioning of the village elders shall offer the necessary legally binding clarity. Yet it is not clarified here in what way a law abusing declaration can be ruled out, nor is it critically questioned whether the statement of the village elder, which most of the time turns on oral tradition, is per se composed specifically enough to function as the foundation for an effective precept towards a third party and therewith another indigenous group. Completely unresolved is furthermore the question: in which cases is it at all permitted for single group members to reveal respective knowledge to third parties? In this context one has to think of 'secret knowledge', for example, which may – due to traditional rites of initiation – only be communicated to a small group of people, who do not need to be identical to the group leadership.

The classification of the postulation by patent law that 'farmers and women', who contributed to the development of new plant varieties, shall be adequately supported is also deficient (Verma 1997: 215). Equally imprecise and unsatisfying is the postulate, raised for instance in the so-called 1993 Mataatua Declaration (Mataatua Declaration: 117), that the commercialisation of traditional plants and medication must only be 'managed' by those indigenous groups who inherited this knowledge.

Generally, the concept of the TRR suffers from a narrowed view on individual interests. Such a view does not do justice to the requirements of the issue. Thus, numerous declared patents will not be retraceable to a single biological substance. If, however, it is about a composition of different substances, which may moreover stem from diverse countries (Sands 2003: 385), it remains totally open in which way there shall be an adequate arrangement between the respectively affected indigenous groups. A sui generis-solution, which neither determines the presuppositions for the conferral of certain rights nor their operating range or which dismisses the relation of the sui generis-system towards the patents and special protective rights, is not workable (Gold and Gallochat 2001: 364).

Moreover, the dual nature of genetic resources, which is in a legal respect attestable, oftentimes remains disregarded. As phenotypes, single plants can undoubtedly fall under an individual power of disposition. However, in view of the genotype, i.e. of the function of plants as bearers of genetic information, collective interests are affected (Straus 1998: XI 101). Against this backdrop it would be inadequate within the scope of a sui generis-patent system to assign rights of disposal on plants as 'information-bearers' only to a single indigenous group. In fact, the protagonists of the 'biopiracy-debate' regularly state that precisely the existing patent law would lead to such a monopolising of knowledge. However, this objection misunderstands that the existence of the research privilege in patent law as well as the opportunity for compulsory licencing leads to the circumstance that – temporally though restricted – patent rights do principally not prevent the appreciation of opposing, justified interests. An exclusive and unrestricted right of disposal of a certain group does not have comparable options.

The proposals for the development of a specific protection of the knowledge of indigenous peoples by patent law can thus not be reconciled with out-of-date standards of a possibly universal comprehension of law – and at times effects the reverse. Thus, in the context of the TRR-discussion, it is considered to prohibit every action that worsens the legal position once granted to the indigenous peoples by patent law, as for instance the assignment transfer or disposal of the law (Singh 1997: 240). By this means it shall be eliminated that members of the respective peoples become outsmarted. The attempt to classify the right to traditional knowledge as 'joint property' apparently rests on the same motivation (Singh 1997: 240). Along these lines every opportunity to an individual right of disposal gets excluded. The protection of the indigenous peoples mutates via this approach to a protection against oneself, thus leading to a partial incapacitation.

Effects on the demarcation between invention/discovery

Besides these specific points of criticism one has to announce concerns of basic nature. The adaptation of the allegedly 'westerly coined' patent law to the needs of the developing countries, especially to the needs of the indigenous peoples living there, rescinds the central distinction between patentable inventions and unpatentable discoveries. This indispensable demarcation of utmost importance especially for the realm of natural substances (Spranger 2000a: 373–80; Spranger 2000b: 170–4) proceeds from the following premises: only the detection of a new feature of a known material or commodity represents a discovery; locating something already existing is mere knowledge (BPatG 1978b: 238–9). Such a discovery is indubitably unpatentable. If, however, a practical utilisation potential can be found for the feature in question, the person concerned then gives an instruction for technical action on the basis of her knowledge and makes it thus count for an invention. In the presence of the other patenting requirements – novelty, inventive step and commercial applicability – this invention is principally patentable (BGH 1975: 430–1; BPatG 1978a: 702f.). The additional contribution to technology – that must stem from the human – thus makes up the essential differentiation criterion (Luttermann 1998: 918).

If one transfers these criteria to the case of medically applicable genetic resources – for instance, in the form of a traditional medical plant – the mere knowledge of the effect of a certain medical plant is certainly no invention, but a mere discovery. The same holds for the case in which the respective plant is not further processed, but is applied in its natural form. If the plant is brewed for tea or gets crushed, pulverised, or the like, it will also lack an instruction to technical action; whereas the prescription for the production of a healing substance in the way of a combination of different plants may possibly be classified as an instruction to technical action that justifies talking of an invention principally worthy of protection.

The thorough inclusion of traditional healing methods into the scope of protection of patent law would consequently lead to a destabilisation if not dissolution of the established line between invention and discovery. This procedure would not have to be further criticised if one understood the issued principles, in line with the proponents of an amendment, simply as a dispensable manifestation of a 'Western apprehension' of patent protection. If one however considers the consequences of such an evaluation, it becomes clear that the differentiation between invention and discovery is about a principle which independently of its origin demands for unconditional and universal validity. Otherwise, also laws of nature – as mere discoveries – could prospectively be patented. Unforeseeable and unbearable relationships of dependence that paralyse economy and massively impair the freedom of each individual could be the consequence.

If one also understands these problems – again in line with the advocates of an amended patent law – merely as specific manifestations of a patent law of Western coinage, the question about sustainable alternatives comes up. Here the only fallback procedure seems to be not to grant a general patent protection, neither to discoveries nor to inventions. An almost entire blocking of technical progress would be the aftermath of such a utopian reorientation. Historical evidence suggests that insufficient protection of intellectual property and technological stagnation are mutually dependent (Kurz 2000: 9). Only the 'reward' of being conferred an exclusive right for a limited period of time constitutes an adequate incentive for the researcher to further advance his works. This conviction already coined a Venetian patent application by Galileo Galilei dated to the year 1593, where he outlines the impact of the protective right as requested by him as follows: 'As a result I will even be more eagerly anxious to create new inventions for the common good [...]'(Kurz 2000: 65).

Intermediate result

The principles of the TRIPS Agreement are not open to essential amendments, not only with regard to the political landscape. Maintenance of these instruments also benefits so-called developing countries if they are applied in a clever way. After the patent system has been rejected here for a long time (Stevenson 2000: 1126), in many states of the southern hemisphere – as for instance in Brazil and Columbia – the interest in the activation of instruments from patent law, to promote own concerns, grows (Stone 1993: 35). The regime of standards of world trade law in this area is increasingly judged to be positive. In addition to the possibility of protecting one's own position, attracting technical knowledge and foreign investment capital for building up the own economy is of great relevance (Stevenson 2000: 1127). With the suggested amendment of the fundamental principles of International Patent Law, developing countries would thus deprive themselves of the opportunity to an extensive economical utilisation of resources.

Further, the TRIPS Agreement only in a strict controlled extent contains variations from the principle of a widespread patent grant. Thus, Art. 30 TRIPS defines that the exceptions may not inadequately contradict the ordinary usage of patents and that the justified interests of patent holders must not inadequately be impaired, whereas the justified interests of third parties have to be considered as well. One can say that the interests of indigenous peoples in the utilisation of their traditional knowledge are certainly substantial. Furthermore it resists the general sense of justice that, for example, traditional healing methods serve as a basis for the patenting of inventions and the bearers of this knowledge cannot at all profit from this economic usage. Nevertheless this circumstance alone may not be sufficient to justify an exception from patent protection according to Art. 30 TRIPS. On the one hand the standard constellation of the utilisation of a patent would thus be impaired in a fundamental way. On the other hand considerations of justice would find entrance into patent law, which are of a kind that can hardly be marked-off against other – ethical, moral or political – motives (Spranger 2000a: 373).

Therewith it remains to be recorded: Art. 16 of the Convention on Biological Diversity does not aim at a fundamental modification of the intellectual property rights. In the case of a different interpretation massive resistance has always been announced on the part of the United States. The European Union, who is the contracting party of the Convention on Biological Diversity, has also pointed out explicitly that rights of access and transfer according to Art. 16 of the Convention may only be granted in accordance with the principles and rules of intellectual property (93/626/EEC). One can respond to the accusation that the regulation of Art. 16 Para. 3 would get robbed of its central regulatory content with reference to the retention of the general claim for the generation of a facilitated opportunity of access to technology (Burhenne-Guilmin and Casey-Lefkowitz 1992: 55). For the goal achievement there is, however, no need for a modification of international patent law.

The relevance of the common concern-concept

In the following, the relevance of the classification of genetic resources by the Convention on Biological Diversity, as an element of the 'common concern of mankind', shall be scrutinised in the context of the set of issues at hand. The attempt to declare the diversity of species and their genes to 'common heritage of mankind' (Ossorio 2007: 425–39) – thus assigning it to an acknowledged utilisation-category of international law – could not be realised politically (Guruswamy and Hendricks 2007: 94; Beyerlin 2000: no. 390). Instead, biological diversity is henceforth qualified as 'common concern of mankind'. Under international law, the concept of 'common concern' is indeed closely related to the category of 'common heritage' (Biermann 1996: 426–31). In contrast to this, it is however characterised by a less marked institutionalisation of the power of utilisation and disposition (Herdegen 2007: §5 no. 14). In the foreseeable future there will thus not be a constellation comparable to something like the seabed-regime (Herdegen 2000: 633) for the realm of genetic resources (Proelß 2007: 650–6).

From the viewpoint of developing countries, which originally contributed in an essential way to the development of a common heritage-principle as driving forces, its practice is not desirable, because it would imply a loss of control over their genetic resources (Beyerlin 2000: 390). In fact the fear of a loss of controlling opportunities is the cause for the circumstance that southern countries consider at the most a restrictive handling of the 'common heritage'-principle to be practicable in the field of genetic resources (Hunter, Salzman and Zaelke 1998: 343). At this point, the certainly understandable and common basic trend manifests itself anew, to only claim compliance with internationally acknowledged principles if this corresponds to one's own advantage. Thus the qualification of genetic resources as 'common concern of mankind' has no impact on the assessment stated above.

Proposal for solution: individual contracts within legal frameworks

The solution concepts discussed so far are, as proven, insufficient. But how can access and benefit interests be brought to unison with rights of intellectual property? One possible solution could be the arrangement of individual contracts. The treaties could still be set within a legal framework, which provides, inter alia, a system of basic requirements as well as sanctions. Within the specific treaties general provisions would then be put into concrete terms. Thus they can, other than general regulations, react to specific or individual needs and circumstances. Art. 19 Para. 1 of the Nagoya Protocol partly brings this concept to realisation already: 'Each party shall encourage, as appropriate, the development, update and use of sectoral and cross-sectoral model contractual clauses for mutually agreed terms.'

Of course, the concept of drafting an individual contract is lengthy and certainly difficult. But experience shows that it can work. The Merck-InBio contract (Henne 1998: 61) is the most prominent example of how access and benefit interests on one hand and interests of commercial utilisation on the other can be balanced by individual terms. The National Biodiversity Institute of Costa Rica (InBio) collected and processed plant, insect and soil samples. Against the payment of 1 million dollars, Merck, a pharmaceutical company from the United States, was supplied with such samples and granted the possibility to conduct pharmaceutical research with the genetic resources. In case of the development of a new drug from these genetic resources, Merck was granted the right of patenting the drug while on the other hand InBio was granted a profit share. The profit InBio made by this arrangement was used for purchasing better laboratory equipment as well as being invested in Costa Rica's national parks. It is, however, questionable whether a deal like this can be repeated. The problem is that comparable treaties call for a strong scientific base and a considerable degree of political stability. Only if these two parameters are given can the concept of individual treaties work. Furthermore it can be said that on the contractual level at least six basic problems have to be faced:

- First of all, a general question has to be discussed: is the application of 'Western law' the right approach?
- Disregarding which law will be applied in the future, it secondly has to be cleared how substantial benefit sharing can be guaranteed.
- If the system of arranging individual contracts is applied, then who of the indigenous groups should be entitled to negotiate on the local or community level? Should it be individuals or the whole community, a single chief, a 'traditional healer' or a council of elders?
- Subsequently it has to be asked, who is competent or qualified to support local or indigenous communities? Is it the communities themselves or should state authorities, lawyers, IOs or NGOs interact?
- All approaches to solving the conflict could further be diminished if one party refuses to use contractual elements as such ('two to tango').
- Finally, it should be taken into account that a large variety of contracts brings a variety of possible contractual situations, which could be a hazard to legal certainty and legal clarity. As positive as the arrangement of individual contracts might seem, a 'battle of the contracts' is undesirable.

These issues should not be misunderstood as a list of paralysing obstacles. Instead, similar problems have to be solved – and are solved – in every bi- or multilateral contract. To put it in a nutshell: if they are adequately used to concretise more general rights and obligations enshrined in international treaties or domestic laws, contractual instruments are the right tool for the fine-tuning of access and benefit sharing projects.

Note

1 Cf. the Nagoya Protocol Signature List. Online. Available HTTP: <http://www.cbd.int/abs/nagoya-protocol/signatories> (accessed 13 June 2013).

References

Andean Community (1996) *Cartagena Agreement. Andean Subregional Integration Agreement.* Online. Available HTTP: <http://www.comunidadandina.org/en/TreatiesLegislation.aspx?id=16&title=cartagena-agreement&accion=detalle&cat=1&tipo=TL> (accessed 13 June 2013).

Beyerlin, U. (2000) *Umweltvölkerrecht*, München: C.H. Beck.

BGH (Bundesgerichtshof) (1975), 'Bäckerhefe', *GRUR*: 430–1.

Biermann, F. (1996) '"Common Concern of Humankind": The Emergence of a New Concept of International Environmental Law', *Archiv des Völkerrechts* 34: 426–31.

Blakeney, M. (1997) 'Access to Genetic Resources: The View from the South', *Bio-Science Law Review* 2: 94–100.

Blakeney, M. (1999) 'Biotechnology, TRIPS and the Convention on Biological Diversity', *Bio-Science Law Review* 4: 144–50.

Bonn Guidelines – United Nations Environment Programme (UNEP) (2002) *Bonn Guidelines on Access to Genetic Resources and Fair and Equitable Sharing of the Benefits Arising out of their Utilisation*, COP 6 Decision VI/24. Online. Available HTTP: <www.cbd.int/decision/cop/?id=7198> (accessed 13 June 2013).

BPatG (Bundespatentgericht) (1978a) 'Menthonthiole', *GRUR*: 702f.

BPatG (Bundespatentgericht) (1978b) 'Naturstoffe', *GRUR*: 238f.

Burhenne-Guilmin, F., Casey-Lefkowitz, S. (1992) 'The Convention on Biological Diversity: A Hard Won Global Achievement', *Yearbook of International Environmental Law*: 43–59.

Cartagena Protocol – Secretariat of the Convention on Biological Diversity (2000) *Cartagena Protocol on Biosafety to the Convention on Biological Diversity: Text and Annexes.* Online. Available HTTP: <http://bch.cbd.int/protocol/text/> (accessed 13 June 2013).

CBD – United Nations (1992) *Convention on Biological Diversity.* Online. Available HTTP: <http://www.cbd.int/doc/legal/cbd-en.pdf> (accessed 12 June 2013).

Chaves, J. (2004) 'The Andean Pact and Traditional Environmental Knowledge', in N.P. Stoianoff (ed.) *Accessing Biological Resources – Complying with the Convention on Biological Diversity*, The Hague: Kluwer Law International, 223–59.

Chiappetta, V. (2000) 'The Desirability of Agreeing to Disagree: The WTO, TRIPS International IPR Exhaustion and a Few Other Things', *Michigan Journal of International Law*: 333–92.

Cottier, T. (1998) 'The Protection of Genetic Resources and Traditional Knowledge: Towards More Specific Rights and Obligations in World Trade Law', *Journal of International Economic Law*: 555–84.

FAO (Food and Agriculture Organization of the United Nations) (2004) *International Treaty on Plant Genetic Resources for Food and Agriculture.* Online. Available HTTP: <ftp://ftp.fao.org/docrep/fao/011/i0510e/i0510e.pdf> (accessed 13 June 2013).

FAO (Food and Agriculture Organization of the United Nations) (1983) *International Undertaking on Plant Genetic Resources for Food and Agriculture.* Online. Available HTTP: <http://www.fao.org/ag//CGRFA/iu.htm> (accessed 13 June 2013).

Federle, C. (2005) *Biopiraterie und Patentrecht*, Baden-Baden: Nomos.

Footer, M.E. (1999) 'Intellectual Property and Agrobiodiversity: Towards Private Ownership of the Genetic Commons', *Yearbook of International Environmental Law*: 48–81.

Gold, E.R. and Gallochat, A. (2001) 'The European Biotech Directive: Past as Prologue', *European Law Journal*: 331–66.

Götting, H.-P. (2004) 'Biodiversität und Patentrecht', *GRUR Int.*: 731–6.

Grubb, P.W. (2004) *Patents for Chemicals, Pharmaceuticals and Biotechnology*, 4th edn, Oxford: Oxford University Press.

Guruswamy, L.D. and Hendricks, B.R. (eds) (2007) *International Environmental Law in a Nutshell*, St. Paul: Thomson/West.

Hansen, B. (2001) 'Hände weg vom absoluten Stoffschutz – auch bei DNA-Sequenzen', *Mitteilungen der deutschen Patentanwälte*: 477–93.

Henne, G. (1998) *Genetische Vielfalt als Ressource*, Baden-Baden: Nomos.

Herdegen, M. (2000) 'Die Erforschung des Humangenoms als Herausforderung für das Recht', *Juristenzeitung*: 633–41.

Herdegen, M. (2007) *Völkerrecht*, 6th edn, München: Beck.

Hunter, D., Salzman, J. and Zaelke, D. (1998) *International Environmental Law and Policy*, New York: Foundation Press.

Ipsen, K. (2004) *Völkerrecht*, 5th edn, München: Beck.

Kawai, M. (1995) 'Biotechnology and Biological Diversity', in H.-J. Rehm and G. Reed (eds) *Biotechnology, Vol. 12, Legal, Economic and Ethical Dimensions*, 2nd edn, Weinheim: VCH, 434–45.

Kunczik, N. (2007) *Geistiges Eigentum an genetischen Informationen*, Baden-Baden: Nomos.

Kurz, P. (2000) *Weltgeschichte des Erfindungsschutzes*, Köln: Carl Heymanns Verlag.

Levy, C.S. (2000) 'Implementing TRIPS – A Test of Political Will', *Law and Policy in International Business*: 789–95.

Luttermann, C. (1998) 'Patentschutz für Biotechnologie, Die europäische Richtlinie über den rechtlichen Schutz biotechnologischer Erfindungen', *Recht der Internationalen Wirtschaft*: 916–20.

Mataatua Declaration – *Mataatua Declaration on the Cultural and Intellectual Property Rights of Indigenous Peoples', First International Conference on the Cultural & Intellectual Property Rights of Indigenous Peoples, Whakatana, 12–18 June 1993, Aotearoa, New Zealand.* Online. Available HTTP: <www.ankn.uaf.edu/IKS/mataatua.html> (accessed 13 June 2013).

Mes, P. (1997) *Patentgesetz*, München: Beck.

Mes, P (2005) *Patentgesetz*, 2nd edn, München: Beck.

Mishra, J.P. (2000) 'Biodiversity, Biotechnology and Intellectual Property Rights', *The Journal of World Intellectual Property*: 211–24.

Mugabe, J., Barber, C.V., Henne, G., Glowka, L. and La Viña, A. (eds) (1997) *Access to Genetic Resources – Towards Strategies for Sharing Benefits*, Nairobi: ACTS Press.

Nagoya Protocol – Secretariat of the Convention on Biological Diversity (2010) *Nagoya Protocol on Access to Genetic Resources and the Fair and Equitable Sharing of Benefits arising from their Utilization to the Convention on Biological Diversity: Text and Annex, COP 10 Decision X/1.* Online. Available HTTP: <http://www.cbd.int/abs/doc/protocol/nagoya-protocol-en.pdf> (accessed 13 June 2013).

Nair, G., 'European Patent Office revokes WR Grace's Neem Patent', *The Financial Express* 18.05.2000. Online. Available HTTP: <http://www.financialexpress.com/old/fe/daily/20000518/fec18041.html> (accessed 13 June 2013).

Neemfoundation (2008) *Patent on Neem*. Online. Available HTTP: <http://www.neemfoundation.org/neem-articles/patents-on-neem.html> (accessed 13 June 2013).

Normand, V. (2004) 'Access to Genetic Resources and the Fair and Equitable Sharing of Benefits Arising out of Their Utilization: Developments Under the Convention on Biological Diversity', *Journal of International Biotechnology Law*: 133–41.

OECD (Organisation for Economic Co-operation and Development) (1996) *Intellectual Property, Technology Transfer and Genetic Resources. An OECD Survey of Current Practices and*

Policies. Online. Available HTTP: <http://www.oecd.org/science/biotech/1947170.pdf> (accessed 13 June 2013).

Ossorio, P.N. (2007) 'The Human Genome as Common Heritage: Common Sense or Legal Nonsense?', *The Journal of Law, Medicine & Ethics*: 425–39.

Pearce, D. and Puroshothaman, S. (1995) 'The Economic Value of Plant-Based Pharmaceuticals', in T. Swanson (ed.) *Intellectual Property Rights and Biodiversity Conservation*, Cambridge: Cambridge University Press, 127–38.

Posey, D.A. and Dutfield, G. (1998) 'Plants, Patents and Traditional Knowledge: Ethical Concerns of Indigenous and Traditional Peoples', in G. Overwalle (ed.) *Patent Law, Ethics and Biotechnology*, Bruylant: Brussels. 109–32.

Primlane, R. (1999) 'Heureka – Es ist der Neem-Baum', *Gen-ethischer Informationsdienst* 131: 23–6.

Proelß, A. (2007) 'Die Bewirtschaftung der genetischen Ressourcen des Tiefseebodens – Ein neues Seerechtsproblem ?', *Natur und Recht*: 650–6.

Raj, R.D. (2000) 'Biopiracy-friendly Laws Worry Neem Battle Winner', *Asia Times* 27.05. Online. Available HTTP: <http://www.atimes.com/ind-pak/BE27Df01.html> (accessed 13 June 2013).

Ramani, R. (2001) 'Market Realities v. Indigenous Equities', *Brooklyn Journal of International Law*: 1147–76.

Sands, P. (2003) *Principles of International Environmental Law Vol. I – Frameworks, Standards, and Implementation*, 2nd edn, Cambridge: Cambridge University Press.

Senti, R. (2000) *WTO – System und Funktionsweise der Welthandelsordnung*, Zürich: Schulthess Verlag.

Singh, N. (1997) 'Developing a "Rights Regime" in Defence of Biodiversity and Indigenous Knowledge', in J. Mugabe, C.V. Barber, G. Henne, L. Glowka and A. La Viña (eds) *Access to Genetic Resources – Towards Strategies for Sharing Benefits*, Nairobi: ACTS Press, 233–4.

Spranger, T.M. (2000a) 'Ethical Aspects of Patenting Human Genotypes According to EC Biotechnology Directive', *International Review of Industrial Property and Copyright Law*: 373–80.

Spranger, T.M. (2000b) 'Patente auf Leben ?', *Sozialrecht und Praxis*: 170–4.

Spranger, T.M. (2002) 'Stoffschutz für "springende Gene"? – Transposons im Patentrecht', *GRUR*: 399–403.

Spranger, T.M. (2008) *Rechtliche Rahmenbedingungen für Access and Benefit Sharing-Systeme*, Bonn: Institut für Wissenschaft und Ethik.

Spranger, T.M. (2012) 'Case C-34/10, Oliver Brüstle v. Greenpeace e.V., Judgment of the Court (Grand Chamber) of 18 October 2011', *Common Market Law Review* 49: 1197–209.

Stevenson, G.R. (2000) 'Trade Secrets: The Secret to Protecting Indigenous Ethnobiological (Medicinal) Knowledge', *Journal of International Law and Politics*: 1119–74.

Stoll, P.-T. and Schillhorn, K. (1998) 'Das völkerrechtliche Instrumentarium und transnationale Anstöße im Recht der natürlichen Lebenswelt', *Natur und Recht*: 622–5.

Stone, C.D. (1993) *The Gnat is Older than Man*, Princeton, NJ: Princeton University Press.

Straus, J. (1998) 'Biodiversity and Intellectual Property', *AIPPI Yearbook* IX: 99–119.

Subramanian, A. and Watal, J. (2000) 'Can TRIPS Serve as an Enforcement Device for Developing Countries in the WTO ?', *Journal of International Economic Law*: 403–16.

TRIPS – World Trade Organization (WTO) (1994) *Agreement on Trade-Related Aspects of Intellectual Property Rights*. Online. Available HTTP: <http://www.wto.org/english/tratop_e/trips_e/t_agm0_e.htm> (accessed 13 June 2013).

UNEP (United Nations Environment Programme) (1992) *Rio Declaration on Environment and Development.* Online. Available HTTP: <http://www.unep.org/Documents.Multilingual/Default.Print.asp?DocumentID=78&ArticleID=1163&l=en> (accessed 13 June 2013).

United Nations (1992a) *Forest Principles Declaration.* Online. Available HTTP: <http://www.un.org/documents/ga/conf151/aconf15126–3annex3.htm> (accessed 13 June 2013).

United Nations (1992b) *United Nations Conference on Environment & Development Rio de Janerio, Brazil, 3 to 14 June 1992, AGENDA 21.* Online. Available HTTP: <http://sustainabledevelopment.un.org/content/documents/Agenda21.pdf> (accessed 13 June 2013).

United Nations (1992c) *United Nations Framework Convention on Climate Change.* Online. Available HTTP: <http://unfccc.int/resource/docs/convkp/conveng.pdf> (accessed 13 June 2013).

Verma, S. K. (1997) 'Biodiversity and Intellectual Property Rights', *Journal of the Indian Law Institute*: 203–15.

WIPO (World Intellectual Property Organization) (2000) *Intergovernmental Committee on Intellectual Property and Genetic Resources, Traditional Knowledge and Folklore.* Online. Available HTTP: <http://www.wipo.int/tk/en/igc/> (accessed 13 June 2013).

10 The legal protection of biodiversity in Namibia[1]

Manfred O. Hinz and Oliver C. Ruppel

Introduction

Biological resources can without exaggeration be described as the fundamental building blocks for social and cultural development. They provide the basis for local food sufficiency, and are hence the backbone for many countries' economies (Ruppel 2009a). For millennia, people have relied on ecosystems to meet their basic needs such as food, water and other natural resources. Apart from these, there are a multitude of further benefits of biodiversity – it is regarded as a global asset and expected to benefit people in all parts of the world (McNeely et al. 1990). For instance, a significant proportion of drugs are derived, directly or indirectly, from biological sources. This interest in biological resources is by no means a recent one. As early as the mid-nineteenth century, the Scottish adventurer and missionary David Livingstone brought plants from the African continent, hoping they would serve as a basis for medicinal drugs (Blaikie 2004). Over the last decade, the interest in drugs of plant origins and their use in various diseases has increased in many industrialised countries since plants used in traditional medicine are more likely to yield pharmacologically active compounds (Paing et al. 2006: 1). Indeed, in most cases, it is impossible to synthesise plant-based medicinal drugs in a laboratory setting. Moreover, a wide range of industrial materials are derived directly from biological resources. These include building materials, fibres, dyes, resins, gums, adhesives, rubber and oil. And there are further advantages in biodiversity: increased biodiversity also constrains the spread of contentious diseases as viruses will need to adapt if they are to infect different species. Many people also derive value from biodiversity through leisure activities. And finally, many cultural groups view themselves as an integral part of the natural world and show respect for other living organisms.

Given all these reasons, biological diversity has to be safeguarded and conserved (Groombridge 1992: xvi).[2] Efforts to maintain the diversity of biological resources are urgently required at local, national and international-level southern Africa and, as part of this region, Namibia is no exception. Van Wyk and Gericke introduced their publication, titled *People's Plants*, by stating the following:

> Southern Africa is exceptionally rich in plant diversity with some 30,000 species of flowering plants, accounting for almost 10% of the world's higher plants. The region also has great cultural diversity, with many people still using a wide variety of plants in their daily lives for food, water, shelter, fuel, medicine and the other necessities of life.
>
> (Van Wyk and Gericke 2000: 7)

In the last few decades, however, the southern African region has seen great changes in access to modern health care and education, shifts from rural to urban areas, changes from subsistence farming to cash-crop production, greater flows of migrant labour, and unprecedented environmental degradation. These changes in the socio-cultural and environmental landscape have severely eroded the indigenous knowledge base, and this loss in turn endangered Namibia's biodiversity. At the same time Namibia's legislation can be taken as an example for successful implementation of national and international environmental legislation.

In the following we shall provide a synoptic overview on the legal biodiversity protection efforts in Namibia. Based upon some general remarks on biodiversity and the relevant legislative history of Namibia, the application of international agreements on the protection of biodiversity will be outlined, and detailed information on the most important regulations on the local and the national level which have been developed within the framework of international agreements and Namibia's constitution will be given. Last but not least, information on current political efforts in respect to further developing and implementing environmental protection laws in Namibia will be sketched out.

Biological diversity in Namibia

The coinage of the term biological diversity can be attributed to Lovejoy (1980), Norse and McManus (1980: 32) and Wilson (1985: 400). Lovejoy was probably the first person to use the term in 1980. In the time of its infancy, the concept of biological diversity (now more commonly biodiversity) comprised an estimate of roughly 1.5 million described species living on earth. Today's estimates range wider, largely because we now realise that most living species are micro-organisms and tiny invertebrates. Estimates range from 5 to 30 million species. Roughly 1.75 million species have been formally described and given official names. The number of unclassified species is much higher. Currently, biological diversity tends to be defined as the variability among living organisms from all sources, including terrestrial, marine and freshwater ecosystems, which includes diversity within species, between species, and habitats or ecosystems (CDB: Art. 2). This describes most circumstances and presents a unified view of the traditional three levels at which biodiversity has been identified: genetic diversity, referring to the diversity of genes within a species. There is a genetic variability among the populations and the individuals of the same species. Species diversity means the diversity among species in an ecosystem; and ecosystem diversity describes diversity at a higher level of organisation, the ecosystem. Ecosystem diversity refers to all the various habitants, biological communities and biological processes as well as the variations and interconnections and interrelations between and/or among various ecosystems.

Namibia's biodiversity includes innumerable species of wild plants and animals. Indeed, as little as about 20% of Namibia's wildlife species have been captured scientifically to date. More than 13,000 species have been described, of which almost 19% are endemic or unique to Namibia (Government of the Republic of Namibia 2004a: 164).

Netumbo Nandi-Ndaitwah, the then Minister of Environment and Tourism, once outlined Namibia's large biodiversity endowment as follows:

> Namibia has a large biodiversity endowment, which is of global significance. Although predominantly a semi-arid country, Namibia contains a remarkable variety of ecosystems, ranging from hyper-arid deserts with less than 10mm of rainfall to subtropical wetlands and savannahs receiving over 600mm of precipitation per annum. Four major terrestrial biomes exist, namely: Succulent Karoo, Nama Karoo, Desert and Tree and Shrub Savannah. On a finer scale, 29 different vegetation types are currently recognised, many of which are wholly unique to Namibia or to the southern African sub-continent. These biomes are storehouses of high species richness: the country harbours 4,000 species and subspecies of higher plants and 658 species of birds have been recorded, of which approximately 30% is migrant. 217 species of mammals are found including unique arid varieties of desert-adapted rhino and elephant. This biodiversity richness generates global and national benefits through protecting globally important ecosystems.
>
> (Nandi-Ndaitwah 2010)

By 2006, the International Union for Conservation of Nature (IUCN) had classified 79 species in Namibia as threatened, which includes those species listed as critically endangered, endangered or vulnerable.[3] The 2011 IUCN Red list[4] counts a total of 95 threatened (critically endangered, endangered and vulnerable categories only) species in Namibia (12 mammals, 25 birds, 4 reptiles, 1 amphibian, 27 fishes and 26 plants).

Five major threats have been identified as threats to biodiversity:

- *Habitat loss, alteration, and fragmentation*: mainly through conversion of land for agricultural, aquaculture, industrial or urban use; damming and other changes to river systems for irrigation, hydropower or flow regulation; and damaging fishing activities.
- *Over-exploitation of wild species populations*: harvesting of animals and plants for food, materials or medicine at a rate above the reproductive capacity of the population.
- *Pollution*: mainly from excessive pesticide use in agriculture and aquaculture; urban and industrial effluents; mining waste; and excessive fertiliser use in agriculture.

- *Climate change*: due to rising levels of greenhouse gases in the atmosphere, caused mainly by the burning of fossil fuels, forest clearing and industrial processes.
- *Invasive species*: introduced deliberately or inadvertently to one part of the world from another; they then become competitors, predators or parasites of native species. (WWF 2010: 12)

This list indicates that Namibia's situation is not dissimilar to the situation of other parts of the world. Its example rather allows us to understand which aspects need to be taken care of politically and legislatively in other parts of the world as well.

For most of human history, the natural world has been protected from the most disruptive human influences by relatively humble technology; cultural-ecological factors, such as taboos preventing over-exploitation; inter-tribal peace, maintained by keeping wide areas of wilderness 'buffer zones' between groups; land ownership by ancestors or lineages rather than individuals; relatively sparse human populations; and many other factors (McNeely et al. 1990: 18). Today, most countries rich in biodiversity have national parks and national legislation promoting conservation. Most governments have joined international conservation conventions, and built environmental considerations into the national education system.

In addition, Non-Governmental Organisations (NGOs) are active in promoting public awareness of conservation issues, including those dealing with biological diversity. Naturalists, including interested amateurs and trained biologists, have initiated the conservation movement. Still, devastation continues. Why?

One possible answer is that while their contributions have been fundamental, they are unable to fully address the basic problems of conservation because the problems are not only biological, but rather political, economic, social, and even ethical. Practical decisions affecting the natural environment are prone to pressures. Conservation action, therefore, needs to be based on the best available scientific information and be implemented by development practitioners, engineers, sociologists, anthropologists, agronomists, economists, lawyers and politicians. Local resource users are often the ones who make local-level decisions, and their decisions are, above all, affected by enlightened self-interest. Those seeking to conserve biodiversity need to be able to identify the legitimate self-interest of rural people, and design ways of ensuring that the interest of conservation and community coincides. Considering these points, it becomes comprehensible why biodiversity protection has been given high importance under environmental law in Namibia.

But how can the law or the legal sciences contribute to the conservation of biodiversity in Namibia?

The aim of environmental protection in general and biodiversity maintenance in particular can be achieved by different means (Barnard 1998: 283ff.). Traditional legal methods, *inter alia*, include the establishment of protected areas, regulation of harvesting and trade in certain species or prohibition of the introduction of new, alien or invasive species. Pollution control and the management of hazardous substances are other effective mechanisms to contribute to the preservation of biological diversity. Other innovative regulatory techniques or policies to preserve biological diversity include access to genetic resources and biotechnology as well as access to and transfer of technology. All the aforementioned methods are to a certain extent governed by legal mechanisms, and the success or failure of Namibia's effort to control, manage, and maintain the sustainable use of biodiversity depends to a large extent on the effectiveness of the different legal instruments in place.

Namibia's legislative context[5]

The inhospitable Namib Desert constituted a barrier to European colonisation until the late eighteenth century, when traders and missionaries first explored the area. In 1878, the United Kingdom annexed Walvis Bay on behalf of the Cape Colony, while the rest of south western Africa would soon thereafter fall under German administration, henceforth to be known as German South West Africa. Resulting from the Herero and Nama wars of anti-colonial resistance of 1904–1908 (cf. Ruppel 2009b and Ruppel 2011: 33), Germany consolidated its hold over the colony, and prime grazing land passed to white control. German overlordship ended during World War I in the wake of South Africa's military occupation of the German colony. On 17 December 1920, South Africa took over the administration of South West Africa in terms of Article 22 of the 1919 Peace Treaty of Versailles (which incorporated the Covenant of the League of Nations) and a mandate agreement by the League Council. South Africa was mandated with the power of administration and legislation over the territory.[6] After more than a century of domination by other countries and a long struggle on both diplomatic and military levels, Namibian Independence was achieved and officially declared on 21 March 1990.

Today, Namibian law reflects the country's history. The law in place is the product of different sources: firstly, Roman law; secondly, the fusion of Roman law and Roman-Dutch customary law – hence the term Roman-Dutch law – which came in the wake of Dutch colonisation at the Cape of Good Hope; thirdly, from the early nineteenth century onwards English law asserted itself, leaving deep traces in Roman-Dutch law, after British hegemony in southern Africa had been established; and fourthly, indigenous customary law from time immemorial (Hinz and Santos 2002). After independence a broad spectrum of statutory laws and policies has been developed on the national level and the important role of international law has been firmly anchored in Namibia's constitution.

International environmental law pertinent to biodiversity protection in Namibia

Article 144 of the Namibian constitution incorporates international law explicitly as law of the land; it needs no legislative act to become so. International law is thus *ab initio* integrated into domestic law. An international treaty will be binding in terms of Art. 144, if the relevant international and constitutional requirements have been met in terms of the law of treaties, and the Namibian constitution. International agreements, therefore, will become Namibian law when they come into force for Namibia (Erasmus 1991: 102).

On the global level, several multilateral environmental agreements have been established that directly or indirectly contain provisions relating to the protection of biological diversity. Since the Convention on Biological Diversity (CBD) and the Convention on International Trade in Endangered Species of Wild Fauna and Flora (CITES) are the most relevant international biodiversity-related agreements, we will start our overview on the Namibian environmental legislation by sketching out their history and purpose.[7]

The Convention on Biological Diversity (CBD)

There was no consensus regarding biodiversity, on how to regulate, protect and preserve global environmental resources, among the nations of this world until the 1992 Earth Summit in Rio. It was at this Summit, the first of its kind at international level, where consensus was reached among scientists, policymakers and civil society that humanity was in the process of unconsciously depleting an invaluably important resource central to our food, health and economic security. The consensus reached at the Summit was in the form of a legal instrument, the CBD. The document was signed by Namibia on 12 June 1992 in Rio de Janeiro and ratified on 18 March 1997. Accordingly, Namibia is obliged to ensure that its domestic legislation is in conformity with the objectives and obligations of the CBD. Namibia gives effect to the CBD *inter alia* by implementing the National Biodiversity Strategy and Action Plan and has issued its fourth national report under the CBD (Government of the Republic of Namibia 2010).

The CBD's preamble affirms that biodiversity is humankind's common concern and that it has to be conserved for continued human survival. However, rather than lay down substantive rules, the CBD sets up overall principles, objectives and goals, leaving it up to the contracting states to develop and adopt detailed means to achieve them. I.e., it leaves it up to individual countries to determine exactly how to implement most of its provisions. Thus, major decision-making is placed on the national level. The CBD provides guidelines and directions to state parties as to how they should use these resources in a conservative manner for the benefit of present and coming generations. The objectives of the CBD comprise the conservation of biodiversity, the sustainable use of its components, and the fair and equitable sharing of benefits arising from the use of genetic resources.

Methods applied to ensure the maintenance of biological diversity are *in situ* and *ex situ* conservation. *In situ* conservation is defined as being[8]

> the conservation of ecosystems and natural habitats and the maintenance and recovery of viable populations of species in their natural surroundings and, in the case of domesticated or cultivated species, in the surroundings where they have developed their distinctive properties.
>
> (CBD: Art 2)

Ex situ conservation refers to the conservation of components of biodiversity outside their natural habitats, for example in zoos and aquaria (Glazewski, Kangueehi and Figueira 1998: 281).

The CBD provides that states have and should maintain their sovereign rights over their biological or generic resources, and that they bear the power to determine access to these resources through established mechanisms for the fair and equitable sharing of benefits arising from their use. There was consensus on the need to protect, conserve and sustainably utilise the available biological diversity for the benefit of humanity.

Thus, the CBD becomes the basis of domestic legislation on the promotion, protection and preservation of biological diversity. It gives the green light to states to exercise full control over their natural resources, provided that proper mechanisms protecting biological diversity are in place. Article 8(j) of the CBD provides that a state is obliged,

> subject to its national legislation, [to] respect, preserve and maintain knowledge, innovations and practices of indigenous and local communities embodying traditional lifestyles relevant for the conservation and sustainable use of biological diversity and promote their wider application with the approval and involvement of the holders of such knowledge, innovations and practices and encourage the equitable sharing of benefits arising from the utilisation of such knowledge, innovations and practices.[9]

Although national sovereignty is recognised, states are obliged to conserve biodiversity and regulate the sustainable use of its component resources. They are also urged to cooperate with each other regarding areas beyond national jurisdiction and other matters of mutual interest. Article 5 of the CBD states that contracting parties are obliged to develop and adopt national biodiversity strategies, plans, or programmes, and integrate the conservation of biodiversity and the sustainable use of its components into relevant sectoral or cross-sectoral plans, programmes and policies.

Due to the fact that the trade in wild animals and plants crosses borders between countries, efforts towards its regulation require international cooperation to safeguard certain species from over-exploitation.

The Convention on International Trade in Endangered Species of Wild Fauna and Flora

Aside from the CBD, the Convention on International Trade in Endangered Species of Wild Fauna and Flora (CITES) is considered one of the most relevant international biodiversity-related agreements on biodiversity protection. This convention, too, is legally binding on its parties, and was conceived in the spirit of such cooperation. Today, it accords varying degrees of protection to more than 30,000 species of animals and plants, whether they are traded as live specimens, fur coats or dried herbs. CITES was drafted as a result of a resolution adopted in 1973 at a meeting of members of IUCN and entered in force in 1975. CITES provides a framework to be respected by each party, which has to adopt its own domestic legislation to ensure that CITES is implemented at national level. To date, CITES has 178 parties.[10]

Namibia acceded to the Convention in 1990, and the Convention came into force for Namibia in March 1991.[11] The commercialisation of goods and services derived from native biodiversity, referred to as biotrade, has become reasonably well established in Namibia. In 2008, exports of indigenous natural plants, the devil's claw in particular, were estimated to have contributed N$24 million (ca. US$2.35 million) to GDP (Government of the Republic of Namibia 2012b: 30). These developments in the biotrade sector have prompted the Government to enhance its revenue collection by introducing differentiated rates on the export of all natural resources. The new taxes to be charged on biotic and abiotic natural resources destined for markets outside the country range between 0% and 2% (Government of the Republic of Namibia 2012a: 24).

One current issue in Namibia under the CITES convention is the production of high-value modern jewellery pieces containing traditional ivory amulets, known as *ekipas*. Such items have thus far used antique *ekipas* considered as pre-Convention ivory. Since the supply of antique *ekipas* has become severely limited, the Ministry of Environment and Tourism in collaboration with the jewellery industry of Namibia has designed a control system for worked ivory and the legal production of new *ekipas* in particular. CITES approval was sought for the export of items of modern jewellery of high value, involving *ekipas* permanently mounted in precious metals and other materials and rendered uniquely identifiable through a combination of engraved marks, documentation and a photographic record of each item (Government of the Republic of Namibia 2004b).

Embedded in and complementing these international regulations are a number of particular African legislations, such as the African Union's Convention on the Conservation of Nature and Natural Resources, and sub-regional agreements such as the various protocols under the umbrella of the Southern African Development Community (SADC).

The African Union's Convention on the Conservation of Nature and Natural Resources

Major foundations of biodiversity protection are contained in the African Convention on the Conservation of Nature and Natural Resources. The original African Convention on the Conservation of Nature and Natural Resources was adopted in Algiers, Algeria, in September 1968 and entered into force in June 1969. Of the 53 member states 40, excluding Namibia, have signed the Convention of which 30 have ratified it. Recognising that soil, water, flora and faunal resources constitute resources of vital importance for mankind, the Convention's fundamental principle is that the contracting states shall adopt the necessary measures to ensure conservation, utilisation and development of soil, water, flora and faunal resources in accordance with scientific principles and with due regard to the best interests of the people. The Convention contains several provisions related to the conservation and perpetuation of species. Special provisions as to protected species and trade in specimens are formulated.

The Revised African Convention on the Conservation of Nature and Natural Resources was adopted by the second ordinary session of the Assembly of Heads of States and Government of the African Union in Maputo, Mozambique, in July 2003. It commits parties to manage their natural resources more sustainably. The Convention has however not yet come into force, as the requirements for coming into force have so far not been fulfilled. According to Article 38 the Convention comes into force on the thirtieth day following the date of deposit of the fifteenth instrument of ratification, acceptance, approval or accession with the Depositary. As of May 2013, 39[12] of the 54 member states have signed the Convention, while only eight member states[13] have deposited their instrument of ratification.[14] Provisions directly related to the protection of biodiversity are contained in Article IX on Species and Genetic Diversity; Article X on Protected Species; Article XI on Trade in Specimens and Products thereof; and Article XII on Conservation Areas.

The parties to the Convention shall maintain and enhance species and genetic diversity of plants and animals whether terrestrial, freshwater or marine. They shall for that purpose establish and implement policies for the conservation and sustainable use of such resources. Parties are obliged to undertake to identify the factors that are causing the depletion of animal and plant species threatened by elimination, and accord a special protection to such species. Furthermore, domestic trade in the transport as well as possession of specimens and products must be regulated by the parties' appropriate penal sanctions, including confiscation measures. In order to ensure the long-term conservation of biological diversity, the parties shall establish, maintain and extend conservation areas.

Protocols under the umbrella of the Southern African Development Community

The various protocols under the umbrella of the Southern African Development Community (SADC) constitute sub-regional agreements also relevant for biodiversity protection in Namibia (Ruppel 2012). SADC parties may conclude protocols as may be necessary in each area of cooperation, which shall spell out the objectives and scope of, and institutional mechanisms for, cooperation and integration. SADC protocols of major concern with regard to biodiversity conservation are the Protocols on Wildlife Conservation and Law Enforcement (SADC 1999b), on Shared Watercourse Systems (SADC 2000), on Fisheries (SADC 2001) and on Forestry (SADC 2002). Furthermore, the Regional Indicative Strategic Development Plan (RISDP) of the SADC, developed in 1999, recognises a need for policies and strategies to offset the high rate of natural resource degradation, focusing on biodiversity amongst others (SADC 1999a).

Biodiversity protection under national environmental law

According to Article 1(6), the Constitution of the Republic of Namibia provides the rule of law in Namibia. Therefore all legislations ought to be consistent in its provisions. Although the Constitution so far contains no enforceable environmental law as such, the foundation is laid for all policies and legislation in Namibia (Ruppel 2010). Two key 'environmental clauses' relevant to sustainable use of natural resources are included in the Constitution. On the issue of biological diversity and its protection, the Namibian Constitution is very clear. It is one of the provisions enshrined under the chapter on principles of state policy. The relevant clause is Article 95(l) which stipulates that the state shall actively promote and maintain the welfare of the people by adopting policies which include the

> maintenance of ecosystems, essential ecological processes and biological diversity of Namibia and utilisation of living natural resources on a sustainable basis for the benefits of all Namibians both present and future.

With this particular Article, Namibia is obliged to protect its biological diversity and to promote a sustainable use of its natural resources. Furthermore, Article 91(c) includes in the functions of the Ombudsman[15]

> the duty to investigate complaints concerning the over utilisation of living natural resources, the irrational exploitation of non-renewable resources, the degradation and destruction of ecosystems and failure to protect the beauty and character of Namibia.

In addition to these clauses, Article 100 provides that all natural resources, including water, vest in the state, unless otherwise legally owned.

The Constitution sets the framework and Namibia's independence created the opportunity to revise a wide range of national policies and laws. This emphasis placed on environmental concerns at the Rio Summit in 1992, and the increasing awareness towards environmental issues, triggered widespread legislative reform particularly in terms of natural resource management. Thus, recent policy and legislative reforms have created a unique opportunity for Namibia to incorporate environmental sensitivity into its laws. As a result Namibian legislation is supported by sound policy direction regarding sustainable development and sustainable use of natural resources (Ruppel 2008).

So far, no specific Act dealing with the conservation of biological diversity as a main topic has come into force. However, the Draft Bill on Access to Biological Resources and Associated Traditional Knowledge was formulated in 2000. This draft bill, which is aimed specifically at the protection of biodiversity and traditional knowledge, has not yet been passed in Parliament. The Government has, however, developed Namibia's 10-year National Biodiversity Strategy and

Action Plan for Sustainable Development through Biodiversity Conservation (2001–2010), which has been subject to review. Namibia, through the Ministry of Environment and Tourism, has begun the review process, and the second National Biodiversity Strategy and Action Plan (2011–2020) is being prepared. To this end, the Ministry of Environment and Tourism has established a multi-sectoral committee to facilitate, provide guidance and technical inputs. The group comprises technical staffs from several Ministries,[16] the Office of the Prime Minister (OPM), the National Planning Commission (NPC), the Council of Traditional Authorities, the Namibia Association of Community Based Natural Resources Management Support Organisation (NACSO), the University of Namibia, the Polytechnic of Namibia and the United Nations Development Programme (UNDP). The committee's first meeting took place in May 2012.

The goal of the first Biodiversity Strategic and Action Plan was to protect ecosystems, biological diversity and ecological processes, through conservation and sustainable use, thereby supporting the livelihoods, self-reliance and quality of life of Namibians in perpetuity (Barnard, Shikongo and Zeidler 2000: 13). The action plan attempts to provide a national strategic framework for natural resource management activities involving biological resource management and the natural environment, including trade and economic incentives, and to prioritise, through detailed action plans, activities and measures needed to address this strategy effectively for the next decade, with cost estimates for each. The strategic aims of this document include: conserving biodiversity in priority areas; sustainable use of natural resources; monitoring, predicting and coping with environmental change and threats; sustainable land management; sustainable wetland management; sustainable coastal and marine ecosystem management; integrated planning for biodiversity conservation and sustainable development; Namibia's role in the larger world community; and capacity building for biodiversity management in support of sustainable development.

Sectoral legislation covering the protection of biodiversity is wide-ranging in Namibia (Hinz and Ruppel 2008). A myriad of legislative instruments provide for the equitable use of natural resources for the benefit of all. Only the most relevant legal instruments will be introduced briefly in the following paragraphs.

One of the major biodiversity-related laws in Namibia is the legislation governing the conservation of wildlife, and protected areas, the Nature Conservation Ordinance.[17] The Ordinance was amended by the Nature Conservation Amendment Act.[18] One of its major highlights is the creation of conservancies in communal areas. In terms of the amendment, rural communities have to form a conservancy in order to be able to acquire the use-right over wildlife. Conservancies are defined as land units managed jointly for resource conservation purposes by multiple landholders, with financial and other benefits shared between them. Conservancies occur in both communal and commercial land (Barnard 1998: 45).[19] The Ordinance deals with *in situ* and *ex situ* conservation by providing for the declaration of protected habitats as national parks and reserves, and for the protection of scheduled species. It regulates hunting and harvesting, and possession of and trade in these species. Under the existing laws, Namibia has national park zoos and safari areas to conserve biodiversity. Most people consider regions merely as tourist areas, but in fact they are also of scientific significance as they allow for natural movement of large animals and help to ensure that there is enough space and food for all of the indigenous species.

In addition to the broader national agenda on conservation of biodiversity is the Community-Based Natural Resource Management (CBNRM). This has enabled local communities to do *in situ* conservation of natural resources and hence contribute to biodiversity conservation.

The Environmental Management Act[20] requires adherence to the principle of optimal sustainable yield in the exploitation of all natural resources. The Act gives effect to Article 95(l) of the Constitution by establishing general principles for the management of the environment and natural resources. It promotes the coordinated and integrated management of the environment and sets out responsibilities in this regard. Furthermore, it intends to give statutory effect to Namibia's Environmental Assessment Policy, to enable the minister responsible for the environmental policy to give effect to Namibia's obligations under international environmental conventions, and to provide for associated matters. The Act promotes inter-generational equity in the utilisation of all natural resources. Environmental impact assessments and consultations with communities and relevant regional and local authorities are provided for monitoring the development of projects that potentially have an impact on the environment.

The proposed Parks and Wildlife Management Act aims at the protection of all indigenous species and provides a tool for controlling the exploitation of all plants and wildlife. The preamble clearly states that the Bill is intended to give effect to paragraph (l) of Article 95 of the Constitution by establishing a legal framework to provide for and promote the maintenance of ecosystems, essential ecological processes and the biological diversity of Namibia and to promote the mutually beneficial co-existence of humans with wildlife, to give effect to Namibia's obligations under relevant international legal instruments including CBD and CITES. In keeping with the constitution, the principles underlying the draft act are directed at the maintenance of essential ecological processes and life support systems and, in effect, biological diversity. If the proposed act comes into force, it repeals the Nature Conservation Ordinance.[21]

Legislation on forests and forestry is one further important mosaic in the legal system of biodiversity conservation in Namibia. In 2005 almost 7.7 million hectares of Namibia's land were covered by forests, corresponding to 9.3% of the total land area (FAO 2005). Major threats to forests in Namibia include the expansion of land for agriculture, the use of fuel wood and charcoal for domestic use, tobacco curing, land clearing for infrastructure development, uncontrolled wild fires, selective logging through timber concessions and unlicensed curio carving, and habitat destruction by elephants (Zender 2012). The Forest Act[22] consolidates the laws relating to the use and management of forests and forest produce, provides for the control of forest fires and creates a Forestry Council. Protection of the environment is treated in part IV of the act. This part of the act deals with protected areas, protection of natural vegetation and control over afforestation and deforestation. The purpose of the Act is to conserve soil and water resources, maintain biological diversity and to use forest produce in a way that is compatible with the forest's primary role as the protector and enhancer of the natural environment.

Last but not least, the water-related legislation in Namibia is to be mentioned (cf. Ruppel and Bethune 2007). Although the new Water Resources Management Act was approved by Parliament in 2004, the rather out-dated Water Act No. 54 of 1965 remains in force until the new act comes into force upon signature by the minister. The new act is currently being amended to take into account practical aspects of implementing it. The Water Act of 1956 does not directly refer to the protection of biological diversity; however, it contains provisions relating to water quality and conservation which are at least indirectly beneficial for the maintenance of biodiversity.

The Marine Resources Act[23] provides for the conservation of the marine ecosystem and the responsible utilisation, conservation, protection and promotion of marine resources on a sustainable basis. For that purpose, it provides for the exercise of control over marine resources and for matters connected therewith. It replaces the Sea Fisheries Act,[24] which in turn replaced the Sea Fisheries Act.[25]

The Aquaculture Act[26] regulates and controls aquaculture activities and the sustainable development of aquaculture resources (Bethune, Griffin and Joubert 2004). All aquaculture ventures will be subject to strict licensing. Section 27 is of most relevance for the protection of biodiversity. A person may not, without written permission granted by the minister, introduce or cause to be introduced into Namibia or any Namibian waters any species of aquatic organism or any genetically modified aquatic organism or transfer any species of aquatic organism from one aquaculture facility to another or from any location in Namibia to another.

The Inland Fisheries Resources Act[27] deals with the conservation and utilisation of inland fisheries resources and allows for the updating and development of new policies for the conservation and sustainable utilisation of Namibia's inland fisheries. It encourages cooperation with neighbouring countries regarding the management and conservation of shared waterways.

In recognising the worldwide diversity situation, the Government of Namibia enacted the Biosafety Act[28] after having signed the Cartagena Protocol on Biosafety to the CBD, which was adopted in 2000. The act provides for measures to regulate activities involving research, development, production, marketing, transport, application and other uses of genetically modified organisms and to establish a Biosafety Council. The objective of the act is *inter alia* to introduce a system and procedures for the regulation of genetically modified organisms in Namibia in order to provide an adequate level of protection to the conservation and the sustainable use of biological diversity.

Conclusion

Having regard to the above, Namibian environmental law is a complex and interlocking system of statutes, policies, treaties, common, customary and case law with the Constitution as the supreme law of the land and therefore the ultimate source of law in Namibia. However, research done under the Biodiversity Monitoring Transect Analysis in Africa (BIOTA) project administered in the Faculty of Law of the University of Namibia, which has been the source of the presented information, has demonstrated that many obstacles prevent the societally desired degree of implementation. Statutory environmental law meets challenges from customary law (cf. Hinz, Ruppel and Mapaure 2012; Hinz and Mapaure 2010; Ruppel 2009a). Apart from this, environmental policies and their translation into law are, in general (and this applies in all parts of the world), faced with the economic interests of various social groups that are not easy to harmonise with each other.[29]

Notes

1 This chapter is substantially based on Hinz and Ruppel 2013.

2 The term conservation is defined as the management of human use of the biosphere, so that it may produce the greatest sustainable benefit to present generations while maintaining its potential to meet the needs and aspirations of the future generations. Thus, conservation embraces the preservation, maintenance, sustainable utilisation, restoration and enhancement of the natural environment. While ecosystems may be used by present generations for their benefit, they should only be used in a way not depriving future generations of their right to use such ecosystems in the same manner for their survival. The maintenance of biological diversity at all levels is fundamentally the maintenance of viable populations of species or identifiable populations (cf. Groombridge 1992: xvi). The book by M. Wulfmeyer (2006) is an interesting record on how this global task has been incorporated into Namibia's education system.

3 Cf. composition of threatened species: mammals 10; birds 21; reptiles 3; amphibians 1; fish 20; plants 24. Online. Available HTTP: <http://www.iucnredlist.org/info/tables/table5> (accessed 13 June 2013).

4 IUCN Red List version 2011.2: Table 5. Online. Available HTTP: <http://www.iucnredlist.org/documents/summarystatistics/2011_2_RL_Stats_Table5.pdf> (accessed 13 June 2013).

5 This passage is largely taken from Ruppel and Ruppel-Schlichting 2011.

6 Cf. <www.state.gov/r/pa/ei/bgn/5472.htm> (accessed 13 June 2013).

7 Other international agreements which also relate to the protection of biodiversity include the UN Convention to Combat Desertification; the UN Framework Convention on Climate Change; the International Convention for the Protection of New Varieties of Plants (UPOV Convention); international conventions containing fishery provisions, e.g. UN Convention on the Law of the Sea; the Ramsar Convention on Wetlands; and the Global Biodiversity Strategy.

8 Article 2 of the CBD.

9 Cf. here also Articles 10(c), 17(1) and (2), and 18(4): The CBD does not differentiate between *indigenous*, *traditional* and *local*, although the terms may refer to different social situations. For example, compare the use of *indigenous* in the United Nations Declaration on the Rights of Indigenous People (to which we will refer below), which applies to specifically defined groups of people and not to all traditional communities

– and certainly not to all that could be called *local*. For the purpose of this study, the term *traditional* is preferred unless there is a need to differentiate.

10 Cf. for more information on CITES <http://www.cites.org> (accessed 13 June 2013).

11 Cf. <http://www.cites.org/end/disc/parties/alphabet.shtml> (accessed 13 June 2013).

12 The Convention has been signed by Angola, Benin, Burkina Faso, Burundi, Chad, Comoros, the DRC, Congo, Cote d'Ivoire, Djibouti, Equatorial Guinea, Ethiopia, Gambia, Ghana, Guinea, Guinea-Bissau, Kenya, Lesotho, Liberia, Libya, Madagascar, Mali, Mozambique, Namibia, Niger, Nigeria, Rwanda, São Tomé and Príncipe, Senegal, Sierra Leone, Somalia, South Africa, Sudan, Swaziland, Tanzania, Togo, Uganda, Zambia and Zimbabwe.

13 I.e. Burundi, Comoros, Ghana, Lesotho, Libya, Mali, Niger and Rwanda.

14 Cf. <http://www.africaunion.org/root/au/Documents/Treaties/List/Revised%20 Convention%20on%20Nature%20and%20Natural%20Resources.pdf> (accessed 13 June 2013).

15 On the environmental mandate of the Ombudsman cf. Ruppel and Ruppel-Schlichting 2010, 363ff.

16 Including the Ministry of Agriculture, Water and Forestry (MAWF), the Ministry of Regional, Local Government, Housing and Rural Development (MRLGHRD), the Ministry of Mines and Energy (MME), the Ministry of Fisheries and Marine Resource (MFMR), the Ministry of Lands and Resettlement (MLR), and the Ministry of Education (MoE).

17 No. 4 of 1975.

18 No. 5 of 1996.

19 Moreover, Section 1(b) of the Amendment Act defines a conservancy to mean any area declared a conservancy in terms of Section 24A.

20 No. 7 of 2007.

21 No. 4 of 1975.

22 No. 12 of 2001.

23 No. 27 of 2000.

24 No. 29 of 1992.

25 No. 58 of 1973.

26 No. 18 of 2002.

27 No. 1 of 2003.

28 No. 7 of 2006.

29 How to balance environmental policies with economic interests, given the conditions of Namibia, is still an area where more research is needed.

References

African Union Comission (1968) *African Convention on the Conservation of Nature and Natural Resources*. Online. Available HTTP: <http://www.au.int/en/sites/default/files/AFRICAN_CONVENTION_CONSERVATION_NATURE_AND_NATURAL_RESOURCES.pdf> (accessed 13 June 2013).

African Union Comission (2003) *African Convention on the Conservation of Nature and Natural Resources. Revised Version*. Online. Available HTTP: <http://www.au.int/en/sites/default/files/AFRICAN_CONVENTION_CONSERVATION_NATURE_NATURAL_RESOURCES.pdf> (accessed 13 June 2013).

Barnard, P. (ed.) (1998) *Biological Diversity in Namibia – A Country Study in Cooperation with the Namibian National Biodiversity Task Force*, Windhoek: Ministry of Environment and Tourism, Directorate of Environmental Affairs.

Barnard, P., Shikongo S. and Zeidler, J. (2000) *Biodiversity and Development in Namibia: Namibia's Ten-year 2000–2010 Strategic Plan of Action for Sustainable Development Through Biodiversity Conservation*, Windhoek: Ministry of Environment and Tourism.

Bethune, S., Griffin, M. and Joubert, D. (2004) *National Review of Invasive Alien Species – Report for Southern Africa Biodiversity Support Programme (SABSP)*, Windhoek: Ministry of Environment and Tourism, Directorate of Environmental Affairs, 2012. Online. Available HTTP: <http://www.biodiversity.org.na/ias/National%20Review%20of%20Invasive%20Alien%20Species,%20Namibia.pdf> (accessed 13 June 2013).

Blaikie, W.G. (2004) *The Personal Life of David Livingstone*. Online. Available HTTP: <http://www.gutenberg.org/files/13262/13262–h/13262–h.htm> (accessed 13 June 2013).

Cartagena Protocol – Secretariat of the Convention on Biological Diversity (2000) *Cartagena Protocol on Biosafety to the Convention on Biological Diversity: Text and Annexes*. Online. Available HTTP: <http://bch.cbd.int/protocol/text/> (accessed 13 June 2013).

CITES – International Union for Conservation of Nature (IUCN) (1973) *Convention on International Trade in Endangered Species of Wild Fauna and Flora*. Online. Available HTTP: <http://www.cites.org/eng/disc/E-Text.pdf> (accessed 13 June 2013).

Erasmus, M.G. (1991) 'The Namibian Constitution and the application of international law in Namibia', in D. van Wyk et al. (eds) *Namibia – Constitutional and International Law Issues*, Pretoria: VerLoren van Themaat Centre for Public Law Studies.

FAO (Food and Agriculture Organisation of the United Nations) (2005) *Global Forests Resources Assessment 2005*. Online. Available HTTP: <http://www.fao.org/forestry/site/28679/en/> (accessed 13 June 2013).

Glazewski, J., Kangueehi, N.K. and Figueira, M. (1998) 'Legal approaches to protecting biodiversity in Namibia', in P. Barnard (ed.) *Biological Diversity in Namibia: A Country Study*, Windhoek: Namibian National Biodiversity Task Force, 279–97.

Government of the Republic of Namibia (2001) *National Biodiversity Strategy and Action Plan for Sustainable Development through Biodiversity Conservation (2001–2010)*. Online. Available HTTP: <http://www.cbd.int/doc/world/na/na-nbsap-01–en.pdf> (accessed 13 June 2013).

Government of the Republic of Namibia (2004a) *Namibia Vision 2030: Policy Framework for Long-term National Development*, Windhoek: AIM Publications. Online. Available HTTP: <http://www.npc.gov.na/> (accessed 13 June 2013).

Government of the Republic of Namibia (2004b) *Control System for Worked Ivory in Namibia*. Online. Available HTTP: <http.//www.cites.org/common/cop/13/inf/E13i-33.pdf> (accessed 13 June 2013).

Government of the Republic of Namibia (2010) *Fourth National Report to the Convention on Biological Diversity (CBD)*, Windhoek: MET (Ministry of Environment and Tourism). Online. Available HTTP: <http://www.cbd.int/doc/world/na/na-nr-04–en.pdf> (accessed 13 June 2013).

Government of the Republic of Namibia (2012a) *Statement for 2012/2013 Budget* presented by Hon. Saara Kuugongelwa-Amadhila, Minister of Finance. Online. Available HTTP: <http://www.mof.gov.na/Budget%20Documents/Budget%202012/Budget%20new/FINAL%202012%20BUDGET%20STATEMENT%2028%2002%202012%20%2017h00%20(RevI)> (accessed 13 June 2013).

Government of the Republic of Namibia (2012b) *Draft Country Report to the United Nations Commission on Sustainable Development: Namibia*, Windhoek: Ministry of Environment and Tourism.

Groombridge, B. (ed.) (1992) *Global Biodiversity Status of the Earth's Living Resources: A Report Compiled by the World Conservation Monitoring Centre*, London: Chapman and Hall.

Hinz, M.O. and Mapaure, C. (2010) 'Traditional and modern use of biodiversity – customary law and its potential to protect biodiversity', in U. Schmiedel, N. Jürgens and T. Hoffman (eds) *Biodiversity in Southern Africa. Volume 2: Patterns and Processes at Regional Scale*, Göttingen: Klaus Hess Publishers, 195–9.

Hinz, M.O. and Ruppel, O.C. (2008) 'Legal protection of biodiversity in Namibia', in M.O. Hinz and O.C. Ruppel (eds) *Biodiversity and the Ancestors: Challenges to Customary and Environmental Law. Case Studies from Namibia*, Windhoek: Namibia Scientific Society, 3–62.

Hinz, M.O. and Santos, J. (2002) *Customary Law in Namibia: Development and Perspective*, 7th edn, Windhoek: Centre for Applied Social Sciences.

Hinz, M.O., Ruppel, O.C. and Mapaure, C. (2012) *Knowledge Lives in the Lake. Case Studies in Environmental and Customary Law from Southern Africa*, Windhoek: Namibia Scientific Society.

IUCN (International Union for Conservation of Nature) (2006) *IUCN Red List*. Online. Available HTTP: <http://www.iucnredlist.org/info/tables/table5> (accessed 13 June 2013).

IUCN (International Union for Conservation of Nature) (2011) *IUCN Red List. Version 2011.2: Table 5*. Online. Available HTTP: <http://www.iucnredlist.org/documents/summarystatistics/2011_2_RL_Stats_Table5.pdf> (accessed 13 June 2013).

Lovejoy, T.E. (1980) 'Changes in biological diversity', in Council on Environmental Quality and the U.S. Department of State (eds) *The Global 2000 Report to the President, Vol. 2: The Technical Report*, Harmondsworth: Penguin, 327–32.

McNeely, J.A., Miller, K.R., Reid, W.V., Mittermeier, R.A. and Werner, T.B. (1990) *Conserving the World's Biological Diversity*, Washington, D.C.: World Bank Publications.

Nandi-Ndaitwah, N. (2010) *Country Statement on the Occasion of the Fourth Global Environment Facility General Assembly Held in Punta del Este, Uruguay 24–28 May 2010*. Online. Available HTTP: <http://www.thegef.org/gef/sites/thegef.org/files/documents/Namibia.pdf> (accessed 13 June 2013).

Norse, E.A. and McManus, R.E. (1980) 'Ecology and living resources biological diversity', in Council on Environmental Quality (ed.) *The Eleventh Annual Report of the Council on Environmental Quality*, Washington, D.C.: US Government Printing Office, 31–80.

Paing, S., Lwin, T., Chit, K., Zin, T. and Ti, T. (2006) 'The role of traditional medicine in the treatment of multidrug-resistant pulmonary tuberculosis, Myanmar', *Regional Health Forum* 2. Online. Available HTTP: <http://209.61.208.233/LinkFiles/Regional_Health_Forum_CommunicableDiseases.pdf> (accessed 13 June 2013).

Ruppel, O.C. (2008) 'Third-generation human rights and the protection of the environment in Namibia', in N. Horn and A. Bösl (eds) *Human Rights and the Rule of Law in Namibia*, Windhoek: Macmillan Education, 101–20.

Ruppel, O.C. (2009a) 'Local knowledge and legal implications for the conservation of biological diversity in Namibia', in R.N. Pati and D.N. Tewari (eds) *Conservation of Medicinal Plants*, New Dehli: SB Nangia APH Publishing Corporation, 40–72.

Ruppel, O.C. (2009b) 'Koloniale Altlasten in Namibia und die Grenzen des Völkerrechts', *Allgemeine Zeitung*, 12 August.

Ruppel, O.C. (2010) 'Environmental rights and justice in Namibia', in A. Bösl, N. Horn and A. du Pisani (eds) *Constitutional Democracy in Namibia. A Critical Analysis after Two Decades*, Windhoek: Macmillan Education, 323–60.

Ruppel, O.C. (2011) 'International environmental law from a Namibian perspective', in O.C. Ruppel and K. Ruppel-Schlichting (eds) *Environmental Law and Policy in Namibia*, 2nd edn, Windhoek: Orumbonde Press & Welwitschia Verlag Dr. A. Eckl, 33.

Ruppel, O.C. (2012) 'SADC environmental law and the promotion of sustainable development', *SADC Law Journal* 2 (2): 246–80.

Ruppel, O.C. and Bethune, S. (2007) *Review of Namibia's Policy and Legislative Support to the Sustainable Use of Wetlands in the Zambezi Basin, Report for the World Conservation Union*, Harare: IUCN. Online. Available HTTP: <http://www.ramsar.org/wn/w.n.namibia_review.htm> (accessed 13 June 2013).

SADC (Southern African Development Community) (1999a) *Regional Indicative Strategic Development Plan*. Online. Available HTTP: <http://www.sadc.int/files/5713/5292/8372/Regional_Indicative_Strategic_Development_Plan.pdf> (accessed 13 June 2013).

SADC (Southern African Development Community) (1999b) *Protocol on Wildlife Conservation and Law Enforcement*. Online. Available HTTP: <http://www.sadc.int/files/8913/5292/8369/Protocol_on_Wildlife_Conservation_and_Law_Enforcement_1999.pdf> (accessed 13 June 2013).

SADC (Southern African Development Community) (2000) *Revised Protocol on Shared Watercourses*. Online. Available HTTP: <http://www.sadc.int/files/3413/6698/6218/Revised_Protocol_on_Shared_Watercourses_-_2000_-_English.pdf> (accessed 13 June 2013).

SADC (Southern African Development Community) (2001) *Protocol on Fisheries*. Online. Available HTTP: <http://www.sadc.int/files/5613/5292/8363/Protocol_on_Fisheries2001.pdf> (accessed 13 June 2013).

SADC (Southern African Development Community) (2002) *Protocol on Forestry*. Online. Available HTTP: <http://www.sadc.int/files/9813/5292/8364/Protocol_on_Forestry2002.pdf> (accessed 13 June 2013).

The Constitution of the Republic of Namibia. Online. Available HTTP: <http://209.88.21.36/opencms/export/sites/default/grnnet/AboutNamibia2/AboutNamsDocs/Constitution/constitutionDoc.pdf> (accessed 13 June 2013).

van Wyk, B.E. and Gericke, N. (2000) *People's Plants: A Guide to Useful Plants in Southern Africa*, Pretoria: Briza Publications.

Wilson, E.O. (1985) 'The biological diversity crisis', *BioScience* 35: 700–6.

Wulfmeyer, M. (2006) *Bildung für nachhaltige Entwicklung im globalen Kontext. Das Beispiel Namibia*, Frankfurt am Main: IKO – Verlag für Interkulturelle Kommunikation.

WWF (World Wide Fund For Nature/World Wildlife Fund) (2010) *Living Planet Report 2010*. Online. Available HTTP: <http://wwf.panda.org/about_our_earth/all_publications/living_planet_report/> (accessed 13 June 2013).

Zender, A.L. (2012) 'Legal implications of grazing policies on the carrying capacity of rangeland in Namibia', in M.O. Hinz, O.C. Ruppel and C. Mapaure (eds) *Knowledge Lives in the Lake. Case studies in Environmental and Customary Law from Southern Africa*, Windhoek: Namibia Scientific Society, 149–75.

11 The impact of access and benefit sharing programmes on indigenous people in Australia

Karen Bubna-Litic

Introduction

The loss of biodiversity has been recognised as one of the most serious problems our planet faces. Indigenous ecological knowledge (IEK) has been recognised as a valuable tool in biodiversity protection (Commonwealth of Australia 2013; cf. Holcombe 2009) and with this recognition comes a need to acknowledge and respect it in terms of its importance to indigenous people and its current and potential value to industry. The commercialisation of genetic resources has often been protected through patent laws but in most cases this has brought little benefit to the holders of the traditional knowledge on which these patents have been built. The international law regime has attempted, through the Convention of Biological Diversity (CBD), to ensure national governments acknowledge this in a way that respects and includes indigenous people and their knowledge. This is provided through the fair and equitable sharing of benefits with indigenous people for the use of biological resources and IEK. Access and benefit sharing arrangements are crucially important because of the inadequacy of intellectual property rights to deal with IEK.

This chapter will explore the unique relationship between biodiversity, indigenous communities and IEK in Australia. At the outset, it is important to try to grapple with a definition of traditional knowledge (TK) in the Australian context, recognising that traditional societies around the world are all different. The cultural and intellectual property rights in indigenous knowledge have been defined as indigenous heritage rights, heritage consisting of 'the intangible and tangible aspects of the whole body of cultural practices, resources and knowledge systems that have been developed, nurtured and refined (and continue to be developed, nurtured and refined) by Indigenous people and passed on by Indigenous people as part of expressing their cultural identity' (Janke 1998: 11).

Knowledge cannot be separated from the person, so acknowledging and valuing the knowledge holders is as important as protecting the knowledge itself. The dissemination of this knowledge allows the knowledge holders an opportunity to practise their cultures and this can help them exercise both their human rights and cultural responsibilities. This will empower them and in turn their communities. So in determining benefits, it is important to acknowledge indigenous knowledge holders for their intellectual input into projects in addition to any financial reward (Janke 2009).

Even more helpful is to identify some of the unique characteristics of TK in Australia to help establish the framework for this discussion. Five have been suggested. They are: indigenous people often hold communal rights and interests in their knowledge; there is a close interdependence between knowledge, land and spirituality in indigenous societies; knowledge is passed down through generations in indigenous societies; knowledge, innovations and practices are often transmitted orally in accordance with customary rules and principles; and there are rules regarding secrecy and sacredness that govern the management of knowledge (Davis 1999).

This chapter will use the term indigenous ecological knowledge (IEK), indigenous knowledge pertaining to natural resource management (NRM). According to Holcombe, IEK is not simply a subset of indigenous knowledge, because IEK is socially and culturally embedded. This embeddedness means that knowledge and thus intellectual property is collectively held (Holcombe 2009). A definition from the Kimberley Land Council's policy on intellectual property and traditional knowledge is helpful in this context. They define traditional knowledge as,

> the know-how, skills, innovations and practices that form the knowledge systems embodying the traditions, observances, customs, beliefs and lifestyles of Kimberley Aboriginal people, or are contained in knowledge systems passed between generations and continuously developed following any changes in the environment, geographical conditions and other factors. It is not limited to any specific technical field, and may include agricultural, environmental and medicinal knowledge, and any knowledge associated with Traditional Cultural Expressions and Biological Resources
>
> (Kimberley Land Council 2011)

With this background, the next section will begin by examining the state of biodiversity and biodiversity loss in Australia and the link between biodiversity and IEK. The next section of this paper will look at the existing legal framework for the protection of biodiversity, both at the international and national levels. The following section will then examine the inadequacy of intellectual property rights to deal with IEK necessitating the reliance on access and benefit sharing schemes, and the final section will look closely at the impact of access and benefit sharing programmes on indigenous Australians.

Biodiversity in Australia and the link between biodiversity and indigenous ecological knowledge (IEK)

Biodiversity in Australia

Australia is recognised as one of only 17 'mega-diverse' countries in the world and scientists estimate a total of 8.7 million species exist worldwide, with approximately 1 million species inhabiting Australia. There are more than 4,500 species of marine fish, and 55% of the entire family of sharks and rays are unique to Australia. It has 17,580 species of flowering plants, with 91% of its flowering plants being endemic. It has 16 endemic plant families (being the highest in the world) and hosts 57% of the world's mangrove species. South Western Australia has been identified as one of 34 global 'biodiversity hot spots'. Biodiversity loss in Australia is occurring at an alarming rate. The Australian landmass is divided into bioregions, which are large, geographically distinct areas of land with common characteristics such as geology, landform patterns, climate, ecological features, and plant and animal communities. Of Australia's 85 bioregions, 80 bioregions include at least one threatened ecosystem. The National Land and Water Resources Audit (1997–2002) identified almost 2,900 threatened ecosystems and ecological communities across Australia (Yencken, Henry and Rae 2012). The 2011 State of the Environment Report (SOER) identifies climate change and the impact of population and economic growth as the major drivers of risk to Australia's environment and heritage (Commonwealth of Australia 2011).

The World Conservation Union (IUCN) attempts to track global trends in biodiversity loss. The IUCN Redlist identifies extinct or endangered, vulnerable species[1] and tries to determine trends at a global level through a list using consistent categories and criteria. Recently, this technique has been applied to Australian birds, the first time the Red List Index has been applied at a national level (Szabo, Possinghamc and Stephen 2012). For birds found only in Australia, the results suggest only a slight decline over the last two decades, and that Australia is doing better than the rest of the world. There is concern, however, for migratory birds. The numbers of migratory shorebirds that breed in northern Asia and then migrate over winter to our tidal mudflats are plummeting. It has been suggested that the federal government policy to invest in landscape rather than threatened species is a retrograde step and that the recent mammal extinction of the Christmas Island Pipistrelle, the first in 50 years, will be replicated in native birds if governments continue to focus on landscapes instead of species. Clearly, there is a need for Australia to do better in biodiversity conservation and natural resources management.

The link between biodiversity and indigenous ecological knowledge (IEK)

Indigenous Australians make up only 2.5% of the general Australian population (Australian Bureau of Statistics 2006) and 45% of very remote Australia. If you take away mining and service towns, indigenous people are by far the majority across very remote Australia (Bubna-Litic 2009: 355). Indigenous ecological knowledge (IEK) is practised across Australia, mostly in remote and very remote communities, and it is only recently that there has been acknowledgement of the important role that IEK can play in biodiversity conservation and natural resource management. Much work has been done to formally incorporate this knowledge into biodiversity conservation practices in Australia.[2] This includes fire management, digging soakages, maintaining surface waters and water management. It includes stimulating new growth for preferred animal species and increasing the abundance of favoured bush medicine and bush tucker plants, as well as the knowledge of natural resources, plants, animals, and their environments, and the use made of that knowledge. Indigenous Australians now own more than 20% of the Australian land mass, predominantly in both remote and very remote Australia, and the indigenous estate is growing, incorporating both marine and terrestrial environments. This estate includes some of the most biodiverse and ecologically intact parts of Australia.

The link between biodiversity and IEK can be illustrated in the area of depopulated landscapes in Australia. It is argued that a depopulated landscape is a major and growing ecological threat to Australia's biodiversity (Altman et al. 2009). Two examples from the Northern Territory in far north Australia illustrate this. The first concerned the Waanyi/Garawa Aboriginal Land trust, covering 12,000 square kms of high national conservation value. Traditional owners were forced to move off the land trust for the lack of basic services, such as health, housing and education. After their move from this land, this country experienced large-scale late dry season hot fires, covering in excess of 16,000 square kms, beyond the land trust areas. The long-term result of these uncontrolled fires across the land trust has resulted in significant loss of vegetation, resulting in loss of feeding and breeding habitats for endemic threatened species and exposure of skeletal soils to erosion. Without people living on the land and implementing aboriginal fire management, these soils will slowly choke the rivers and impact the habitat of marine species. The fires will also emit additional GHGs. The second example relates to the infestation of mimosa which choke wetlands, thus restricting biodiversity. Indigenous people living on the land have been at the forefront of identifying new infestations and eradicating existing infestations of mimosa. In a depopulated landscape, weeds can quickly take hold, impacting on biodiversity, and increasing the intensity of fire regimes. The benefits from early detection of weed infestations by indigenous people living in very remote areas needs to be recognised (cf. Commonwealth of Australia 2013).

Legal framework for the protection of biodiversity

International level

When a company wishes to patent IEK by isolating and patenting a new substance, generally the holders of this knowledge do not benefit from the resulting revenues as the enforceable right to commercialise rests with the patent holder. This is often referred to as 'biopiracy', 'bioprospecting' or 'biocolonialism'. Prior to the CBD, genetic resources were regarded as the 'common heritage of mankind', allowing for the exploitation of these resources with no acknowledgement of the knowledge of the source of these resources, often held by indigenous communities. An example of this is the anti-rejection drug cyclosporine A, which was extracted from nature reserves in Norway. Its revenue was valued at more than $1.2b in 1997 and Norway received none of the revenue. The international law regime has sought to overcome this problem by highlighting the responsibility of humankind for the preservation of this knowledge through fair access and benefit sharing provisions. A problem within the CBD regime is how access and benefit sharing requirements can be reconciled with the rights of intellectual property holders. Solutions of individual contracts within the existing legal framework or sui generis legislation have been suggested (cf. Stoianoff 2012) but for the moment, the international community has sent this problem off to WIPO.

Australia's obligations in relation to IEK stem from its international obligations. The Convention on Biological Diversity (CBD) is based on three basic principles: the maintenance of biodiversity, the sustainable use of its components, and the equitable sharing of the benefits deriving from the utilisation of genetic resources. It was ratified by Australia in 1993.

Article 8(j) states:

> Each Contracting Party shall, as far as possible and as appropriate:
>
> (j) Subject to its national legislation, respect, preserve and maintain knowledge, innovations and practices of indigenous and local communities embodying traditional lifestyles relevant for the conservation and sustainable use of biological diversity and promote their wider application with the approval and involvement of the holders of such knowledge, innovations and practices and encourage the equitable sharing of the benefits arising from the utilization of such knowledge, innovations and practices.

Article 10(c) states:

> Each Contracting Party shall, as far as possible and as appropriate:
>
> (c) Protect and encourage customary use of biological resources in accordance with traditional cultural practices that are compatible with conservation or sustainable use requirements.

Article 15 of the CBD concerns access to genetic resources and benefit sharing, and is of particular significance to indigenous people and those involved in natural resource management. It provides for access to be on mutually agreed terms and subject to prior informed consent (CBD: Art. 15.4, 15.5), as well as fair and equitable benefit sharing (CBD: Art. 15.7). Article 15 specifically recognises that national governments have the authority to determine access to genetic resources occurring in their countries.

Despite the focus on access and benefit sharing in the CBD, 20 years after it came into force, there is still no international regime protecting IEK and providing for a share of the benefits flowing from the commercialisation of this knowledge. The CBD has been criticised for its soft language and absence of a 'prior informed consent' provision in relation to protection of IEK and its slow progress in moving towards effective access and benefit sharing (ABS) programmes (Koutouki and Rogalla von Bieberstein 2012). The recognition of this lack of progress led to the adoption of the non-binding Bonn Guidelines in 2002. These guidelines have been criticised on a number of grounds, not only because of their voluntary nature but also on the inability of the holders of IEK to enforce their rights (Koutouki and Rogalla von Bieberstein 2012: 523). They focus too much on the access side, ignoring the fact that the benefit-triggering moment often takes place in another jurisdiction and so there is a need to have user-country measures that comply with ABS laws of the provider country as well as measures to monitor compliance with any agreements (Koutouki and Rogalla von Bieberstein 2012). This prompted the negotiation of a legally binding protocol on ABS.

The Nagoya Protocol was adopted in October 2010.[3] Its preamble emphasises the importance of ABS as an incentive for the conservation of biological diversity (Nagoya Protocol: Preamble) and the importance of traditional knowledge for the conservation of biological diversity. It recognises the diversity of ownership of IEK by indigenous communities and gives these communities the right to identify who, amongst their community, are the rightful owners of their IEK. An expanded definition of the utilisation of genetic resources in article 2 creates more legal certainty by including biochemical compositions within the scope of ABS, resulting in drugs based on the extraction of chemicals being subject to benefit sharing (Koutouki and Rogalla von Bieberstein 2012: 527). It also deals with the problem of user-country compliance with ABS (that was a criticism of the Bonn Guidelines) in articles 15 and 16, but commentators have argued that these articles rely on contractual means rather than specified obligations within the Protocol, which is the weaker position of the two (Kamau, Fedder and Winter 2010).

One of the strongest criticisms of the Nagoya Protocol is that it undermines indigenous people's rights in favour of state sovereignty. Using the terms 'in accordance with domestic law' throughout the Protocol provides for indigenous people's inherent right to genetic resources to be deemed to be contingent upon recognition by national legislation in each State. Clearly this is problematic but perhaps inevitable when protection of IEK is attempted to be done within a regime underlined by state sovereignty. In the process, issues of intellectual property were deferred to WIPO, even though WIPO's mandate does not cover the protection of IEK. Unsurprisingly, there has been a call for a sui generis system, acknowledging indigenous peoples as rightful owners of their IEK (cf. North American Indigenous Organisations 2010).

There are some as yet unresolved issues with the Nagoya Protocol which need to be addressed. The first is the timeline of IEK that will be covered by the Protocol. Will it apply to IEK acquired before the Protocol comes into force? What about IEK that has already been acquired and stored (Morse 2011: 4)? There is uncertainty as to whether the definition of traditional knowledge will exclude that which is publicly available. Arguably, this is the area where protection is most needed as secret knowledge will have the greater potential to be protected through intellectual property rights.

There are other important international statements of principle relating to indigenous rights in their traditional ecological knowledge. One example is the United Nations Declaration on the Rights of Indigenous Peoples (UNDRIP, United Nations 2007). Article 29 states:

> Indigenous peoples are entitled to the recognition of the full ownership, control and protection of their cultural and intellectual property. They have the right to special measures to control, develop and protect their sciences, technologies and cultural manifestations, including human and other genetic resources, seeds medicines, knowledge of the properties of fauna and flora, oral traditions, literatures, designs and visual and performing arts.

Article 24 states:

> Indigenous peoples have the right to their traditional medicines and health practices, including the right to the protection of vital medicinal plants, animals and minerals.

The Nagoya Protocol specifically refers to the UNDRIP in its preamble, where it is noted. Canada wanted no reference to UNDRIP and only agreed to this weakened version after international criticisms of its position (North American Indigenous Organisations 2010).

There is an international standard promoting ethical research into the relationships between plants, animals, the environment and indigenous people. This is the ISE Code of Ethics developed by the International Society for Ethnobiology (2006). With respect to attribution, the Code promotes a principle of acknowledgement and due credit, which recognises that indigenous peoples, traditional societies and local communities must be acknowledged in accordance with their preference and given due credit in all agreed publications and other forms of dissemination for their tangible and intangible contributions to research activities. Co-authorship should be considered when appropriate. Acknowledgement and due credit to indigenous peoples, traditional societies and local communities extend equally to secondary or downstream uses and applications, and researchers are to act in good faith to ensure the connections to original sources of knowledge and resources are maintained in the public record (International Society of Ethnobiology 2006: Principle 17).

National level

Australia ratified the CBD in 1993 and has implemented its international obligations through the Environmental Protection and Biodiversity Conservation (EPBC) Act 1999 (Cth).

Interests of indigenous Australians are included in the objects of the EPBC Act:

> Section 3. Objects:
>
> (h) to recognise the role of indigenous people in the conservation and ecologically sustainable use of Australia's biodiversity; and
>
> (g) to promote the use of indigenous peoples' knowledge of biodiversity with the involvement of, and in co-operation with, the owners of the knowledge.

In order to achieve these objectives, the EPBC Act promotes a partnership approach to environment protection and biodiversity conservation through recognising and promoting indigenous peoples' role in, and knowledge of, the conservation and ecologically sustainable use of biological resources.[4] The EPBC Act provides for the introduction of regulations regarding control of and access to biological resources in Commonwealth areas and the implementation of a benefit sharing scheme.[5]

Pursuant to s301, in 2005, part 8A of the Regulations was introduced. The purpose of part 8A is to be achieved to some extent by recognising the special knowledge held by indigenous persons about biological resources.[6] Anybody wanting to access biological resources in Commonwealth areas for research and product development must apply to the relevant National Parks office for a permit.[7] If access to biological resources is for a commercial (or potential commercial) purpose, the permit will only be granted if the applicant has entered into a benefit sharing agreement with the relevant access provider, which includes Aboriginal owners and native title holders.

A benefit sharing agreement must provide for reasonable benefit sharing arrangements, including protection for, recognition of and valuing of any indigenous people's knowledge to be used, and must include

> (h) a statement regarding any use of indigenous people's knowledge, including details of the source of the knowledge, such as, for example, whether the knowledge was obtained from scientific or other public documents, from the access provider or from another group of indigenous persons;
>
> (i) a statement regarding benefits to be provided or any agreed commitments given in return for the use of the indigenous people's knowledge;
>
> (j) if any indigenous people's knowledge of the access provider, or other group of indigenous persons, is to be used, a copy of the agreement regarding use of the knowledge (if there is a written document), or the terms of any oral agreement, regarding the use of the knowledge.[8]

The regulations also provide for informed consent by owners of TK to a benefit sharing agreement.[9] In determining whether informed consent has been given, the Minister must consider:

- the access provider's knowledge of the regulations
- whether reasonable negotiations took place
- the views of the relevant land council or traditional owners, and
- whether there was provision of independent legal advice.[10]

The requirement of informed consent means indigenous access providers must be properly informed as well as consulted before they permit access to biological resources and use of their knowledge for commercial purposes on their land. It would also suggest that indigenous people in these situations can deny interested parties access if they are unwilling to give proper consideration and value to indigenous knowledge (Janke 2009).

These provisions seem to be in line with Australia's international obligations pursuant to the CBD but have attracted criticism (Holcombe, Rimmer and Janke 2009). There are limitations to these provisions in that the jurisdiction of the EPBC Act is limited to actions within Commonwealth lands and actions taken outside Commonwealth land which have, will have or are likely to have a significant impact on the environment on the Commonwealth land.[11] The general limitation of the EPBC Act is that it will only apply to matters that are likely to have a significant impact on a matter of national environmental significance and these matters are rather limited.[12] In other cases, the responsibility for the environment falls to State and Territory regimes.

In 2010, Australia prepared a 20-year Biodiversity Conservation Strategy for 2010–2030 (Commonwealth of Australia 2010). Its first national biodiversity strategy, developed in 1996 (Commonwealth of Australia 1996), was reviewed in 2001 and 2006. One of the main shortcomings was the inability to get agreement on conservation objectives and targets[13] and the difficulty of measuring performance against the qualitative objectives in the strategy. The 2010 strategy emphasises the need for public awareness and involvement and requires measurable targets. It increases the focus on managing biodiversity on a regional or landscape focus with a whole of the ecosystem approach, including connectivity and corridors; and it recognises the vulnerability of biodiversity to rapid climate change, resulting in the need for new approaches. It focuses on the threats to biodiversity loss to develop programmes to address these threats. These threats have been identified as habitat loss, degradation and fragmentation; invasive species; unsustainable use and management of natural resources; changes to the aquatic environment and water flows; changing fire regimes; and climate change (Commonwealth of Australia 2010: 22). In dealing with these threats, the strategy has identified three action priorities. The first is to engage all Australians. The second is to build ecosystem resilience in a changing climate, and the third is to achieve measurable results. It contains 10 measurable national targets for 2015.[14] There is now much more emphasis on implementing robust national monitoring and evaluation and connecting science to policy. There has been a move from a focus on the protection of single species and iconic areas with a static approach to biodiversity conservation, to re-establishing ecosystem functions and reducing threats to biodiversity. Similarly, the earlier focus on government having sole responsibility for biodiversity conservation with limited acknowledgement of indigenous conservation skills has changed to mainstreaming biodiversity, increasing indigenous involvement and enhancing strategic investments and partnerships.

The strategy acknowledges indigenous participation in biodiversity conservation as a sub-priority in its first priority of engaging all Australians. The sub-priority of increasing indigenous engagement has three outcomes of increasing the employment and participation of indigenous peoples in biodiversity conservation activities, increasing the use of indigenous knowledge in biodiversity conservation decision-making, and increasing the extent of land managed by indigenous people for biodiversity conservation (Commonwealth of Australia 2010: 40). Despite this, indigenous groups have been scathing in their criticism of the lack of recognition of current indigenous programmes in biodiversity protection as well as the lack of implementing protection of indigenous knowledge in the strategy. The EPBC regulations have also come under attack for the lack of strong enforcement measures for non-compliance and for inadequately recognising IEK (Holcombe, Rimmer and Janke 2009: 6). The submission by the Northern Australian Indigenous Land and Sea Management Alliance (NAILSMA) criticises the strategy as understating the existing role of indigenous land managers (Morrison, James and Michael 2009)[15] and failing to recognise current research and monitoring by indigenous groups using innovative technologies. It also fails to appreciate the bio-cultural diversity and has limited practical understanding of the region and language-specific nature of much of the traditional ecological knowledge. They argue that this is a failure to implement its commitment under articles 8(j) and 10(c) of the CBD. The submission by the Centre for Aboriginal Economic Policy Research (CAEPR) argues that the draft strategy fails to recognise the role that indigenous biodiversity management plays in conserving biodiversity, particularly through the formalised indigenous land and sea management programmes. This submission also called for greater support of indigenous people's sustainable customary and commercial use of biodiversity (Altman et al. 2009). Principle 7 underlying the strategy's vision was regarded as weak and this submission called for a recognition that 'management that acknowledges and respects indigenous cultural values, innovations, practices and knowledge is fundamental to long-term biodiversity management'. This recommendation was not accepted. This submission also criticised the strategy for not giving enough attention to the rights of traditional owners of IEK and falling short of its obligations under Art 8(j) CBD. They recommended an objective that, 'the use of indigenous knowledge in the scientific, commercial and public domains proceeds only with the cooperation and control of the traditional owners of that knowledge and ensure that the use and collection of such knowledge results in social and economic benefits to the traditional owners'. A third submission argued for a regulatory framework providing for benefit sharing regimes, free prior informed consent, management of intellectual property, and storage and repatriation of research materials (Holcombe, Rimmer and Janke 2009: 3). They stressed that any national framework needs to be consistent with the existing protocols and management tools already developed by indigenous organisations. Indigenous knowledge management guidelines were recommended based on the principle of 'active protection', recognising that IEK is as much about practice as content. Sui generis legislation was recommended as a way forward (Holcombe, Rimmer and Janke 2009: 4–6).

Can IEK be protected through patents?

Under Article 27 of TRIPS, there is an acknowledgment of the patentability of

> any inventions, whether products or processes, in all fields of technology, provided they are new, involve an inventive step and are capable of industrial application.

This raises one of the difficulties around the use of patents to protect IEK and that is the question of novelty. Unfortunately, the TRIPS agreement does not specifically address IEK. The requirement of novelty may not be satisfied because the traditional knowledge may be widely known throughout the community. There also needs to be an inventive step and something may not be patentable if it is knowledge of a naturally occurring product – so much IEK may not be regarded as sufficiently inventive. There are other reasons why patents are not appropriate to be applied to IEK. The cost is prohibitive,[16] and the maximum time limit of 20 years seems incompatible with indigenous notions of responsibility for and ownership of ecological knowledge. A final hurdle is that the patent system operates on the protection of individual rights, and in indigenous communities the knowledge is held by a community or a section of the community (Janke 2009: 40).

The nub of the problem is that where IEK has been well known within an indigenous community to have particular qualities and then subsequently researchers come along and identify the active chemical, isolating it and then synthesising it, they argue that this new product is patentable, if shown to be novel. The traditional owners can argue that but for their IEK, the researchers would never have identified the plant. All would agree that the traditional owners should be compensated but the question is what is the most appropriate manner?

There has been a lot of discussion about whether it would help if a disclosure requirement was introduced into patent legislation, disclosing the origins of the biological material. Many countries have introduced such a provision.[17] Australia has not. Arguments for disclosure include that by increasing transparency, disclosure can help track the commercial exploitation of IEK for ABS. Some say that disclosure is a necessary and effective incentive measure to spur patent applicants to comply with PIC and ABS and that disclosure would also prevent misappropriation and biopiracy and ensure compliance with the CBD (Henniger 2009). Those opposing the introduction of disclosure mechanisms argue that the purpose of patents is to reward innovation, not to protect biodiversity. Perhaps a more practical view is that these disclosure mechanisms should be flexible and only necessary where the source of the material is closely linked to the invention, i.e. it is based on IEK (Henniger 2009).

In most cases, patent law would not be appropriate because of the reasons mentioned above. When the knowledge is secret, the application of patent law is more promising because the requirement of novelty may be maintained. There are examples of where the providers of that knowledge can claim as joint owners of the ensuing patent. For example, the Chuulangan Aboriginal Corporation joint venture with the University of South Australia (cf. University of South Australia 2012). Considering the recognition of the difficulty of patenting IEK and in the absence of sui generis legislation in Australia, Australia relies on a variety of access and benefit sharing schemes to protect IEK. The next section of this chapter assesses the effectiveness of these schemes.

Impact of access and benefit sharing programmes on indigenous Australians

There are a number of important issues relating particularly to access and benefit sharing arrangements with indigenous Australians. The first relates to who, amongst the indigenous community, has the authority to negotiate on behalf of that community. Another relates to how substantial benefit sharing will be guaranteed. Another issue that may arise is to do with the requirement for legal certainty and consistency. If these arrangements are negotiated individually, then legal certainty and consistency may be a casualty of this process. In addition to these is the question of how the knowledge and how the mutually agreed terms regarding benefit sharing are to be determined. Knowledge use must be respectful of indigenous values and custodianship needs to be recognised. To this end, prior informed consent is essential.

This section will consider the existing national and state-based schemes that protect IEK through access and benefit sharing programmes, and the impact of these access and benefit sharing programmes on the holders of this knowledge. Australia signed the Nagoya Protocol in January 2012 and is currently developing its approach to implementation and ratification. It already has extensive legislative provisions in place with regard to access and benefit sharing provisions, and national competent authorities have been established in all Australian jurisdictions.[18]

Legislation

Section 301 of the EPBC Act provides access and benefit sharing arrangements for IEK in Commonwealth areas. To ensure consistency amongst all Australian jurisdictions, in 2002, all Australian governments came together and developed the *Nationally Consistent Approach for Access to and the Utilisation of Australia's Native Genetic and Biochemical Resources.*[19] This was pursuant to Objective 2.8 of the 1996 Biodiversity Strategy. The intention was to provide consistency in the regulation and management of access to genetic resources and it set out some general principles to be applied in all Australian jurisdictions.

Under the Commonwealth legislation, if there is a request for access to genetic resources on indigenous land, then the terms of access must be negotiated with the indigenous owners and the benefits must go to the indigenous community under mutually agreed terms. These may include direct benefits, local priorities and options that reinforce self-determination. As an example, see the Desert Knowledge CRC Protocol for Aboriginal Knowledge and Intellectual Property (Desert Knowledge Cooperative Research Centre). The role of the Commonwealth government is to verify the content of the benefit sharing arrangement, to ensure negotiations are fair and that prior informed consent of the knowledge holder has been obtained.

A number of Australian states have been progressive in legislating for access and benefit sharing in their jurisdictions. The first was the *Biodiscovery Act* Qld (2004) but this legislation did not include any reference to TK in its access and benefit sharing provisions. This legislation is supplemented by the Queensland Biotechnology Code of Ethics which provides for benefit sharing from the use of traditional knowledge (Queensland Government 2001: Art. 11). This code has limited mandatory application, only applying to three types of organisations, known as Queensland Biotechnology Organisations (QBOs). These are:

a Queensland Government agencies, research centres, laboratories and public hospitals that conduct biotechnology activities;
b Private sector companies, academic institutions and research bodies that receive financial assistance from the Queensland government to undertake biotechnological activities;
c Cooperative Research Centres (CRCs) that receive financial assistance from the Queensland Government to undertake biotechnology activities, and all CRCs that conduct biotechnological activities that have a Queensland government body or officer as a participating member. (Stoianoff 2012: 36)

Other organisations can voluntarily subscribe to the Code but so far no major pharmaceutical company has become a voluntary subscriber.[20] As Stoianoff has pointed out, article 11 deals with ABS with regard to genetic resources in Queensland but it is too vague with no details on what will satisfy the prior informed consent requirement or what will constitute a 'reasonable benefit-sharing arrangement' (Stoianoff 2012: 37). Arguably, there should be model templates to help with this process, akin to the Desert Knowledge CRC Protocol.

The Northern Territory enacted the *Biological Resources* Act in 2006. This legislation provides for access and benefit sharing agreements when 'bioprospectors' wish to conduct research into genetic resources in the Northern Territory. Biosprospecting is defined as research relating to the genetic resources or biochemical compounds contained in a biological sample such as a plant or an animal.[21] A potential bioprospector needs to obtain a permit[22] and a permit will only be given if the authority is satisfied that prior informed consent has been obtained from the resource access provider,[23] and there is a fair benefit sharing arrangement,[24] including details of the benefits going to the resource access provider.

The benefit sharing agreements must provide for 'reasonable benefit sharing arrangements including protection for, recognition of and valuing of any Indigenous people's knowledge to be used'.[25]

Under s28 (2), in considering whether a resource access provider has given informed consent, the CEO must consider the following matters:

a whether the resource access provider had adequate knowledge of this Act and was able to engage in reasonable negotiations with the applicant for the permit about the benefit-sharing agreement;
b whether the resource access provider was given adequate time:
   i) to consult with relevant people; and
   ii) if the biological resources are in an area that is Aboriginal land and a resource access provider for the resources is a Land Trust – for the responsible Land Council to consult with the traditional owners for the land; and
   iii) to negotiate the benefit-sharing agreement;
c whether the resource access provider has received independent legal advice about the application and requirements of this Act. Details of these arrangements are to be kept in a register.[26]

According to Holcombe and Janke, of the 45 benefit sharing deeds entered into between the Northern Territory (NT) and biodiscovery groups, only three have been completed with indigenous parties. As half of the NT is aboriginal freehold land, this shows a high level of non-compliance with the regime by public and private bioprospecting entities in Australia, resulting in indigenous knowledge holders missing out on benefits that should flow through to them (Holcombe and Janke 2012: 296).

None of the other Australian states have yet enacted legislation to protect IEK, even though there is the '*Nationally Consistent Approach for Access to and the Utilisation of Australia's Native Genetic and Biochemical Resources*'.

Protocols, guidelines and policies

There are a number of protocols, guidelines and policies that have been drafted by indigenous communities.[27] Protocols can provide an example of best practice when dealing with indigenous people and knowledge but they are not legally enforceable unless included or referred to in a contract. Kathy Bowery has described the use of protocols in the protection of traditional knowledge:

> Protocols are prescriptive – in that they prescribe particular types of behaviour. They also have the capacity to convey a mode of behaviour that institutions and individuals are presumed to follow. Protocols prescribe

> modes of conduct through emphasizing or normalizing particular forms of cultural engagement. Whilst this effect is not assured, over time protocols do have the capacity to influence change in ways that differ to stringent bureaucratic or legislative programs. Protocols are part and parcel of repositioning certain agendas. They are ostensibly based in choice and therefore less than law as command. It is true that an individual, or even an institution either chooses to follow them or not. But this is also true of positive law … The challenge is of allowing for productive 'private' negotiations that can present as alternative and distinctive, without allowing the formal legal order to presume that the indigenous case always falls outside of its categories and power.
>
> (Bowery 2006 as cited in Janke 2009: 107)

The protocols perhaps offer the best opportunity to protect IEK by setting out detailed approaches and guidance to third parties wanting to access indigenous knowledge. They have been developed with extensive indigenous participation and each of them focuses on indigenous participation, detailed prior informed consent and benefit sharing provisions. For example, in terms of free prior informed consent, the Desert Knowledge CRC Protocol says this means,

- Aboriginal participants in the project have been fully informed about the project, and have a clear understanding of the purpose, methodology, and intended outcomes of the research, including potential risks, uses and possible commercialisation options
- Adequate opportunities and timeframes have been provided for Aboriginal participants to make their own decisions about the research and whether they will participate. This may be either as individuals or through their communities and organisations
- Consent is an ongoing engagement between the community and the researcher. Subject to local circumstances, it can be suspended or withdrawn.

The Kimberley Land Council (WA) protocol requires acknowledgement of the value of the traditional knowledge in the research and requires the free, prior and informed consent of traditional owners at every stage of the project, including at the concept stage in the project methodology (Kimberley Land Council 2011: paragraph 4.1). It requires a benefit sharing agreement for access to genetic resources, this being defined as,

> a written agreement providing for the fair and equitable sharing of benefits derived from the use of Biological Resources and any associated Traditional Knowledge with the persons, or groups (or suitable representative) providing access to the Biological Resources and Traditional Knowledge. Such an agreement must be established upon mutually agreed terms with the Free Prior Informed Consent of the relevant Indigenous person(s) or group(s) and should include, inter alia

2.3.1. A dispute settlement clause;
2.3.2. Terms on benefit-sharing, including in relation to intellectual property rights;
2.3.3. Terms on subsequent third-party use, if any; and
2.3.4. Terms on changes of intent, where applicable.

(Kimberley Land Council 2011: paragraph 2.3)

The Aboriginal and Torres Strait Islander library and information resource network has developed a series of protocols to guide libraries, archives and information services interacting with Aboriginal and Torres Strait Islander peoples in the communities which the organisations serve, and to handle materials with Aboriginal and Torres Strait Islander content.[28] These protocols cover topics such as governance and management, intellectual property, secret or sacred materials, and offensive materials.

Australian case studies

There are a number of case studies in all Australian states which we will now consider and attempt to draw some important lessons regarding the impact of access and benefit sharing arrangements.

The smoke bush case study from WA illustrates many of the problems that can come with the patenting of IEK. Indigenous people from Geraldton and Esperance in Western Australia have traditionally used smokebush for healing. In the 1960s, the WA government gave the US National Cancer Institute (NCI) a licence to gather plants to screen for the presence of cancer-fighting properties. They were unsuccessful and were stored for 20 years when they were retested and found to contain 'conocurvone', an active ingredient with the potential to destroy the HIV virus. This discovery was then patented by NCI which tried to leave Australia with more plant samples, amid claims of biopiracy. The NCI then granted a Victorian company, Amrad, exclusive rights to smokebush for research purposes and licence to develop Conocurvone. With the licence came a provision for royalties to the WA government of around $1.5 million. Successful commercialisation of Conocurvone would have resulted in royalties of more than $100 million. These agreements made no provision for indigenous groups to receive any benefits. Development of a drug from Conocurvone was suspended in 2001, although Amrad still holds the licence to do so (Janke 2009: 42).

Consider two more positive examples of patenting medical inventions from indigenous knowledge where true partnerships have been set up and the indigenous holders of this knowledge have been included as inventors.[29]

The first concerns the Jarlmadangah Burru people of the Kimberley, where a community elder found the marjala tree to have healing and pain relief properties. He used the plant to numb pain after losing a finger when crocodile hunting. He then took a sample of the plant to Professor Ron Quinn from Griffith University and explained how he used the bark from the tree as a dressing on his wound (Brisbane Courier Mail 2008). Griffith University and

Jarlmadangah Burru Aboriginal Corporation (JBAC) have filed several patents in relation to the healing properties of the plant as joint applicants. Patent No. 2004293125 for 'novel anaelgesic compounds, extracts containing same and methods of preparation' was sealed in 2010. As has been noted, the patent system has limited value in protecting traditional knowledge as this knowledge will become publicly available after 20 years. In October 2008, the JBAC and Griffith University entered an agreement with Avexis (a biotechnology company) to commercialise the plant in return for sharing of benefits – firstly by the income from sales, and secondly from the opportunity to cultivate and supply the plant in sufficient quantities to make the product marketable. Horticulture, however, is not conducive to the intergenerational transfer of indigenous knowledge and there has been local resistance and a preference for bush harvesting (Holcombe and Janke 2012: 313). The marjala plant bark remedy constitutes secret indigenous knowledge and this case study shows that co-ownership of the patent application with benefit sharing provisions on commercialisation can develop a community supply chain resulting in local economic empowerment and lead to cultural enforcement through harvesting (Holcombe and Janke 2012: 315).

The second example is the collaborative research project between the Chuulangan Aboriginal Corporation from Cape York, Queensland, and researchers from the Quality Use of Medicines and Pharmacy Research centre at the University of South Australia (University of South Australia 2012). The Corporation is a joint applicant with UniSA on a patent concerning the anti-inflammatory components of the plant extract from the native Australian plant, Dodonea polyandra. The indigenous elder, David Claudie, is included as an inventor on the patent application[30] and co-author on numerous academic applications (Simpson et al. 2010, 2011a, 2011b, 2012; Claudie et al. 2012).

The Plants for People project in Titjikala (Tapatjatjaka)[31] was instigated to try to document traditional knowledge in order to support the intergenerational transference of this knowledge, through developing best practice approaches and using that knowledge to advance the livelihoods of aboriginal people. This project was conducted under the Desert Knowledge CRC protocols. Community elders had noticed that traditional learning environments in the bush had been reduced with the impact of Western commercialisation from modern media and they wanted to create community awareness of the importance of retaining and transferring IEK through the generations (Evans et al. 2010: 5). The project began by collecting data on culturally significant plant species and these were stored on a password protected electronic database, called the Tapatjatjaka Plant Database. Fifty-three plant species relating to food, medicine, tools and mythology are kept in text, digital images and audio files in both English and Pitjantjatjara language (Evans et al. 2010: 7). Community children were included in the videos resulting in increased usage and cultural information transference. Some valuable lessons were learned, particularly in regards to engaging the community, drafting agreements that would provide benefits to the community, and protecting the IEK that would be recorded.

The case study contains the following observations:

- protecting IEK of plants and their uses is most likely to be achieved through the development of effective protocols for preserving and recording IEK and the use of contract law in commercial applications of that knowledge (Evans et al. 2010: 7)
- relationship building and a partnership approach based on trust and mutual respect were found to be of fundamental importance
- IEK holders are faced with the dilemma of how to commercialise their knowledge but still remain faithful to their ancestors and to their spiritual beliefs
- a key concept articulated and later formalised as part of the research agreement was that the process should be one of partnership to achieve a common goal
- there was a need to balance the risk of failing to protect the IEK rights of the Titjikala community if commercialisation of a medicinal product did eventuate with the risk of the research not proceeding due to the need to comply with an excessively complicated process
- consultations involving all or most of the community is a more effective approach to agreement making than a process involving a representative person or representative structure.

This approach was used in the Titjikala case study, with recommendations being developed by the research team, endorsed by the Council of Elders and the TCGC and then finally endorsed at community gatherings to which all community members were invited. However, this decision making process was lengthy and involved high transaction costs.

> Questions have been raised regarding the issue of exclusivity of this contract and possible financial benefits to the Titjikala community, given that some of the plants studied occur over wide geographic regions and have cultural significance to many different Aboriginal groups.
>
> (Janke 1998)

As a result of this project, a written agreement was reached between the Tapatjatjaka Community Government Council and Curtin University. The clauses relating to IP state that:

- IP developed in the project will be shared between the participants according to a negotiated agreement, and
- Ongoing indigenous ownership of the cultural and IPR in the material on which the research is based should be acknowledged. (Evans et al. 2010: 140)

The Ara Irititja project in SA has a demonstration video documenting the Anangu's knowledge through a digital library.[32] Developed in order to give the Pitjantjatjara people access to all of the information and records that have been gathered over the past 60 years, one of the important issues is to be able to restrict access, where appropriate. There are restrictions for sorrow and for any embarrassing materials. Pitjantjatjara language is used wherever possible. Although it is a private, secure database, this project is helping other organisations build their own archives using their software.

The Indigenous Knowledge in Water Planning, Management and Policy case study (Roberts 2012) in Cape York sought to obtain indigenous perspectives on how to incorporate indigenous knowledge in water planning, management and policy, as required under the National Water Initiative.[33] This report recognised that the lack of appreciation of indigenous water values and lack of meaningful consultation with the indigenous community had resulted in a disconnect between processes and expectations (Roberts 2012: 3). One of the objectives was to 'Do something positive for the Traditional Owner groups concerned and provide Indigenous knowledge (IK) amenable to being integrated with Western science' (Roberts 2012: 6). Two issues arose in the two case studies involved in this project. In the first, the indigenous knowledge holders were involved near the end of the process. In the second, the indigenous owners only spoke generally and not site-specifically because of their reluctance to share their sacred sites, water places and knowledge with outsiders. A formal benefit sharing agreement may have alleviated their concerns. In interviews with the owners, they seemed to be asking for prior informed consent to be a part of the process (Roberts 2012: 13). Ideally, traditional owners should understand the origin of the research project and the motivation for water reform and be able to generate the project themselves. This would involve at least a day's exercise. There were a number of findings from this case study:

1. Lead times for indigenous research projects must be substantial for optimal results. In many cases, Traditional Owners will not be rushed or unduly stressed. They would sooner let the opportunity pass, as they see additional stresses as unnecessary and bad for their wellbeing.
2. An effort should be made to establish indigenous research priorities from the user and grass roots level, and then broader programme logic developed from responses. Indigenous support organisations should be used to facilitate this approach.
3. Research opportunities could be put directly to Traditional Owner groups, who could be assisted by regional indigenous organisations who are more informed about existing capabilities and with the ability to value-add and cross-link projects. Traditional Owner groups could then make their own judgements about preferred partners if any and the utility of the opportunity.
4. Resources should be provided for an internal planning process where knowledge is freely discussed. Then the group can decide what can be shared and by whom, and information that cannot be shared can be converted to a visual 'use overlay' which simply allows or disallows certain activities. NAILSMA has commissioned a useful schematic of what categories of

knowledge might exist in the indigenous data domain. This is part of their data-sharing agreements relating to I-tracker data collected by local ranger groups, where the local groups are empowered to place restrictions they see fit on different types of information.

5 To blend IK and Western science, the worldviews and different spiritualities of each must be respected.
6 Feelings, emotions and perceptions which are too often regarded as irrelevant to scientific research, management, use and policy can create conflict. Some conflict resolution may be necessary.

This case study illustrates that a focus on Western science has the potential to leave out the cultural and spiritual dimension of IEK in ABS arrangements. Not acknowledging or respecting that indigenous people 'believe in' the spiritual elements of their country is disrespectful. This can lead to a refusal to engage on the part of indigenous groups. The development of protocols for free prior informed consent and access and benefit provisions must be developed by the indigenous communities themselves. They need time to understand the context of the research and acknowledgement that indigenous knowledge cannot be separated from the person.

Conclusion

The problem that access and benefit sharing arrangements have tried to overcome is where traditional knowledge of genetic resources has been exploited without any benefits flowing to indigenous holders of this knowledge, generally through the patent system. Various commentators have called for sui generis as a national regime to protect indigenous knowledge of genetic resources (Stoianoff 2012: 38; Janke 2009: 161; Holcombe, Rimmer and Janke 2009: 5). In Australia there is a need for sui generis legislation to cover the field and avoid the piecemeal results from other forms of legislation. However, the Australian government has rejected this approach, preferring to move towards non-binding measures and 'non-intellectual based elements of a sui generis system' (Holcombe and Janke 2012: 300).

In the absence of sui generis legislation, we have the patent system and the provisions of access and benefit sharing schemes under the CBD. Despite the criticisms and some of the problems with the Nagoya Protocol, an international regime on access and benefit sharing would provide a consistent international framework requiring consultation with holders of IEK regarding the use of that knowledge. The Protocol could draw on existing templates for ABS arrangements from various jurisdictions and collate, draw lessons from experience and disseminate best practice examples and clauses. It is recommended that standard terms and conditions be developed with information for IEK holders to engage commercially and negotiate uses.

Similarly, despite the many problems that patents have in protecting IEK, true partnership-type, collaborative scientific and other research projects may lead to the development of patentable inventions based on IEK. As the case studies have shown, in these circumstances, it is important that researchers discuss the potential patent application with indigenous people, initially. If a patent is filed, the indigenous knowledge holders should benefit from any commercialisation of the resulting product. One option is to include the indigenous knowledge holder as a joint applicant in the patent application. Alternatively, or additionally, the researcher or institution applying for the patent could enter into a benefit sharing agreement with the relevant knowledge holder. Benefit sharing agreements with indigenous access providers are now legally required under the NT *Biological Resources Act* (2006) and the Commonwealth *Environment Protection and Biodiversity Conservation Regulations*, but as we have seen, there seems to be large non-compliance with the NT legislation.

At a local level, implementing protocols and guidelines for NRM practitioners in various fields should be encouraged. ABS agreements have the potential to protect IEK and so impact on indigenous Australians in a positive way. This is the most practical way of ensuring IEK is dealt with appropriately in the immediate future. A number of issues relating to ABS agreements can impact on indigenous knowledge holders and these need to be addressed, most likely through protocols. There are a number of regional and local protocols already developed by indigenous communities. Some of this is already happening but a lot more work needs to be done to try to have consistency through the numerous indigenous groups working across this area. One of the most crucial issues borne out by this research is the loss of intergenerational transfer of IEK. There is a need to continue to promote a living tradition of intergenerational knowledge by putting people into their country. Knowledge centres, such as have been developed with the *Ara Irititja* database software, may also be a way forward as a way of allowing users to access the content, but still limiting access where this is necessary. These knowledge databases would be one way of holding this knowledge for future generations. Another issue is that of free and informed proper consent. This must be a collaborative and ongoing process that should enhance local self-determination (Holcombe and Janke 2012: 302). Finally, these protocols must be enforced through contractual clauses.

Notes

1 Cf. <http://www.iucn.org/about/work/programmes/species/our_work/the_iucn_red_list/> (accessed 13 June 2013).
2 For example, Uluru Kata Tjuta Park note on fire management. Online. Available HTTP: <http://www.environment.gov.au/parks/publications/uluru/pubs/pn-fire-management.pdf > (accessed 13 June 2013). Cf. Commonwealth of Australia 2013.
3 This protocol will come into force on the 90th day following the 50th ratification. As at 28 March 2013, there were 15 ratifications.
4 *Environmental Protection and Biodiversity Conservation* Act 1999 (Cth) s 3(2)(g)(iii).
5 *Environmental Protection and Biodiversity Conservation* Act 1999 (Cth) s 301.
6 *Environmental Protection and Biodiversity Conservation Regulations* (Cth) reg 8A.01(c).
7 *Environmental Protection and Biodiversity Conservation Regulations* (Cth) reg 8A.06.
8 *Environmental Protection and Biodiversity Conservation Regulations* (Cth) reg 8A.08.

9 *Environmental Protection and Biodiversity Conservation Regulations* (Cth) reg 8A.10.
10 *Environmental Protection and Biodiversity Conservation Regulations* (Cth) reg 8A.09(2).
11 *Environmental Protection and Biodiversity Conservation* Act 1999 (Cth) s26 (2).
12 There are eight matters of national environmental significance. They are world heritage properties, national heritage places, wetlands of international importance (listed under the Ramsar Convention, UNESCO 1971), listed threatened species and ecological communities, migratory species protected under international agreements, Commonwealth marine areas, the Great Barrier Reef Marine Park, and nuclear actions (including uranium mines).
13 The National Objectives and Targets for Biodiversity Conservation 2001–2005 (Commonwealth of Australia 2001) were produced but were not agreed to by all jurisdictions.
14 They are, by 2015: to achieve a 25% increase in the number of Australians and public and private organisations participating in biodiversity conservation activities; achieve a 25% increase in employment and participation of indigenous peoples in biodiversity conservation; achieve a doubling of the value of complementary markets for ecosystem services; achieve a national increase of 600,000 km^2 of native habitat managed primarily for biodiversity conservation across terrestrial, aquatic and marine environments; restore 1,000 km^2 of fragmented landscapes and aquatic systems to improve ecological connectivity; reduce the impact of alien species on threatened species by at least 10%; research activities to be guided by nationally agreed science and knowledge priorities for biodiversity conservation; all jurisdictions to align legislation, policies and programmes with this strategy; establish a national long-term biodiversity monitoring and reporting system.
15 Also confirmed by CAEPR's submission. Cf. Altman et al. 2009.
16 Estimated cost of set up is between $6,000–$10,000 and maintaining it for 20 years, another $8,000.
17 Model Law of the Andean Community, 18 Belgium, 19 Bolivia, 20 Brazil, 21 China, 22 Colombia, 23 Costa Rica, Denmark, Ecuador, Egypt, the European Community (EC), and most European countries, India, the Kyrgyz Republic, New Zealand, Norway, Panama, Peru, the Philippines, Portugal, Romania, South Africa, Sweden, Switzerland, Thailand, Venezuela.
18 Cf. the List of National Competant Authorities. Online. Available HTTP: <http://www.environment.gov.au/biodiversity/science/access/contacts/act/national.html> (accessed 13 June 2013).
19 Cf. <http://www.environment.gov.au/biodiversity/publications/nca/index.html> (accessed 13 June 2013).
20 Cf. the list of QBOs and subscribing organisations. Online. Available HTTP: <http://www.sd.qld.gov.au/dsdweb/v3/documents/objdirctrled/nonsecure/pdf/3150.pdf> (accessed 13 June 2013).
21 *Biological Resources Act* 2006 (NT) s5(1).
22 *Biological Resources Act* 2006 (NT) s11.
23 *Biological Resources Act* 2006 (NT) s28.
24 *Biological Resources Act* 2006 (NT) s29.
25 *Biological Resources Act* 2006 (NT) s29.
26 *Biological Resources Act* 2006 (NT) s33.
27 For example, the Desert Knowledge CRC Protocol for Aboriginal Knowledge and Intellectual Property, Northern Australian Indigenous Land and Sea Management Alliance Guidelines and Protocols for Research, AIATSIS Guidelines for Ethical Research, Aboriginal and Torres Strait Islander Protocols for libraries, archives and information services, Kimberley Land Council IP and Traditional Knowledge Policy.
28 Cf. <http://aiatsis.gov.au/atsilirn/docs/ProtocolBrochure2012.pdf> (accessed 13 June 2013).

29 Cf. <http://www.itek.com.au/portfolio/index.php?option=com_zooandtask=itemanditem_id=145andItemid=124> (accessed 13 June 2013).
30 Cf. the patent application 'Anti-inflammatory compounds' PCT/AU2010/001502. Online. Available HTTP: <http://patentscope.wipo.int/search/en/detail.jsf?docId=WO2011057332andrecNum=138anddocAn=AU2010001502andqueryString=aquacultureandmaxRec=2052> (accessed 13 June 2013).
31 A joint project between the Tapatjatjaka Community Government Council and Curtin University.
32 Cf. <http://irititja.com/the_archive/demo/demo.html> (accessed 13 June 2013).
33 Cf. <http://nwc.gov.au/nwi> (accessed 13 June 2013).

References

Altman, J., Kerins, S., Ens, E.-J., Buchanan, G. and May, K. (2009) 'Submission to the Review of the National Biodiversity Strategy: Indigenous People's Involvement in Conserving Australia's Biodiversity', *CAEPR Topical Issue* 08/2009.

Australian Bureau of Statistics (2006) *4705.0 – Population Distribution, Aboriginal and Torres Strait Islander Australians.* Online. Available HTTP: <http://www.abs.gov.au/ausstats/abs@.nsf/mf/4705.0> (accessed 13 June 2013).

Bonn Guidelines – United Nations Environment Programme (UNEP) (2002) *Bonn Guidelines on Access to Genetic Resources and Fair and Equitable Sharing of the Benefits Arising out of their Utilisation,* COP 6 Decision VI/24. Online. Available HTTP: <www.cbd.int/decision/cop/?id=7198> (accessed 13 June 2013).

Bowery, K. (2006) 'Alternative Intellectual Property? Indigenous Protocols, Copyleft and New Juridifications of Customary Practices', *Macquarie Law Journal* 6: 65–95.

Brisbane Courier Mail (2008) *Aborigine, Scientist Find Pain Relief in Marjarla Tree.* Online. Available HTTP: <http://www.couriermail.com.au/news/queensland/native-tree-bark-eases-pain/story-e6freoof-1111117814741> (accessed 13 June 2013).

Bubna-Litic, K. (2009) 'The Impacts of Carbon Pricing on Indigenous Communities: A Comparison of New Zealand and Australia', in L.-H. Lye, J. Milne, H. Ashiabor, K. Deketelaere and L. Kreiser (eds) *Critical Issues in Environmental Taxation, International and Comparative Perspectives. Vol. VII*, New York: Oxford University Press, 349–74.

CBD – United Nations (1992) *Convention on Biological Diversity*. Online. Available HTTP: <http://www.cbd.int/doc/legal/cbd-en.pdf> (accessed 12 June 2013).

Claudie, D.J., Semple, S.J., Smith, N.M. and Simpson, B.S. (2012) 'Ancient but New: Developing Locally-driven Enterprises Based on Traditional Medicines in "Kuuku I'Yu" (Northern Kaanju Homelands, Cape York, Queensland, Australia)', in P. Drahos and S. Frankel (eds) *Indigenous Peoples' Innovation: IP Pathways to Development*, Canberra: ANU epress, 29–56.

Commonwealth of Australia (1996) *The National Strategy for the Conservation of Australia's Biological Diversity.* Online. Available HTTP: <http://www.environment.gov.au/archive/biodiversity/publications/strategy/pubs/national-strategy-96.pdf> (accessed 13 June 2013).

Commonwealth of Australia (2001) *The National Objectives and Targets for Biodiversity Conservation 2001–2005.* Online. Available HTTP: <http://www.environment.gov.au/biodiversity/publications/objectives/pubs/objectives.pdf> (accessed 13 June 2013).

Commonwealth of Australia (2010) *Australia's Biodiversity Conservation Strategy 2010–2030. Prepared by the National Biodiversity Strategy Review Task Group convened under the Natural Resource Management Ministerial Council.* Online. Available HTTP: <http://www.environment.gov.

au/biodiversity/publications/strategy-2010-30/pubs/biodiversity-strategy-2010.pdf> (accessed 13 June 2013).

Commonwealth of Australia (2011) *State of the Environment Report 2011, Ch. 8, Biodiversity, The Importance of Biodiversity*. Online. Available HTTP: <http://www.environment.gov.au/soe/2011/report/biodiversity/1-1-the-importance-of-biodiversity.html#s1-1> (accessed 13 June 2013).

Commonwealth of Australia (2013) *Caring for Country – Outcomes 2008–2013*. Online. Available HTTP: <http://fedpub.ris.environment.gov.au/fedora/objects/mql:1887/methods/c4oc-sDef:Document/getPDF pp 21> (accessed 2 April 2013).

Davis, M. (1999) 'Indigenous Rights and Biological Diversity – Approaches to Protection', *AILR* 4: 1.

Desert Knowledge Cooperative Research Centre (no year) *Protocol for Aboriginal Knowledge and Intellectual Property*. Online. Available HTTP: <http://www.desertknowledgecrc.com.au/resource/DKCRC-Aboriginal-Intellectual-Property-Protocol.pdf> (accessed 13 June 2013).

Evans, L., Cheers, B., Fernando, D., Gibbs, J., Miller, P., Muir, K., Ridley, P., Scott, H., Singleton, G., Sparrow, S. and Briscoe, J. (2010) *Plants for People: Case Study Report. DKCRC Report 55*, Alice Springs: Desert Knowledge Cooperative Research Centre. Online. Available HTTP: <http://www.nintione.com.au/resource/NintiOneResearchReport_55_PlantsForPeopleCaseStudyReport.pdf> (accessed 13 June 2013).

Henniger, T. (2009) *Disclosure Requirements in Patent Law and Related Measures. A Comparative Overview of Existing National and Regional Legislation on IP and Biodiversity*. Online. Available HTTP: <http://ictsd.org/downloads/2009/11/henninger-biodiversity-ip-think-piece-final.pdf> (accessed 13 June 2013).

Holcombe, S. (2009) 'Indigenous Ecological Knowledge and Natural Resources in the Northern Territory: Guidelines for Indigenous Ecological Knowledge Management. A Report Commissioned by the Natural Resources Management Board (NT) Component 1 (of 3)', *ANU College of Law Research Paper*. 10–26.

Holcombe, S. and Janke, T. (2012) 'Patenting the Kakadu Plum and the Marjala Tree: Biodiscovery Intellectual Property and Indigenous Knowledge', in M. Rimmer and A. McLennan (eds) *Intellectual Property and Emerging Technologies: The New Biology*, Cheltenham: Edward Elgar Publishing, 293–319.

Holcombe, S., Rimmer, M. and Janke, T. (2009) 'Australia's Biodiversity Conservation Strategy 2010–2020 Submission to the National Biodiversity Strategy Review Taskforce', *ANU College of Law Research Paper*. 10–24. Online. Available HTTP: <http://papers.ssrn.com/sol3/papers.cfm?abstract_id=1630068> (accessed 13 June 2013).

International Society of Ethnobiology (2006) *International Society of Ethnobiology Code of Ethics (with 2008 additions)*. Online. Available HTTP: <http://ethnobiology.net/code-of-ethics/> (accessed 13 June 2013).

Janke, T. (1998) *Our Culture: Our Future Report on Australian Indigenous Cultural and Intellectual Property Rights. Prepared for Australian Institute of Aboriginal and Torres Strait Islander Studies and the Aboriginal and Torres Strait Islander Commission*. Online. Available HTTP: <http://www.frankellawyers.com.au/media/report/culture.pdf> (accessed 13 June 2013).

Janke, T. (2009) 'Report on the Current Status of Indigenous Intellectual Property Component 3 (of 3)', *ANU College of Law Research Paper*. 10–27.

Kamau, E., Fedder, B. and Winter, G. (2010) 'The Nagoya Protocol on Access to Genetic Resources and Benefit Sharing: What is New and What are the Implications for Provider and User Countries and the Scientific Community?', *Law, Environment and Development*

Journal 6 (3): 246. Online. Available HTTP: <www.lead-journal.org/content/10246.pdf> (accessed 13 June 2013).

Kimberley Land Council (2011) *Intellectual Property and Traditional Knowledge Policy.* Online. Available HTTP: <http://uploads.klc.org.au/2012/05/KLC_IP_TK_Policy_V1_final1.pdf> (accessed 13 June 2013).

Koutouki, K. and Rogalla von Bieberstein, K. (2012) 'The Nagoya Protocol: Sustainable Access and Benefits-Sharing for Indigenous and Local Communities', *Vermont Journal of Environmental Law* 13: 513–36.

Morrison, J., James, G. and Michael, C. (2009) *Northern Australian Indigenous Land and Sea Management Alliance, Submission to the Review of the National Biodiversity Strategy*. Online. Available HTTP: <http://www.environment.gov.au/biodiversity/strategy/review-submissions/pubs/id916-nailsma-05062009.pdf> (accessed 13 June 2013).

Morse, J. (2011) 'Nurturing Nature, Nurturing Knowledge: The Nagoya Protocol on Access and Benefit Sharing', *Indigenous Law Bulletin* 7 (24): 3–6.

Nagoya Protocol – Secretariat of the Convention on Biological Diversity (2010) *Nagoya Protocol on Access to Genetic Resources and the Fair and Equitable Sharing of Benefits Arising from their Utilization to the Convention on Biological Diversity: Text and Annex, COP 10 Decision X/1*. Online. Available HTTP: <http://www.cbd.int/abs/doc/protocol/nagoya-protocol-en.pdf> (accessed 13 June 2013).

North American Indigenous Organisations (2010) *Joint Statement of North American Indigenous Organisations on the Nagoya ABS Protocol of the Convention on Biological Diversity.* Online. Available HTTP: <http://indigenouspeoplesissues.com/index.php?option=com_contentandview=articleandid=8084:international-joint-statement-of-north-american-indigenous-organizations-on-the-nagoya-abs-protocol-of-the-convention-on-biological-diversityandcatid=65:indigenous-peoples-generalandItemid=92> (accessed 13 June 2013).

Queensland Government (2001) *Code for Ethical Practice in Biotechnology in Queensland.* Online. Available HTTP: <http://www.sd.qld.gov.au/dsdweb/v3/documents/objdirctrled/nonsecure/pdf/4130.pdf> (accessed 13 June 2013).

Roberts, C. (2012) *Indigenous Knowledge in Water Planning, Management and Policy – Cape York Peninsula, Qld. Case Studies. NAILSMA Knowledge Series 10/2012*, Darwin: North Australian Indigenous Land and Sea Management Alliance Ltd.

Simpson, B., Claudie, D., Smith, N., Wang, J., McKinnon, R. and Semple, S.J. (2010) 'Evaluation of the Anti-inflammatory Properties of Dodonaea Polyandra, a Kaanju Traditional Medicine', *Journal of Ethnopharmacology* 132: 340–3.

Simpson, B., Claudie, D., Gerber, J., Pyke, S., Wang, J., McKinnon, R. and Semple, S.J. (2011b) 'In Vivo Activity of Benzoyl Ester Clerodane Diterpenoid Derivatives from Dodonaea Polyandra', *Journal of Natural Products* 74: 650–7.

Simpson, B.S., Claudie, D.J., Smith, N.M., McKinnon, R.A. and Semple, S.J. (2012) 'Rare, Seven-membered Cyclic Ether Labdane Diterpenoid from Dodonaea Polyandra', *Phytochemistry* 84: 141–6.

Simpson, B.S., Claudie, D.J., Smith, N.M., Gerber, J.P., McKinnon, R.A. and Semple, S.J. (2011a) 'Flavonoids from the Leaves and Stems of Dodonaea Polyandra: A Northern Kaanju Medicinal Plant', *Phytochemistry* 72: 1883–8.

Stoianoff, N. (2012) 'Navigating the Landscape of Indigenous Knowledge – A Legal Perspective', *Biotechnology and the Law* 90: 23.

Szabo, J., Possinghamc, H.P. and Stephen, T.G. (2012) 'Adapting Global Biodiversity Indicators to the National Scale: A Red List Index for Australian Birds', *Biological Conservation* 148 (1): 61–8.

TRIPS – World Trade Organization (WTO) (1994) *Agreement on Trade-Related Aspects of Intellectual Property Rights.* Online. Available HTTP: <http://www.wto.org/english/tratop_e/trips_e/t_agm0_e.htm> (accessed 13 June 2013).

UNESCO (United Nations Educational, Scientific and Cultural Organization) (1971) *Convention on Wetlands of International Importance especially as Waterfowl Habitat. Ramsar (Iran), 2 February 1971. UN Treaty Series No. 14583. As Amended by the Paris Protocol, 3 December 1982, and Regina Amendments, 28 May 1987.* Online. Available HTTP: <http://www.ramsar.org/cda/en/ramsar-documents-texts/main/ramsar/1-31-38_4000_0__> (accessed 13 June 2013).

United Nations (2007) *Declaration on the Rights of Indigenous Peoples, Resolution Adopted by the General Assembly. 61/295. 107th Plenary Meeting 13/9/2007.* Online. Available HTTP: <http://www.un.org/esa/socdev/unpfii/documents/DRIPS_en.pdf> (accessed 13 June 2013).

University of South Australia (2012) *Developing Aboriginal Medicines to Fight Inflammation.* Online. Available HTTP: <http://www.unisa.edu.au/media-centre/releases/280812/> (accessed 13 June 2013).

Yencken, D., Henry, N. and Rae, J. (2012) *Biodiversity in Australia*, Victoria: The Australian Collaboration. Online. Available HTTP: <http://www.australiancollaboration.com.au/pdf/FactSheets/Biodiversity-FactSheet.pdf> (accessed 13 June 2013).

12 Reconstructing the 'biopiracy' debate from a justice perspective

Klara H. Stumpf

Introduction

One of the most important treaties regulating the conservation and use of biodiversity on the international level is the Convention on Biological Diversity (CBD) of 1992.[1] It differs from other international environmental treaties in a remarkable way: questions of fairness, equity and justice play an explicit and prominent role.[2]

In the Access and Benefit-Sharing (ABS) framework established by the CBD and the Nagoya Protocol,[3] the main requirements are that benefits shall be granted in exchange for access to genetic resources and traditional knowledge (*benefit-sharing*, CBD: Art. 1 and 15.7; Nagoya Protocol: Art. 3 and 6), that prior informed consent (*PIC*) shall be obtained before access (CBD: Art. 15.5; Nagoya Protocol: Art. 6), and that mutually agreed terms (*MAT*) for the benefit-sharing shall be negotiated (CBD: Art. 15.4 and 15.7; Nagoya Protocol: Art. 5, 6, 7 and 18, inter alia). The ABS framework is thus based on an *exchange perspective.*

But is this perspective sufficient? To answer this question, it is helpful to examine why the ABS framework was established. It emerged as a response to the practice of *bioprospecting* and connected allegations of *'biopiracy'*. Bioprospecting is the search for biological and genetic materials and the investigation of the properties these materials have for practical applications in science and industry (cf. Lößner 2005: 36). Many plants and microorganisms have a high potential to provide prototypes for new pharmaceuticals, agrochemicals and other goods (WBGU 2000: 69). In this process of bioprospecting, the so-called traditional knowledge of indigenous and local communities can play an important role by enhancing the probability of success (to find a new lead substance, e.g. for pharmaceuticals) up to four times (Shiva 2002: 85). The term 'biopiracy' has come up as a political term connected to this practice of bioprospecting and to the ABS framework. As no general definition of the term 'biopiracy' exists, different authors use it in different ways, emphasising different aspects.[4] But still, we can say that the term is generally associated with some kind of offence to developing countries or indigenous and local populations regarding the utilisation of genetic resources and traditional knowledge for (potentially patented) inventions (cf. Federle 2005: 25).

As the variety of understandings of 'biopiracy' indicates, the narrow perspective of the ABS framework may not be sufficient to capture all aspects of the justice problems arising when genetic resources and traditional knowledge are made use of. This chapter therefore reconstructs the debate from the perspective of the concept of justice. The lead question is: *Which problems of justice arise regarding the utilisation of genetic resources and traditional knowledge,*[5] *especially if associated with patenting?* My hypothesis is that the perspective of justice-in-exchange (*sensu* Aristotle), as underlying the ABS requirements of the CBD and the Nagoya Protocol, is too narrow to understand all claims of (in)justice in the 'biopiracy' debate.

The aim of this study is thus to contribute to the clarification of the problems of justice concerning biodiversity, genetic resources and traditional knowledge, and to highlight some implications for a step-by-step improvement of justice.

The analysis focuses on arguments which either explicitly use the term 'biopiracy' or refer explicitly to claims of (in)justice connected to patenting, genetic resources and traditional knowledge. To limit the scope of the discussion, only claims which concern the relationship between humans are analysed.[6]

In the following, I start by laying some theoretical foundations – first, by discussing the legal, philosophical and economic background of patenting, and second, by presenting the philosophical framework for the analysis centred on the concept of justice. Building on this theoretical and mainly philosophical framework, I examine claims of (in)justice frequently put forward in the 'biopiracy' debate, in order to show that the perspective of justice-in-exchange is too narrow and that at least some of the rationally reconstructable claims made in the debate belong to other domains. In a subsequent step, I summarise the discussion and end by presenting some conclusions.

Theoretical foundations

Legal, philosophical and economic background of patenting

From a legal point of view, an invention needs to fulfil the criteria of novelty, inventive step (or non-obviousness) and utility (or industrial applicability) in order to be patentable (cf. e.g. Federle 2005: 33, 44). These criteria are roughly the same in all major national and international patent laws.[7] Briefly put, inventions should be patentable, while discoveries should not.[8] One important difference between the US-American and the European system is the understanding of novelty – while the European system defines novelty as 'absolute novelty' (cf. Lößner 2005: 50), in the US-American system only certain instances are detrimental to the claim of novelty, such as prior publication or patenting in the US or another country, or prior public use or recognition *inside the United States* (cf. Federle 2005: 48) ('relative novelty').

From a philosophical and economic perspective, Fritz Machlup (1958) identified four lines of reasoning or theses for the general justification of patents (i.e. why there should be any patents at all):

1. The 'natural law' thesis, stating that there is a natural (human) property right of the inventor to his or her invention,
2. the 'reward-by-monopoly' thesis, stating that the inventor deserves some kind of reward for the favour done by him or her to society by the invention – and this reward can take the form of the monopoly granted by the patent (or some other form),
3. the 'monopoly-profit-incentive' thesis, stating that the profit extractable from the monopoly position granted is necessary to recoup the costs of research, so that the prospect of obtaining a patent is necessary as an incentive for the inventor to even start with his research, and
4. the 'exchange-for-secrets' thesis, stating that the patent is necessary as an incentive for the inventor to disclose his or her invention to the public (as he or she has to describe it in detail in the patent application documents) when he would otherwise try to keep its underlying principles secret for as long as possible.

Theses one and two are essentially 'justice arguments', as they are about individual rights and merit. Theses three and four can be labelled 'incentive arguments', as they are about furthering the common good by giving individual inventors incentives to invest in research or to disclose their knowledge to the public, thus stimulating further research (even if the invention itself cannot be used by others before the end of the patent term).

Natural law arguments (best known from John Locke's theory of property), which typically justify property by referring to first occupancy or extended personality, create serious problems when applied in order to justify *intellectual* property in inventions or in traditional knowledge. Apart from the fact that the soundness of these natural law approaches to the justification of property in general can be questioned (cf. Becker 1980), their applicability to abstract objects (like knowledge) should especially be doubted (Drahos 1996: 52). Therefore, the 'natural law' thesis is treated as rejected in this chapter.

In contrast to the natural law thesis, none of the three remaining theses can be totally rejected and all three have some validity. I therefore argue that justifications of patents must take the form of a 'mixed justification' relying mainly on arguments of societal interest (i.e. incentives for disclosure of knowledge and incentives for innovation), but also of individual merit. The patent system should thus be judged according to how it lives up to these justifications, i.e. how it provides for individual rewards and for incentives to contribute to societal welfare.[9]

Philosophical framework: justice

The philosophical analysis of this chapter centres on the concept of *justice*. Justice is a central and contested concept in philosophy (cf. Mazouz 2006: 371; Dobson 1998: 5; Höffe 2004: 9; more generally for different conceptions of justice cf. Aristotle 1998; Dworkin 2000; Nozick 1974; Nussbaum 2000; Pogge 2008; Rawls 1971; Risse 2012; Sen 2009, among others). However, a kind of 'minimum definition' could be that justice refers to the mutual (intersubjective) claims and obligations that members of the community of justice could agree on from the standpoint of impartiality and equal consideration (cf. Gosepath 2007: 82). In the following I present a way to systematise claims and conceptions of justice employing a 'conceptual structure of justice', and different 'domains of justice'. I furthermore propose to use a comparative approach to assessments of justice.

Conceptual structure of justice

In order to describe claims and/or conceptions of justice in a comprehensive way, we have to specify a number of elements which are part of what I call the 'conceptual structure of justice'.[10] These include the following (cf. also Box 1):

The *judicandum*: That which is to be judged just or unjust (actors, actions, institutions or states of affairs) (Pogge 2006: 863). The assessment can be taken from a *process and/or outcome* perspective.

The *metric* for the judgement: This includes the *informational base* (Sen 1990: 111) on which the judgement is based and the *principle(s)* to be applied to this base.[11] *Informational bases*[12] are mostly discussed for the domain of distributive justice; candidates given in the literature are, for example, capabilities (Sen 1990; Nussbaum 2000), primary goods (Rawls 1971), or utility.[13] H. Peyton Young identifies three broad *principles*[14] to be applied to this informational base: equality (or parity), proportionality, and priority (cf. Young 1994: 8), which might need to be further specified (e.g. proportionality according to merit, need, ability to pay; priority according to being the worst-off or according to being able to make the most out of some good; etc.). A further principle is that of sufficiency, meaning that everyone should have enough of some good without necessarily involving a comparison to what others have (Frankfurt 1987: 22; Krebs 2003: 237).[15]

The *community of justice*:[16] This includes *claim holders* (holding particular claims) and *claim addressees* (responsible for the fulfilment of the claims).[17]

The *claim(s)*: The notion of a claim is central to the concept of justice (Ott and Döring 2008: 47). Therefore, the content of the claims held by the claim holders should be specified.

On a more practical level, it is also important to define the *instruments of justice* (that which is to be used to satisfy the legitimate claims of justice).[18]

Finally, the *ethical foundation*[19] of a conception (or claim) of justice influences what is specified as the judicandum, which metric is seen as appropriate, which claims are seen as legitimate, and who belongs to the community of justice. I am

Box 12.1 Conceptual structure of justice

Judicandum
Process/outcome perspective
Metric
- Informational base
- Principles

Community of justice
- Claim holders
- Claim addressees

Claim (content)
Instrument of justice
Ethical foundation

not subscribing to or advocating any particular ethical foundation here. Rather, I will assess the possibility to rationally reconstruct the claims brought forward in the 'biopiracy' debate resorting to different lines of ethical reasoning.

The conceptual structure can be used in two different ways. It can be employed to describe a *full conception of justice* (in which case all of the above elements should be specified, with 'claim addressee' and 'claim holder' referring to general groups of potential claim addressees/claim holders), and it can be employed to describe *single claims about (in)justice* (in which case primarily the elements of claim content, claim holder and claim addressee should be specified).[20] I will use it in the second way in this study.

Domains of justice

Claims of justice can be classified into different fields of application or 'domains' of justice (Pogge 2006), based on Aristotle's fruitful and still very influential distinction between general justice and particular justice. While *general justice* is about the lawful, that which creates and maintains the good for the community, *particular justice* has different forms, referring to different fields of application for more specific demands of justice. It can be divided into distributive justice, justice-in-exchange and corrective justice (Aristotle 1998).[21] The domains of justice can be associated with different kinds of social interaction (Koller 2007): *Distributive justice* is the domain of justice concerning community relationships, i.e. relationships between persons who have common claims for certain goods or share the obligation to carry certain burdens (Koller 2007: 8). While the claim holder can be a single person, the claim addressee of such a kind of claim is the community as a whole, or some representative of the community, e.g. a central authority like the state (cf. Petersen 2009a: 25). In the domain of distributive justice, different principles of distribution can apply. For Aristotle, it was the *axa* (dignity) of the person determining her share (Petersen 2009a: 25f.). In modern conceptions of distributive justice, it requires *equal* consideration and treatment, unless there are reasons for unequal consideration which are acceptable for all (cf. Koller 2007: 9). These acceptable reasons usually refer to contributions or merits, (basic) needs, or legitimate expectations (cf. Engisch 1971: 152f.; Koller 2007: 9; Ladwig 2004: 130f.; Petersen 2009a: 26).

Justice-in-exchange is applicable to exchange relationships, i.e. whenever two or more persons interact with each other voluntarily to mutually confer certain goods and services on each other (Koller 2007). The claim addressees and claim holders of the resulting claims are the respective partners in such an exchange. The principle of justice-in-exchange (in an *outcome* perspective) is equivalence (cf. Petersen 2009a: 26). In the – nowadays more prominent – *process* perspective on justice-in-exchange an exchange is just if the partners have fair negotiation positions (cf. Koller 2007: 9; Mazouz 2006: 372; Wagner 1990: 277).

Corrective justice is required whenever a member of the community breaks the rules of the community, injures the rights of others or does not fulfil his or her obligations towards others. Corrective justice can be further subdivided into restitutory justice dealing with remedies for harm done, and retributive justice dealing with punishments (cf. Koller 2007: 10; for a more detailed account cf. Lumer 1999: 466). The question of who is claim addressee and who is claim holder is more complicated in this domain. First, (corrective) claims are addressed against the rule-breaking person. But claims can also be addressed to the community or its representatives (e.g. the judiciary) to persecute the wrongdoer and enforce the corrective claims.[22] The claim holder of corrective claims is generally the harmed party (in the case of restitution), but it can also be the general public who want to see the wrongdoer punished for breaking the rules (retributive justice).

Regarding the background conditions of the mentioned domains of justice, Malte Faber and Thomas Petersen (2008) coin the term *Ordnungsgerechtigkeit* (*structural justice*), describing the moral quality of the setup of a community regarding its judicial and political institutions and moral principles (Faber and Petersen 2008: 411). The justice of the order of a community can be measured by the degree to which it enables its members to lead a good life (Faber and Petersen 2008: 412). Thus, the members of the community (as claim holders) can be said to hold a claim against the structure or setup of the community (or those responsible for that structure) that the preconditions of a good life be provided. Faber's and Petersen's notion of structural justice builds on Aristotle's category of general justice (cf. Petersen 2009a: 25).[23]

Comparative approach

I propose to use a comparative approach to assessments of justice, aiming at enhancing justice case by case rather than achieving perfect justice. This has been advocated by several authors. Amartya Sen (2009) claims in *The Idea of Justice* that an analysis of justice related to practical politics 'cannot be but about comparisons' (Sen 2009: 400), e.g. by comparing the justice of different states of affairs before and after a proposed reform. Thomas Pogge (1991) also argues that for a gradual improvement of the justice of the global order, it is necessary to compare between the 'feasible and morally accessible avenues of institutional change' (Pogge 1991: 260).[24] Referring to the CBD, Doris Schroeder and Thomas Pogge write: '[T]he CBD framework should be assessed by reference to the common good of humankind. In making this assessment, one must consider the effects of the CBD relative to those of its politically available alternatives' (Schroeder and Pogge 2009: 273; cf. also Schroeder 2009). One example of a justice assessment on the basis of a comparative approach would be to use John Rawls' (1971) difference principle in a modified form – by evaluating judicanda in terms of their impact on the worst-off (cf. Byström, Einarsson and Nycander 1999: 23).

Analysis of selected claims from the 'biopiracy' debate

Equipped with this theoretical and especially the philosophical framework, I will now examine some of the statements about (in)justice frequently put forward in the 'biopiracy' debate, in order to answer the question posed in the introduction: *Which problems of justice arise regarding the utilisation of genetic resources and traditional knowledge, especially if associated with patenting?* Provided that my hypothesis is that the perspective of justice-in-exchange, as underlying the ABS requirements of the CBD and the Nagoya Protocol, is too narrow to understand all claims of (in)justice in the 'biopiracy' debate, each claim will first be described using the domains of justice and the conceptual structure, and will then be discussed. I will start by examining one of the most common claims, to respect the requirements of the CBD, then clarify the allegations connected to the term 'bad patents', furthermore look at claims of moral desert, reciprocity, human rights, and – somewhat outside the moral sphere – incentives, as well as conservation financing, and finally touch upon further issues of structural justice.

'Respect the requirements of the CBD' – claims for PIC, MAT and benefit-sharing

One very common concern in the 'biopiracy' debate is that bioprospecting companies or other bioprospecting actors (as claim addressees) should, in accordance with the requirements of the CBD, obtain 'prior informed consent' (PIC) from and reach 'mutually agreed terms' (MAT) (claim contents) with countries of origin or local and indigenous communities (as claim holders). Another demand (claim content) is that they should share benefits from bioprospecting with countries and communities of origin.[25] In short, bioprospectors are asked to comply with the requirements of the CBD (cf. for example FUE 2002: 16 as cited in Lößner 2005: 36; Nilles 2003: 216; Baumgartner and Mieth 2003: 319). These claims fall in the domain of justice-in-exchange with the corresponding principles of equivalence and fairness. PIC and MAT refer to the procedural dimension, as these requirements aim to establish fair negotiation positions. Benefit-sharing refers to the outcome dimension.

These claims show that (of course) the emphasis of the ABS regulations on exchange is not totally misplaced. In their justification, these claims refer to the precondition of all claims of justice-in-exchange: an act of exchange. Thus, as soon as an access to genetic resources or traditional knowledge takes place, the reference to claims of justice-in-exchange in general is justified (but not necessarily to particular specifications of these claims, e.g. regarding the amount of benefit-sharing[26]) – provided that rights (of property, or merit) of countries of origin or of indigenous or local communities to the resources in question are acknowledged. This means that such rights need to be clearly delimited (although not necessarily in legal terms) before it is possible to make conclusive statements about justice-in-exchange.[27] However, PIC and MAT may not be sufficient to reach fair negotiation positions – a substantial capacity building might be required.

Bad patents

The term 'bad patents' (Federle, 2005; cf. also Hamilton 2006: 170) is used to describe patents on inventions based on traditional knowledge or genetic resources which do not satisfy the criteria of inventive step, novelty and/or utility.

In terms of the conceptual structure, different kinds of justice claims can be identified here. To start with, an (illegitimate) claim held by a biotechnology firm (as – illegitimate – claim holder) against the general public (as claim addressee) to accept the property claim on the invention (i.e. the patent claims as claim content). Then, we can identify corrective claims which can be further subdivided. First, restitutory claims directed towards the bioprospecting firm (the claim addressee) to give up the patent and to compensate those harmed (claim contents), the claim holder being the general public (harmed by the illegitimate monopoly and the reduction of the public domain), but also groups more directly harmed[28] by the patent. Second, retributive claims like punishments (claim content) could be brought up against the biotech firm (claim addressee) by the judiciary (claim holder). As a second-order claim addressee of the corrective claims, the judiciary can be demanded to 'take action' on the issue. In sum, 'bad patents' touch upon both the domain of distributive justice (when constituting illegitimate claims of merit; and when having adverse distributive effects) and the domain of corrective justice.

The fact that 'bad patents' are granted although they do not fulfil the criteria for a patentable invention (novelty, inventive step, utility) is attributed on the one hand to the 'unscrupulous users' of the patent system, and on the other hand to a failure of the patent examination process (Dutfield 2004: 50f.). The corrective claims against these illegitimate patents are justified by the harm done by them, but also by their illegitimacy itself. The claims against 'bad patents' also hint at the broader 'justice effects' of the patent system: if existing inequalities are reinforced by the patent system, it fails to pass the comparative 'minimum test' (Byström, Einarsson and Nycander 1999) of not worsening the situation of the worst-off, creating a problem of structural justice.

Claims of moral desert

Claims of moral desert are raised in regard to the conservation and development of genetic resources and with respect to the production of useful knowledge by local and indigenous communities (as claim holders, against the 'general public' as claim addressee) (cf. for example Wullweber 2004: 88; Klaffenböck and Lachkovics 2001: 134). These claims belong to the domain of distributive justice (because they allude to community relations, target the 'general public' as claim addressee, and refer to the contribution to a community good for their justification).

The validity of these claims is difficult to establish. First, it of course depends on whether the deserving action was actually undertaken by the supposed claim holder. So, did local communities actually develop genetic resources, produce useful knowledge, etc.? How to distinguish between desert and contingent factors (factors beyond human control, e.g. 'natural nature')?[29] At least for some cases, a specific, 'deserving' human action can be confirmed (cf. e.g. Dutfield (2004: 99) referring to genetic resources and traditional resource management, Castle and Gold (2008: 73) regarding the utility of traditional knowledge). However, claims of desert are normally attributed to individual performance – how to deal with claims of collective or inherited desert? Which claims can today's generations infer from the achievements of their ancestors? As a 'causal' approach doesn't work (the presently living cannot claim credit for achievements of the past), a different approach could be to use an analogy: just as the inhabitants of industrial countries benefit from their countries' industrial developments in the past, members of indigenous and local communities should benefit from the achievements of their predecessors. Finally, the claim content is not quite clear – it has to be discussed what kind of reward is expected for the deserving action.

Arguments from reciprocity

The utilisation of traditional knowledge or genetic resources in bioprospecting and biotechnology may serve as a proof of existing social cooperation (*sensu* Rawls) between the holders of such knowledge (or owners of such resources) and the users of those resources. The existence of such a community of social cooperation then establishes legitimate claims of indigenous and local communities (as claim holders) for some of the fruits of the cooperation (claim content) against the community (claim addressee).[30] These claims belong to the domain of distributive justice.[31]

David Castle and Richard E. Gold argue that the utility of traditional knowledge is made clear by the fact the bioprospectors are interested in it (Castle and Gold 2008: 73), and this shows, one could add, that a community of social cooperation indeed exists. Thus, there is some plausibility to these claims, even if they are not very specific in terms of the claim content. How much of the fruits of the cooperation can be claimed by whom? Wilfried Hinsch (2005: 59f.) argues for 'prima facie equal claims to share the fruits of social cooperation'. At least, the comparative minimum test should be applied.

Arguments from human rights

From a human rights point of view, benefit-sharing can be seen as a mechanism that should contribute to the fulfilment of these human rights (Castle and Gold 2008; de Jonge and Korthals 2006). The proponents of this view claim that benefit-sharing should be directed first towards fulfilment of basic needs (de Jonge and Korthals 2006: 151f.; Schroeder and Pogge 2009: 277f.). Such claims can be taken to belong to the domain of distributive justice (on a sufficientarian account), but also to structural justice (referring to an absolute standard that needs to be guaranteed by those responsible for the institutional order). This line of argument breaks the link between the possession of traditional knowledge (or genetic resources) and being eligible for benefit-sharing which becomes 'a mechanism to ensure the equitable distribution of both scientific research capacity and gains arising from scientific research, at least in the health care and agricultural fields' (Castle and Gold 2008: 75). Consequently, anyone in need could be seen as a claim holder, irrespective of their possession of traditional knowledge or genetic resources (but this could of course still include local and indigenous communities).

The first responsibility to ensure these claims lies with the states (Castle and Gold 2008: 74f.). Still, it is also possible to connect the obligation to share benefits (i.e. the question of who is claim addressee) to the utilisation of genetic resources and traditional knowledge, i.e. to construct an obligation of bioprospecting actors to share benefits.

As long as human rights are taken as justified,[32] these claims are quite plausible; however, their justification (ethical foundation) has only a very loose connection to the utilisation of genetic resources and traditional knowledge.[33]

Non-moral 'entitlements to legitimate expectations': incentives

Another possible line of argument does not refer to moral claims, but to incentives (or 'entitlements to legitimate expectations' *sensu* Rawls, as claim content), for example for local communities (as claim holders) to disclose (traditional) knowledge, to produce such knowledge, and to protect biodiversity (cf. for example de Jonge and Korthals 2006: 147; Cunningham 1991).[34]

Whether (monetary) incentives can reasonably be expected to achieve their desired effect depends on the kind of desired activity. They might work for the conservation of biodiversity (insofar as it is a matter of choice about alternative land uses), but might not work for, say, the production of traditional knowledge grounded in social practices which do not react to, and might even be disturbed by, those incentives (for the last point cf. Mulligan 1999: 47; Wullweber 2004: 130f.; but cf. also Byström, Einarsson and Nycander 1999: 51, who indicate that bioprospecting agreements of the International Cooperative Biodiversity Group (ICGB) in Nigeria and Suriname were 'said to have reinforced cultural values related to traditional use of medicinal plants'). The success of incentives for conservation also presupposes that local communities have full control over their territories and that there are no harmful outsiders like oil companies, gold mining, etc. destroying nature in the area.

Benefit-sharing as a means of funding biodiversity conservation

Benefit-sharing can also be seen as a means to contribute to an internationally shared funding of biodiversity conservation activities, i.e. as a requirement of distributive justice. The claim content in this case is the sharing of the burdens of biodiversity conservation, the claim holders are biodiversity-rich countries and communities, and the claim addressee can be various actors depending on the way this claim is justified.

Such a common funding might be in order for different reasons, e.g. because the costs of a public good should be distributed fairly (de Jonge and Korthals 2006), because funding should come from those who benefit from biological resource use (Cunningham 1991) or because of the historical responsibility of industrialised countries (Frein and Meyer 2008).

Further claims of structural justice

Finally, as already hinted at above, the term 'biopiracy' also encompasses statements that criticise unequal access to the patent system (cf. for example Byström, Einarsson and Nycander 1999: 49; Hamilton 2006: 169) or its corrective mechanisms (Frein and Meyer 2008: 141; Hamilton 2006: 169 referring to the costs of patent challenges), as well as the negation of custom-based regulations (Dutfield 2000: 63f.). These concerns touch upon the moral quality of the international order regulating intellectual property, genetic resources and traditional knowledge.[35] The judicandum is thus to be found in the structures, rules and institutions underlying the global patent system and other regulations for intellectual property, genetic resources and traditional knowledge, and the domain is that of structural justice. The main criterion and claim content in this domain is the degree to which the members of an order (as claim holders) are enabled to lead a good life (Faber and Petersen 2008: 412). The claim addressees are those responsible for that order and/or able to influence it.

Does the international order regulating intellectual property, genetic resources and traditional knowledge satisfy the formal criterion of equality? Does it enable its members, i.e. the persons affected by it, to lead a good life? At least it should not worsen their situation. There is certainly room for reform here. As patents are no natural rights and are thus modifiable, they should be designed so as to achieve their 'mixed' goals (giving incentives for innovation and disclosure of knowledge and rewarding inventors), while also being measured by the standards of structural justice.

Discussion

The analysis of the 'biopiracy' debate shows that the perspective of justice-in-exchange cannot cover all the claims of (in)justice regarding the utilisation of genetic resources and traditional knowledge. At least some of the other claims made in the debate can be rationally reconstructed, and they belong to the domains of distributive justice, corrective justice and structural justice. The analysis of the selected claims also shows the fruitfulness of classifying claims of justice according to the domains and the conceptual structure of justice.

What does this analysis imply for step-by-step justice improvements (as called for under a comparative approach to justice) and for further analysis? Two relatively practical implications can be named. First, the analysis of the claims of justice-in-exchange implies that indeed in specific cases of bioprospecting or 'biopiracy' *standards of justice-in-exchange* should be enabled and enforced (i.e., PIC, MAT and benefit-sharing as minimum requirements, complemented by a substantial capacity building) – the Nagoya Protocol could be an important step in this direction. Second, the analysis regarding bad patents and some of the structural injustice claims imply that the *international order regulating intellectual property, genetic resources and traditional knowledge should be reformed*, e.g. by posing a strict emphasis on the distinction between invention and discovery and by applying a standard of absolute novelty. Additionally, other forms of rewards and incentives other than intellectual property should be considered.

The analysis of the arguments regarding moral desert, reciprocity, incentives, and conservation funding furthermore shows that *distributive issues* arise that go beyond genetic resources and traditional knowledge, but concern the sharing of the benefits and burdens (according to different principles of justice) of use and conservation of *biodiversity more generally*.

The analysis of the *human rights claims* implies that, as these rights are fundamental (referring to the background conditions of other justice claims), it is legitimate to ask for their recognition also in the benefit-sharing context. However, benefit-sharing also needs to take account of concerns of, e.g., environmental effectiveness, efficiency and merit. This points to the conclusion that human rights claims (as well as other claims of structural justice) might not be fully satisfiable in the CBD/ABS framework; but have to be addressed taking into account a broader number of institutions and their interplay.

Concluding remarks

As a general conclusion, the question how institutions and structures shape (perceptions of) justice deserves more attention. For example, it became clear that the question how to define (property) rights to genetic resources, but also to biodiversity and nature more generally, is far from settled. This also touches upon the questions of land rights so important for indigenous communities, but merely mentioned in the 'biopiracy' debate. This analysis should also include local conceptions of justice and of the human relationship to biodiversity. Another open question for implementation is how to coordinate international and national regulations – or, in other words, which claims of justice should be directed towards the national level, and which to the international level?

Finally, policymakers, activists and scientists involved in the 'biopiracy' debate alike should broaden their perspectives on how the issues of justice that come up when genetic resources, traditional knowledge and more generally biodiversity are used. The concern for justice in the CBD should be more broadly conceived, concerning all levels of biodiversity including local and global ecosystem services. A further analysis should also include aspects such as justice towards nature (which was deliberately beyond the scope of this chapter) or justice towards future generations. These aspects of justice are especially important when thinking about biodiversity policy in the context of *sustainability*.

The development and implementation of (more) just institutions of biodiversity conservation and use, embedded into an account of other institutions and structures which influence justice, is one of the big tasks of sustainability policy and politics on all levels, by governments, social movements, individuals and organisations.

Acknowledgements

I would like to thank Stefan Baumgärtner, Christian Becker, Stefanie Glotzbach and Konrad Ott for fruitful discussions of earlier versions or parts of this study; and an anonymous reviewer for constructive comments. I would also like to thank the participants at the workshop 'Biodiversity – Concept and Value', Bonn, 2011, at the International Conference of the European Society for Ecological Economics, Istanbul, 2011, and members of the research project EIGEN for discussion. Financial support from the German Federal Ministry of Education and Research (BMBF) under grant 01UN1011A is gratefully acknowledged.

Notes

1 The Convention gives a well-known definition of biodiversity in its Article 2: '"Biological diversity" means the variability among living organisms from all sources including, inter alia, terrestrial, marine and other aquatic ecosystems and the ecological complexes of which they are part; this includes diversity within species, between species and of ecosystems.'

2 One of the three objectives of the Convention is 'the fair and equitable sharing of benefits arising out of the utilization of genetic resources' (CBD: Art. 1), and it also calls for benefit-sharing in respect to 'traditional knowledge' (CBD: Art. 8j).

3 The Nagoya Protocol on Access and Benefit-Sharing of 2010, a protocol to the CBD, further develops the requirements and recommendations of the CBD for benefit-sharing regarding genetic resources and traditional knowledge (Nagoya Protocol: Art. 3).

4 That is why Chris Hamilton (2006: 173) calls it an 'index for a number of different concerns'. For example, David Castle and Richard E. Gold describe 'biopiracy' using the following 'wrongful exploitation scenario': 'An indigenous group has traditional knowledge. Another group, typically but not necessarily members of an industrialized country, recognizes the potential utility of the knowledge and exploits it. When the latter does so, it gains access to and control over the benefits arising from the knowledge to the exclusion of the indigenous group' (Castle and Gold 2008: 67). Vandana Shiva defines 'biopiracy' as 'the process through which the rights of indigenous cultures to these resources and knowledge are erased and replaced by monopoly rights for those who have exploited indigenous knowledge and biodiversity' (Shiva 1997: 31, as cited in Gehl Sampath 2003: 22). By referring to 'monopoly rights', Shiva touches upon the problem of patenting. Gertrude Klaffenböck and Eva Lachkovics (2001) similarly describe 'biopiracy' as referring to a situation in which biological resources are acquired from local communities or indigenous peoples and are patented – or a direct derivative is patented – while the resulting profits do not benefit the communities having originated, utilised and disclosed the resources (and their properties) (Klaffenböck and Lachkovics 2001: 134). Others, e.g. the Friends of the Earth, refer to the standards of the CBD – for them, 'biopiracy' occurs when biological or genetic resources and/or knowledge of indigenous or local groups about their utilisation are appropriated without adhering to the standards of the CBD (FUE 2002: 16, as cited in Lößner 2005: 36; Wullweber 2004: 88).

5 'Utilisation of genetic resources' is understood in this chapter as the utilisation of the species-specific information contained in the functional units of heredity, including the genetic and biochemical composition of a species (cf. e.g., Byström, Einarsson and Nycander 1999: 13, Nagoya Protocol, Art. 2d). 'Traditional knowledge' is understood as locally adapted practical ecological knowledge (cf. e.g. Federle 2005: 20f.; Dutfield 2004: 91).

6 Resource-based patenting also raises claims regarding the relationship between humans and their non-human environment. For example, 'patents on life' (i.e. patents on plants, animals or microorganisms or on organic compounds or parts thereof, cf. Baumgartner and Mieth 2003: 11) are often rejected for reasons of the inherent value or dignity of these organisms.

7 The international system of patent law is constituted by national patent laws, the European Patent Convention, several international treaties on intellectual property rights under the World Intellectual Property Organization, as well as diverse regional treaties and directives of the EU (Dutfield 2000: 8). In the context of biodiversity and the CBD, the Agreement on Trade-Related Aspects of Intellectual Property Rights (TRIPS) is the most important international treaty on intellectual property rights (Dutfield 2000: 8).

8 The European patent system makes this distinction explicitly, while the American patent system reaches a similar effect with the 'laws of nature doctrine' (cf. Federle 2005: 60). For further details and discussion cf. also Spranger in this volume.

9 As any structure or order, it should also be judged according to its broader structural justice effects. For example, if existing inequalities are reinforced by the patent system, it fails to pass the comparative 'minimum test' (Byström, Einarsson and Nycander 1999) of not worsening the situation of the worst-off.

10 This 'conceptual structure' has been developed in the Sustainability Economics Group at Leuphana University of Lüneburg, inter alia for a course on 'Sustainability Governance'. A detailed discussion of an earlier version can be found in my diploma thesis, Gerechtigkeit in der Biodiversitäts-Konvention (2010), Leuphana University of Lüneburg, which is available from the author. Cf. also Stefan Baumgärtner et al. (2012).

11 This corresponds to questions 3 and 4 in the conceptual analysis of the notion of justice by Simon Caney (2005: 103).

12 Also called 'currency' of justice or advantage/benefit by some authors (Kersting 2000: 35; Page 2007: 1). Mathias Risse (2012: 4) speaks of '*distribuendum*, *metric*, or *currency* of justice'.

13 For other domains of justice, and for other judicanda and perspectives, other informational bases will be important. For instance, in the domain of justice-in-exchange the information might refer to the existence of fair bargaining positions characterised by sufficient information, rational self-control and the absence of force and fraud (procedural perspective) or to the value of the exchanged goods, which should be equivalent (outcome perspective) (cf. Koller 2007: 9). Equivalence can be seen as a special case of the principle of proportionality (where the ratio is 1:1).

14 Also called 'pattern of justice' (Page 2007: 1) or 'principle of distribution' (Dobson 1998: 63).

15 As a special case in situations of scarcity, the principle of sufficiency can lead to a priority rank-ordering of whose claims need to be satisfied first (Krebs 2003: 241).

16 Also called 'scope of justice' (Caney 2005: 103; Page 2007: 1) or 'relevant population' (Risse 2012: 4). Cf. also Andrew Dobson (1998: 64ff.), who uses the term 'community of justice', but, referring only to *distributive* justice, speaks of 'recipients' and 'dispensers' instead of claim holders and claim addressees.

17 There can be claim addressees on different levels of consideration: First-level claim addressees are those persons directly addressed. Second-order claim addressees are those who are responsible for the enforcement, enabling, etc. of certain claims.

18 Cf. e.g. Anand and Sen (2000). Dobson asks 'What is distributed?' (Dobson 1998: 73ff). However, instruments of justice could take many forms. They could refer to a (re)distribution of particular goods such as income or primary goods, but also to institutional reform or other measures which lead to a better 'score' on the justice metric.

19 Also called 'grounds of justice' (Risse 2012). Dobson (1998: Chapter 3) distinguishes *universal and particular norms* underpinning theories of justice.

20 The two levels will often interact – the conception of justice may inform about possible justifications of single claims; the analysis of single claims may test if the conception of justice really yields results which are consistent with our moral convictions, etc.
21 Justice-in-exchange and corrective justice are both forms of commutative justice, which is characterised by an interaction between single members of the community (as opposed to distributive justice which is characterised by community relationships). Justice-in-exchange refers to voluntary exchanges, while corrective justice refers to involuntary 'exchanges' (Kersting 2000: 42; Mazouz 2006: 372). The domains of justice can be further subdivided into more sophisticated domains, but I resign from doing that here. Furthermore, procedural justice is often mentioned as a category of justice in its own right. I don't distinguish a separate domain of procedural justice as procedural aspects play a role in all domains.
22 The judiciary is thus a second-order claim addressee.
23 The domain of structural justice could be interpreted as a special case of distributive justice, where the judicandum and claim addressee is an institution or structure (or those responsible for these institutions and structures), and the principle is one of sufficiency (having enough to lead a good life). However, I argue that it is worthwhile to keep this as a distinct category of justice, as it includes an emphasis on the importance of a certain quality and stability of the addressed order, institution or structure. Petersen (2009b: 92) connects this domain of justice to an absolute standard of justice (such as the one by Konrad Ott and Ralf Döring (2008), who in turn build on the capability approach by Sen and Nussbaum) and argues that such an absolute standard requires a certain quality of the order guaranteeing this standard. Similarly, Krebs (2003: 243) implies that the guaranty principles of the sufficientarian approach belong to the category of general justice and are *prior* to the principles of particular justice (including distributive justice).
24 Sen and Pogge follow different lines of reasoning, though. Sen argues that there can be a plurality of principles of justice (Sen 2009: x), while Pogge speaks of *one* global principle of justice (Pogge 1991: 260).
25 In this part of the 'biopiracy' debate, the legitimacy of patents or compliance with patent criteria are not questioned – Christina Federle therefore designates the connected patents as the 'good patents' of biopiracy (Federle 2005: 107).
26 The question of a 'just' amount of benefit-sharing is indeed not trivial. One approach could be to share benefits proportional to contribution, but the quantification of these contributions can turn out to be very difficult (cf. Mulligan 1999: 38; Brush 1993: 661; Byström, Einarsson and Nycander 1999: 22; Dutfield 2000: 54).
27 In this sense, structural or distributive justice have to precede justice-in-exchange.
28 Not all bad patents are directly harmful, but 'some are potentially harmful and others are actually harmful' (Dutfield 2004: 50), e.g. through exclusion from a previous utilisation of the erroneously patented invention (cf. Dutfield 2004: 50ff., 119; Federle 2005: 24; Byström, Einarsson and Nycander 1999: 48; Frein and Meyer 2008: 116). A kind of harm to the general public is the so-called 'tragedy of the anticommons' (Heller and Eisenberg 1998).
29 For a general discussion of the problem to distinguish between desert and contingent factors cf. Wolfgang Kersting (2000: 249), Wilfried Hinsch (2005: 71).
30 Cf. Hinsch (2005: 59f.) for a general description of Rawls' argument and development of the 'community of cooperation' argument for the international level (without reference to bioprospecting or 'biopiracy').
31 This is because this argument refers to cooperation within the community at large, not cooperation between two specific members (which would hint at the domain of justice-in-exchange).
32 Justifications for human rights themselves can be, e.g., constructed from a legal point of view, i.e. in terms of obligations from international human rights agreements (Universal Declaration of Human Rights, International Covenant on Economic,

Social and Cultural Rights), or from an ethical point of view, e.g. in terms of the generalisability and feasibility of consensus in the sense of discourse ethics (following Ott and Döring (2008: 83f.), who even justify the more demanding humanitarian base this way). Cf. also Risse (2012), who discusses different justifications for human rights and argues for a pluralistic account.

33 Following this argument, what matters are a person's human rights, not their possession or utilisation of genetic resources or traditional knowledge. So, the claim holders are not primarily determined with reference to genetic resources or traditional knowledge. But of course, the satisfaction of their claims stemming from human rights can be more or less tightly connected to these resources: for instance, thinking of agricultural biodiversity, these resources might be one *instrument* to satisfy the claims.

34 The claim addressee would thus be the community/a state authority which is responsible to assure the 'entitlements to legitimate expectations'.

35 The latter concern – negation of custom-based regulations – can also be taken to be a matter of missing *recognition*, a category emphasised, e.g., by David Schlosberg (2004).

References

Anand, S. and Sen, A.K. (2000) 'Human Development and Economic Sustainability', *World Development* 28: 2029–49.

Aristotle (1998) *The Nicomachean Ethics*, trans. W.D. Ross, rev. J.L. Ackrill and J.O. Urmson, New York: Oxford University Press.

Baumgartner, C. and Mieth, D. (eds) (2003) *Patente am Leben? Ethische, rechtliche und politische Aspekte der Biopatentierung*, Paderborn: Mentis.

Baumgärtner, S., Glotzbach, S., Hoberg, N., Quaas, M.F. and Stumpf, K.H. (2012) 'Economic Analysis of Trade-offs Between Justices', *Intergenerational Justice Review* 1/2012: 4–9.

Becker, L.C. (1980) *Property rights. Philosophic foundations*, Boston, MA: Routledge and Kegan Paul.

Brush, S.B. (1993) 'Indigenous Knowledge of Biological Resources and Intellectual Property Rights: The Role of Anthropology', *American Anthropologist* 95: 653–71.

Byström, M., Einarsson, P. and Nycander, G.A. (1999) *Fair and equitable. Sharing the benefits from the use of genetic resources and traditional knowledge*, Uppsala: Swedish Scientific Council on Biological Diversity. Online. Available HTTP: <http://www.grain.org/docs/fairandequitable.pdf> (accessed 13 June 2013).

Caney, S. (2005) *Justice beyond borders: a global political theory*, Oxford: Oxford University Press.

Castle, D. and Gold, E.R. (2008) 'Traditional Knowledge and Benefit Sharing: From Compensation to Transaction', in W.B. Peter and C.B. Onwuekwe (eds) *Accessing and sharing the benefits of the genomics revolution*, Dordrecht: Springer, 65–79.

CBD – United Nations (1992) *Convention on biological diversity*. Online. Available HTTP: <http://www.cbd.int/doc/legal/cbd-en.pdf> (accessed 12 June 2013).

Cunningham, A.B. (1991) 'Indigenous Knowledge and Biodiversity. Global Commons or Regional Heritage?', *Cultural Survival Quarterly* 15 (3). Online. Available HTTP: <http://www.culturalsurvival.org/ourpublications/csq/article/indigenous-knowledge-and-biodiversity-global-commons-or-regional-heritag> (accessed 13 June 2013).

de Jonge, B. and Korthals, M. (2006) 'Vicissitudes of Benefit Sharing of Crop Genetic Resources: Downstream and Upstream', *Developing World Bioethics* 6: 144–57.

Dobson, A. (1998) *Justice and the environment. Conceptions of environmental sustainability and theories of distributive justice*, Oxford: Oxford University Press.

Drahos, P. (1996) *A philosophy of intellectual property*, Aldershot: Dartmouth.

Dutfield, G. (2000) *Intellectual property rights, trade and biodiversity. Seeds and plant varieties*, London: Earthscan.

Dutfield, G. (2004) *Intellectual property, biogenetic resources, and traditional knowledge*, London: Earthscan.

Dworkin, R. (2000) *Sovereign virtue. The theory and practice of equality*, Cambridge, MA: Harvard University Press.

Engisch, K. (1971) *Auf der Suche nach der Gerechtigkeit. Hauptthemen der Rechtsphilosophie*, München: Piper.

Faber, M. and Petersen, T. (2008) 'Gerechtigkeit und Marktwirtschaft – das Problem der Arbeitslosigkeit', *Perspektiven der Wirtschaftspolitik* 9: 405–23.

Federle, C. (2005) *Biopiraterie und Patentrecht*, Baden-Baden: Nomos.

Frankfurt, H.G. (1987) 'Equality as a Moral Ideal', *Ethics* 98: 21–43.

Frein, M. and Meyer, H. (2008) *Die Biopiraten. Milliardengeschäfte der Pharmaindustrie mit dem Bauplan der Natur*, Berlin: Econ.

Gehl Sampath, P. (2003) 'Biodiversity Prospecting Contracts for Pharmaceutical Research. Institutional and Organisational Issues in Access and Benefit-sharing', dissertation, University Hamburg.

Gosepath, S. (2007) 'Gerechtigkeit', in D. Fuchs and E. Roller (eds) *Lexikon Politik. Hundert Grundbegriffe*, Stuttgart: Reclam, 82–5.

Hamilton, C. (2006) 'Biodiversity, Biopiracy and Benefits. What Allegations of Biopiracy Tell Us about Intellectual Property', *Developing World Bioethics* 6: 158–73.

Heller, M.A. and Eisenberg, R.S. (1998) 'Can Patents Deter Innovation? The Anticommons in Biomedical Research', *Science* 280: 698–701.

Hinsch, W. (2005) 'Global Distributive Justice', in T.W. Pogge (ed.) *Global justice*, Malden, MA: Blackwell, 55–75.

Höffe, O. (2004) *Gerechtigkeit. Eine philosophische Einführung*, 2nd edn, München: Beck.

Kersting, W. (2000) *Theorien der sozialen Gerechtigkeit*, Stuttgart: Metzler.

Klaffenböck, G. and Lachkovics, E. (eds) (2001) *Biologische Vielfalt. Wer kontrolliert die globalen genetischen Ressourcen?*, Wien: Brandes and Apsel/Südwind-Agentur.

Koller, P. (2007) 'Der Begriff der Gerechtigkeit', *Generationengerechtigkeit!* 4: 7–11. Online. Available HTTP: <http://www.generationengerechtigkeit.de/images/stories/Publikationen/gg_25!_web.pdf > (accessed 13 June 2013).

Krebs, A. (2003) 'Warum Gerechtigkeit nicht als Gleichheit zu begreifen ist', *Deutsche Zeitschrift für Philosophie* 51: 235–53.

Ladwig, B. (2004) 'Gerechtigkeit', in G.Göhler, M. Iser and I. Kerner (eds) *Politische Theorie. 22 umkämpfte Begriffe zur Einführung*, Wiesbaden: VS Verlag für Sozialwissenschaften, 119–36.

Lößner, M. (2005) *Nutzung der Biodiversität. Eine Analyse struktureller Umsetzungsmodelle der CBD*, Aachen: Shaker.

Lumer, C. (1999) 'Gerechtigkeit', in H.J. Sandkühler and D. Pätzold (eds) *Enzyklopädie Philosophie*, Hamburg: Meiner, 464–70.

Machlup, F. (1958) *An economic review of the patent system, study of the subcommittee on patents, trademarks, and copyrights of the committee on the judiciary, United States senate, eighty-fifth congress, second session, pursuant to S. Res. 236: Vol. 15*, Washington, D.C.: Government Printing Office.

Mazouz, N. (2006) 'Gerechtigkeit', in M. Düwell, C. Hübenthal and M.H. Werner (eds) *Handbuch Ethik*, 2nd edn, Stuttgart: Metzler, 371–6.

Mulligan, S. (1999) 'For Whose Benefit? Limits to Sharing in the Bioprospecting "Regime"', *Environmental Politics* 8 (4): 35–65.

Nagoya Protocol – Secretariat of the Convention on Biological Diversity (2010) *Nagoya protocol on access to genetic resources and the fair and equitable sharing of benefits arising from their utilization to the convention on biological diversity: text and annex, COP 10 Decision X/1.* Online. Available HTTP: <http://www.cbd.int/abs/doc/protocol/nagoya-protocol-en.pdf> (accessed 13 June 2013).

Nilles, B. (2003) 'Biopatente aus entwicklungspolitischer Perspektive', in C. Baumgartner and D. Mieth (eds) *Patente am Leben? Ethische, rechtliche und politische Aspekte der Biopatentierung*, Paderborn: Mentis, 213–27.

Nozick, R. (1974) *Anarchy, state, and utopia*, New York: Basic Books.

Nussbaum, M.C. (2000) *Women and human development. The capabilities approach*, Cambridge: Cambridge University Press.

Ott, K. and Döring, R. (2008) *Theorie und Praxis starker Nachhaltigkeit*, 2nd edn, Marburg: Metropolis-Verlag.

Page, E.A. (2007) 'Intergenerational Justice of What: Welfare, Resources or Capabilities?', *Environmental Politics* 16: 453–69.

Petersen, T. (2009a) 'Der Gerechtigkeitsbegriff in der philosophischen Diskussion', in M. Schmidt, T. Beschorner, C. Schank and K. Vorbohle (eds) *Diversität und Gerechtigkeit*, München: Rainer Hampp Verlag, 23–35.

Petersen, T. (2009b) 'Nachhaltigkeit und Verteilungsgerechtigkeit', in T. v. Egan-Krieger, J. Schultz, P. Pratap Thapa and L. Voget (eds) *Die Greifswalder Theorie starker Nachhaltigkeit. Ausbau, Anwendung und Kritik*, Marburg: Metropolis-Verlag, 69–84.

Pogge, T.W. (1991) *Realizing Rawls*, Ithaca, NY: Cornell University Press.

Pogge, T.W. (2006) 'Justice', in D.M. Borchert (ed.) *Encyclopedia of philosophy*, 2nd edn, Detroit, MI: Macmillan Reference USA, 862–70.

Pogge, T.W. (2008) *World poverty and human rights*, 2nd edn, Cambridge: Polity.

Rawls, J. (1971) *A theory of justice*, Cambridge, MA: Belknap Press.

Risse, M. (2012) *On global justice*, Princeton, NJ: Princeton University Press.

Schlosberg, D. (2004) 'Reconceiving Environmental Justice. Global Movements and Political Theories', *Environmental Politics* 13 (3): 517–40.

Schroeder, D. (2009) 'Justice and Benefit Sharing', in R. Wynberg, D. Schroeder and R. Chennells (eds) *Indigenous peoples, consent and benefit sharing. Lessons from the San-Hoodia case*, Dordrecht: Springer, 11–26.

Schroeder, D. and Pogge, T.W. (2009) 'Justice and the Convention on Biological Diversity', *Ethics and International Affairs* 23 (3): 267–80.

Sen, A.K. (1990) 'Justice. Means versus Freedoms', *Philosophy and Public Affairs* 19 (2): 111–21.

Sen, A.K. (2009) *The idea of justice*, London: Allen Lane.

Shiva, V. (2002) *Biopiraterie. Kolonialismus des 21. Jahrhunderts – Eine Einführung*, Münster: Unrast.

TRIPS – World Trade Organization (WTO) (1994) *Agreement on trade-related aspects of intellectual property rights.* Online. Available HTTP: <http://www.wto.org/english/tratop_e/trips_e/t_agm0_e.htm> (accessed 13 June 2013).

Young, H.P. (1994) *Equity in theory and practice*, Princeton, NJ: Princeton University Press.

Wagner, H. (1990) 'Gerechtigkeit. Historische Entwicklung', in H.J. Sandkühler (ed.) *Europäische Enzyklopädie zu Philosophie und Wissenschaften*, Hamburg: Meiner, 276.

WBGU (Wissenschaftlicher Beirat der Bundesregierung Globale Umweltveränderungen) (2000) *Erhaltung und nachhaltige Nutzung der Biosphäre*, Berlin: Springer.

Wullweber, J. (2004) *Das grüne Gold der Gene. Globale Konflikte und Biopiraterie*, Münster: Westfälisches Dampfboot.

Part III

Challenges and chances

A conceptual and normative clarification of biodiversity is not being based on theoretical reflections only. Since the concept of biodiversity is connected with its way of utilisation and the value of biodiversity is imbedded in our ideas of a certain quality of life, rather all theoretical discussions must be rooted in a process of practical implementation and need to be monitored by practical experiences. Therefore the third part of this book raises a number of specific and typical practical problems of the use of information on biodiversity and discusses challenges and chances which are linked with the protection of species and their natural habitats. As the theoretical reflection should lead to good practice, the practical experiences should inspire the conceptual and ethical debate.

Many conservation policies focus on the protection of biodiversity hotspots. Those hotspots indicate biogeographic regions with a significant natural potential of biodiversity and a large number of endemic species. The success and longevity of the biodiversity hotspot concept arises, as Fred Van Dyke points out, from its ability to combine a scientific assessment of biological diversity with clarification of goals and targets which appeal to conservation organisations and their supporters. But the conflation of semantic assumptions inherent in the hotspot concept carries the seeds of potential confusion that may reduce effectiveness in efforts of the protection of biodiversity. Biodiversity hotspots feature exceptional concentrations of endemic species and experiencing exceptional loss of habitats. Due to their limited natural range, endemic species are endangered and very vulnerable to extinction. Due to their small habitats and vulnerability, endemic species are quite often considered as indicator species for conservation efforts. But it is not always clear – as Ines Bruchmann argues in her chapter – how the concept works in conservation practice. Not only those species which are limited to a small geographical unit raise problems in conservation policy; also those species which continuously cross geographical and political borderlines cause challenges for those who want to protect them. The more stakeholders are involved and the more value systems for biodiversity apply, the more challenging sustainable management becomes. The vulnerability of migratory species crossing different countries on their annual journeys is to a large extent caused by such complex governance structures. The chapter by Aline Kühl-Stenzel illustrates how international legal instruments, such as the Convention on Migratory Species, can play an important role in facilitating coordination and regulation amongst different countries sharing a biological resource.

More and more it becomes evident that we learn from the diversity of organismic construction and the various physiological processes how different species solve problems to survive in reaction on particular environmental conditions. This knowledge can be considered as a chance and an enormous inspiration for challenges in engineering. This particular field of bionics is being picked up by Wilhelm Barthlott and M. Daud Rafiqpoor. The natural treasures are not only valuable for the modern art of engineering. Since human beings exist they use the knowledge of therapeutically effective substances produced by plants and animals to cure diseases and to relieve pain. Nowadays research in biodiversity becomes very successful in detecting even more effectual substances for therapeutic purposes. The knowledge on biodiversity develops into a particular knowledge about an almost infinite natural source of pharmaceutical substances: ecosystems become natural pharmacies. The scientific background of this potential is described in the chapter by Tobias A.M. Gulder and René Richarz. Those chances and challenges in applied life sciences lead to the question what kind of knowledge the knowledge on biodiversity is and whether the protection of biodiversity could be argumentatively based on the values of pure scientific purposes, which looks at biodiversity as being mainly an object of science. The argument of the pure scientific value of biodiversity is discussed by Gesine Schepers. She critically reflects the widespread, and usually undisputed, belief that biodiversity is to be protected because of its scientific value. According to Schepers' analysis, biodiversity as such might have to be protectable for a number of specific reasons but not for the sake of science as such.

13 Biodiversity hotspots

Concepts, applications and challenges

Fred Van Dyke

What is biodiversity and what is a hotspot?

'Biodiversity' is one of those modern contractions, like 'cyberspace', that has become so familiar, oft-used and well worn that we no longer recognize what it is a contraction of. The term is a shortening of the phrase 'biological diversity', and, according to conservation biologist Stuart Pimm, first appeared in a rather obscure US government report authored by Elliot Norse in 1980 (Pimm 2001). Norse was ahead of his time. As a term, the word 'biodiversity' did not attain common use in science until after the American National Forum on Biodiversity in 1986 (Thompson and Starzomski 2006) and through Harvard zoologist E.O. Wilson's extensive use of the term in his own writings, beginning with his book, *Biodiversity* (Wilson 1988). It has now been defined by a multitude of authors and agencies, but not always consistently. To those engaged in the study of natural history, biodiversity represents the biotic elements of nature that can be described and classified. To environmental activists, biodiversity is an intrinsic value-laden quality of natural systems that should be preserved for its own sake. To ecologists and conservation biologists, biodiversity is a measurable parameter relevant to an understanding of community structure, environmental processes, and ecosystem functions (Mayer 2006). All of these 'thought styles' of defining and describing biodiversity are embedded and conflated in the concept of 'biodiversity hotspots'.

In conservation, the term 'hotspot' has been with us, in scientific literature, for the last three decades, with roughly the same period of active use as 'biodiversity'. The use of the word 'hotspot' first appears in a seminal work by the noted conservation scientist and activist Norman Myers in an article entitled 'Threatened Biotas: Hotspots in Tropical Forests', published in *The Environmentalist* in 1988 (Myers 1988). Here Myers was concerned with saving endemic plant species unique to tropical forests, not biodiversity *per se*, but he makes the case that densities of endemic species are particularly high in tropical forests, and, therefore, such sites should be given priority in conservation efforts. In this first attempt, Myers identified ten specific regions of tropical forests he designated as hotspots which possessed exceptionally high levels of endemic plants and were threatened with significant habitat loss. Here Myers treats biodiversity as a quality of intrinsic value associated with these areas, something that ought to be protected for its own sake. Further, the hotspot concept he joins to it is itself a conflation of three different and more or less independent concepts: the concept of endemism-dependent *rarity* (a concept focusing on individual endemics), the concept of biological *diversity* (the number of species per unit area) and the concept of *vulnerability* (the prospect of imminent loss without immediate conservation action). It would be unfair to criticize Myers for not achieving complete clarity in a new concept the first time it was used. Unfortunately, 33 years later, the conflation of rarity, diversity and vulnerability in setting conservation priorities is still with us.

Conservation International (CI), one of the world's largest and most effective non-governmental conservation organizations, has perhaps been the best at developing, refining, and using the hotspot concept in practical ways in the actual work of conservation, and so represents a recognized authority for its definition. According to CI, a region qualifies for the designation of 'hotspot' if it contains at least 1,500 species of vascular plants (or > 0.5% of the world's total vascular plant species) and if it has lost at least 70% of its original habitat (Conservation International 2011). A general but more functional criterion is the arbitrary cutoff that designates the most species-rich 2.5% of any given ecoregion as that ecoregion's 'hotspot' (Guilhaumon et al. 2008). A further criterion, unstated but necessarily and practically assumed, is that the area meeting these three criteria must be sufficiently definable to render it a feasible target for protection and management. There are, in fact, many areas of the world which, if we only drew their boundaries the right way, could contain 1,500 species of vascular plants, but most of them would be so large that no feasible conservation strategy would be possible to protect them, short of international treaties and conventions, instruments far beyond the scope of even the largest conservation organizations. Hotspots, then, must be areas that can be defined by physical location, possess some potential for common or cooperative jurisdiction, and provide the possibility for practical management and restoration efforts that could be applied to and actually affect their entire area.

CI currently lists 34 hotspots grouped in five continental regions, North and Central America, South America, Europe and Central Asia, Africa, and (southeast) Asia-Pacific. Although the specific regions within these areas that contain the most biodiversity (52% of all known vascular plants and 42% of all known terrestrial vertebrates), they comprise, in an absolute sense, large areas. Nevertheless, large as they are, all retain the essential features mentioned previously: floristic diversity coupled with significant current and potential loss of habitat, but generally definable boundaries and potential for effective administrative jurisdiction. It is important to realize, from the start, that biodiversity hotspots of this scale are not preserves, and that no *designation* of a biodiversity hotspot ever *preserved* anything. Hotspots of this scale are intended to identify broad regional and continental areas where well-placed reserves might be disproportionately effective at preserving species, but actual and potential protected areas within them are usually small. Eighty-five percent of over 100,000 areas recorded in the 2003 UN List of Protected Areas are <100 km^2 in area, and almost 60% are <10 km^2 (Grenyer et al. 2006). Problems of scale and conflation of concepts in the hotspot approach are troubling when we try to explain exactly what hotspots protect, yet hotspot strategies have proved adaptable to conservation in many contexts. Why has this concept worked so well for so long?

Why the hotspot concept works

Fundamentally, the hotspot concept in conservation is foundational to area-specific conservation strategy. As such, a hotspot approach naturally suggests practical algorithms for prioritizing conservation targets. Most conservation organizations operate, explicitly or implicitly, with the goal of protecting as many species as possible. But, unless the organization is building a zoo, the species themselves cannot be protected by direct market purchase. The primary means of protection is *land acquisition*, either through direct purchase, or purchase of property rights that govern land use (conservation easements), which might prevent things like development, livestock grazing, farming, hunting, logging, mining, or any of a myriad of activities that could diminish the number of species on the protected site. Constrained by a limited budget, conservation organizations find themselves more often faced with economic decisions than ecological ones in determining conservation strategy. They must determine *the optimal allocation of expenditures that will protect the greatest number of species*. Because species cannot be protected directly but only through the purchase of the land on which they occur or the property rights thereof, this means that *optimal land and property rights purchases are those in which the number of species per area is highest per unit of land purchased.* Given an array of potential land purchases, the economically rational decision is to purchase land with the highest per unit species richness.

The hotspot approach adapts itself well not only in suggesting decision rules for prioritizing conservation targets by optimizing the ratio of species protected per unit of land acquired, but also in the more basic, and more essential, requirement of organizational survival. All conservation organizations have explicit, well-defined, conservation missions, such as The Nature Conservancy's memorable aspiration of *saving the last great places*. When an organization can effectively define hotspots, it not only generates decision rules for its conservation strategies, it also generates *easily identifiable funding objectives that attract and retain major donors and public support.* A hotspot approach permits a conservation organization to put the targets of its funding appeals on the map, literally. When such appeals are presented in a tangible way, the positive response to those appeals rises dramatically.

It is not only the concreteness of the hotspot approach that contributes to its success. From its first use in 1988, the hotspot approach has not only been tied to rarity and diversity, but also to *urgency and vulnerability* (Figure 13.1). In considering these additional dimensions of conservation concern, the hotspot concept again creates relevance to potential donors. If an organization focuses its definition of hotspots on *vulnerability*, it conveys the message, 'Act now, for tomorrow will be too late.' If the organization focuses on the *irreplaceability* of a targeted hotspot, it conveys the message, 'Save species here, for there is nowhere else on Earth these species can be saved.' And, when these strategies can be combined, when a

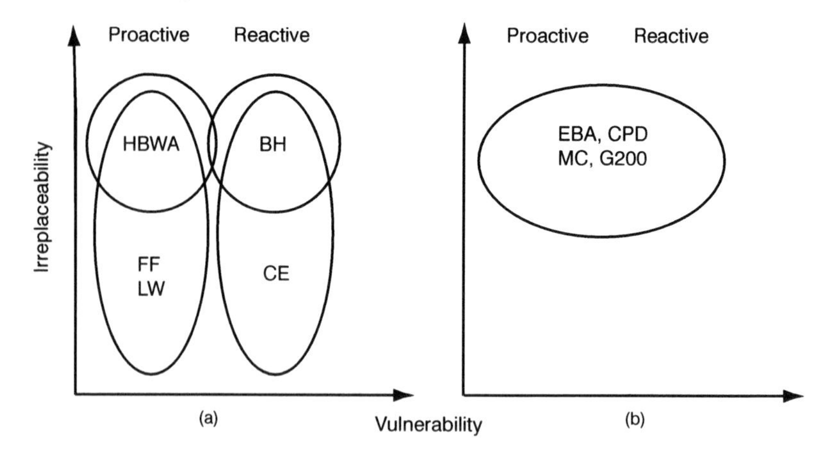

Figure 13.1 Global biodiversity conservation priority templates of High Biodiversity Wilderness Areas (HBWA), Biodiversity Hotspots (BH), Forever Forests (FF), Last of the Wild (LW), Crisis Ecoregion (CE), Endemic Bird Areas (EBA), Centers of Plant Diversity (CPD), Megadiversity Countries (MC), and Global 200 Ecoregions (G200) ranked according to criteria of irreplaceability and vulnerability. After Brooks et al. 2006.

tangible target with a unique species association also a *threatened target* with urgent need for protection, appeals for funds are likely to be highly effective, placed within the conceptual framework of irreplaceability and vulnerability. (a) Reactive approaches focus on the protection of areas of high vulnerability and immediate threat. Proactive approaches focus on areas with high biodiversity still relatively unaffected by human influence. (b) Four approaches that do not incorporate vulnerability as a criterion, but consider only irreplaceability (uniqueness) in conservation.

Because different organizations stress different combinations of these criteria, the adaptability of the hotspot approach becomes more apparent, as does its effectiveness. Recall that Myers originally defined hotspots in terms of vascular plants, but few vascular plants can match the charisma of large birds and mammals when it comes to popular appeal and effective fund-raising for conservation causes. Similarly, effective fund-raising in conservation is largely a matter of clearly and correctly defining specific interest-based constituency groups. Some of these may be plant lovers, but appeals that generate more response are likely to be about birds and mammals, and, to a lesser extent, amphibians, reptiles, fish, and invertebrates (especially butterflies). Thus, the hotspot approach can and has been successfully adapted from plants to Endemic Bird Areas (the primary conservation target of the conservation organization BirdLife), The Wildlife Conservation Society's Last of the Wild areas, which are usually promoted in terms of their protection of large mammalian megafauna species, or the World Wildlife Foundation's Global 200 Ecoregions (Brooks et al. 2006). In these and other examples, the concept's adaptability has made it a favorite of conservation organizations. The relative ease with which the hotspot concept, properly adapted, can define attractive targets for donors in any interest-specific conservation constituency makes it a powerful tool in generating constituent and public support. The hotspot approach, in this regard, not only has proven effective at the 'front end' of fund-raising efforts and donor attraction, but at the 'back end' as well. When an organization can provide its constituency with a simple, non-technical summary of the number of species of plants, birds, mammals, or other groups protected by its work, it projects the image of a cost-effective organization that produces real results for every dollar contributed. When different criteria are used to designate hotspots, different hotspots get designated. This means that more areas are *targeted* for protection, but not necessarily that more areas *will be protected,* because now already *limited* financial resources also become more *diluted*. Such divergent criteria also undermine the sense of coherency or integration of a supposedly global biodiversity conservation strategy. And when, in the public's eyes, that sense of *coherency* is absent, the sense of *credibility* can also suffer with it. When this happens, the public asks, not without warrant, do you conservationists really know what you are doing?

These points are not made to suggest criticism or cynicism toward conservation organizations or government agencies that use the hotspot concept effectively to achieve conservation goals, or to attract and keep public support for those efforts. The success and longevity of the hotspot concept, however, is not because there are not alternative approaches. It arises, rather, from its ability to combine a scientific assessment of biological diversity with clarification of goals and targets which appeal to conservation organizations and their supporters. But the conflation of different concepts and assumptions inherent in the hotspot concept carry the seeds of potential confusion that can reduce effectiveness in conservation efforts. We can better appreciate these problems with more detailed analysis.

Underlying assumptions of the hotspot concept and their problems

Assumptions of Measurement – At one level, biodiversity is a measurable scientific parameter providing information about the structure of biological communities, but there is no one measure or metric that can fully capture it. Yet, if the conservation community hopes to speak accurately about where biological hotspots are, it needs such single estimate measures, and it must be able to say what the conservation of that measure would actually conserve. The most common measure of diversity in a hotspot, or in any defined area, is species richness, the total number of species present. Zoogeographers have known for centuries that there exists a positive relationship between species richness and area (Figure 13.2), first documented in island flora and fauna (Darlington 1957), environments that provided areas of varying sizes and well-defined boundaries that could be compared because of similar conditions. The theory of island biogeography (MacArthur and Wilson 1967) generalized this species-area rule as a power function through the equation $S = cA^z$, where S represents species, A area, and c is a taxon-specific constant and z is the slope of the line, the so-called 'extinction coefficient', because, as we move down and to the left along the line, we should be able to predict how many species

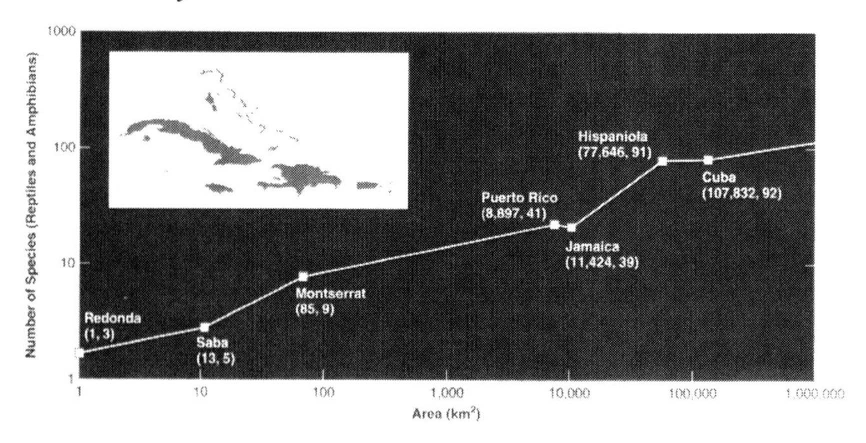

Figure 13.2 A general species-area relationship among some Caribbean islands. First number in parenthesis indicates area, second number indicates number of species. Note that species richness on islands increases with increasing area of the island. After Darlington 1957.

will be lost as area decreases. Hotspots are assumed to hold disproportionately more species per unit area than non-hotspots, and this is what gives them enhanced value and higher conservation priority. But to refine our understanding, we must complement the concept of species richness with three related concepts: species *diversity*, species *endemism*, and species *rarity*.

Species *diversity* differs from *richness* in that diversity adds the element of *equitability* as a method of *weighting* species in calculating *community* diversity. For example, one of the most common measures of diversity is the Shannon Index (Shannon and Weaver 1949),

$$H' = -\sum p_i(lnp_i)\ .$$

That is, diversity (H') is the sum of the product of the proportional abundance of each species (the 'ith' species, represented by p_i) times the natural log of that value (lnp_i). Because the natural log of the proportion will decrease as the proportional value (p_i) gets bigger, valuing diversity using the Shannon Index has the effect of giving higher values to communities that not only have many species, but those in which the species have relatively equal abundance. To see the effect, consider the most extreme case. In a community consisting of a single species, the value of p_i is 1, and the value of lnp_i is 0, resulting in a value of 0 for H'. A community in which every individual is the same has no diversity. The value of H' then rises from zero diversity to higher values if 1) more species are present, and 2) if the proportional abundance of each species becomes increasingly similar, or both. Consider two tallgrass prairies of relatively equal size in the Midwest United States, and the bird communities that reside in them (Table 13.1, data based on Van Dyke et al. 2007). Site A differs from Site B in both characteristics. It has three fewer species than Site B, and is dominated by a single species, the Common Yellowthroat, whereas no single species dominates

Table 13.1 Differences in species richness, proportional abundance, and their effect on an estimate of community diversity from two sites in tallgrass prairie remnants at DeSoto National Wildlife Refuge, Iowa, USA. Values for species reflect density of singing males/ha

Species	*Site A*	*Site B*
Common yellowthroat	8.24	1.21
Field sparrow	2.94	2.84
Dickcissel	1.18	2.23
Red-winged blackbird	0.29	0.81
Brown-headed cowbird	2.06	1.82
American goldfinch	1.47	1.02
Ringneck pheasant	0.59	1.63
Mourning dove	1.18	0.61
Eastern kingbird	–	1.60
Grasshopper sparrow	–	4.48
Northern bobwhite	–	2.64
Shannon diversity (H′)	1.64	2.25

Notes: As measured by the Shannon Index (Shannon and Weaver 1949), Site B is more diverse than Site A because of its greater species richness and more equitable relative abundance of each species. Data derived from Van Dyke et al. 2007.

the bird community on Site B. That is, on Site B, there is less chance of sampling or observing the same species in consecutive samples or observations. The Shannon Indices of these sites tell us that Site B is more diverse than Site A.

Just as it takes longer to explain species diversity than species richness, it takes longer to calculate it in the field. Species richness can be estimated with rapid assessment techniques, but species diversity requires some measure of the relative abundance of every species, and therefore an investment of time and effort that most conservation organizations do not have the luxury to make. But without an estimate of species diversity, sites with equal species richness may in fact harbor communities with very different levels of diversity. Preserving a community dominated by one or a few species is not the same as preserving a community with many equitably distributed species, and therefore does not have the same value for conservation. The greater the number of species, and the more even their abundance, the greater the relative *rarity* of each species. Areas of high species diversity are also areas in which many, or even most of the species, may be relatively rare. If preventing extinction is as important as preserving biodiversity, we are now faced with a potentially problematic choice in the application of the conservation strategy. That is the choice between preserving diversity, endemism, and rarity. Can we preserve all three at the same time?

Endemism predisposes a species to rarity because it makes the species more vulnerable to area-specific disturbance. The Haleakala silversword (Figure 13.3) is a plant species found only in the Hawaiian Islands. In fact, it is endemic to a single

Hawaiian island, Maui. And it is not endemic to all of Maui, but only to the crater, rim, and surrounding slopes of a single volcano, the Haleakala Volcano, which, fortunately for the silversword, lies within Haleakala National Park.

The silversword is an extreme case, but it illustrates why endemic species can be vulnerable. Conservationist Stuart Pimm illustrated the problem with his now famous 'cookie cutter' model. Imagine a disturbance of a certain size, perhaps a logging sale, represented by the 'cookie cutter' in Figure 13.4 (Pimm 1998). Now imagine that each shape represents a different species, and the size of the shape is proportional to the species' range. The populations on the right, with larger ranges, are reduced by the disturbance, but not exterminated. On the left, endemic species with smaller ranges see four of their seven populations become extinct.

As Darwin correctly observed, rarity precedes extinction (Darwin 1859), and high diversity areas may be places where most species are relatively rare. If the rare species are also endemic species, their small populations are located within relatively small areas. We can illustrate this concept with a diagram that categorizes the components of rarity, which are three (Figure 13.5, categories based on Rabinowitz, Cairns and Dillon 1986). A species can have populations in which at least some are large, or where every population is small. This creates rarity associated with low abundance. A species can be tolerant of different habitats, or specialize in one or a few. This creates rarity associated with habitat specialization. And a species can have a broad geographic range, or a very narrow

Figure 13.3 The Haleakala silversword (*Argyroxiphium macrocephalum*), an example of a 'rare' species with a dense population of individuals (50,000) confined to a single site, the crater of Haleakala, a Hawaiian volcano. Photographers: Forest and Kim Starr.

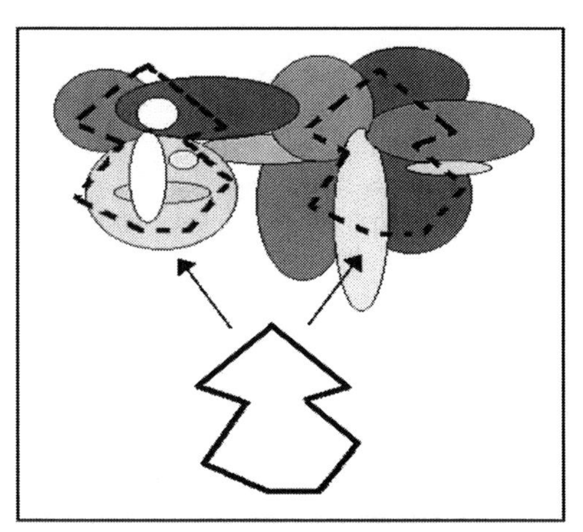

Figure 13.4 The 'cookie cutter' model of the effects of habitat loss on endemic species. Assume each shape represents a different species, and the size of the shape is the size of the species range. Populations on the right, with larger ranges, are reduced by the disturbance, but not exterminated. On the left, more endemic species with smaller ranges see four of their seven populations exterminated. Thus, random habitat loss produces a disproportionately high rate of extinction in endemic species. Based on concepts from Pimm 1998.

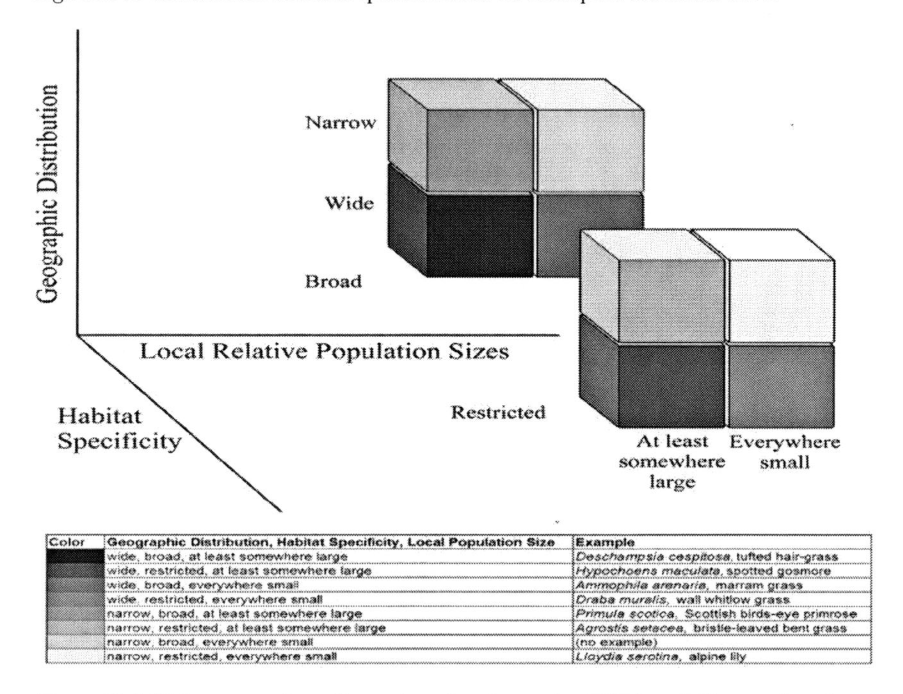

Color	Geographic Distribution, Habitat Specificity, Local Population Size	Example
	wide, broad, at least somewhere large	*Deschampsia cespitosa*, tufted hair-grass
	wide, restricted, at least somewhere large	*Hypochoens maculata*, spotted gosmore
	wide, broad, everywhere small	*Ammophila arenaria*, marram grass
	wide, restricted, everywhere small	*Draba muralis*, wall whitlow grass
	narrow, broad, at least somewhere large	*Primula scotica*, Scottish birds-eye primrose
	narrow, restricted, at least somewhere large	*Agrostis setecea*, bristle-leaved bent grass
	narrow, broad, everywhere small	(no example)
	narrow, restricted, everywhere small	*Lloydia serotina*, alpine lily

Figure 13.5 Eight categories of species abundance in British wildflowers based on geographic range, habitat use, and relative population size. Note that only one category (broad habitat specificity, wide geographic distribution, and large local population) can truly be considered 'common'. Species in all seven other categories are rare in one or more dimensions. Adapted from Rabinowitz et al. 1986.

one. This creates rarity associated with endemism. In various combinations, these factors create eight categories of species abundance, and species in seven of these categories would be rare in at least one way. The Kirtland's warbler (*Dendroica kirtlandii*) (Figure 13.6) is a species which achieves rarity in all three. Geographically, it nested, until recently, in only three counties in the US state of Michigan. In habitat specificity, it will use only young, even-aged stands of pine established by fire. And, not surprisingly, on all its nesting sites, its populations are small. A species with any of these conditions of rarity, whether that rarity is based on populations, distribution, or habitat specialization, is one that can be hard to conserve. When a species is 'rare' in all three, its conservation becomes particularly difficult.

Assumptions of Correlation – Hotspots are designated to preserve biodiversity, but biodiversity of what? Do all taxonomic groups experience the same level of biodiversity protection in every hotspot? In one hotspot, South Africa's Succulent Karoo, scientists found a positive correlation between small mammal species richness and plant species richness, possibly because of the seed dispersing activities of the small mammals (Keller and Schradin 2008). Ecological interactions like those suspected in the Succulent Karoo would create cause-and-effect mechanisms for correlations of species richness in different taxonomic groups, but interactions like this tend to be ecosystem- and site-specific, not something that can be assumed, without investigation, for all hotspots. In general, for the hotspot concept to be effectively applied, four assumptions must be warranted. First, richness at one taxonomic level (for example, species) should be correlated with richness at other levels (for example, genera or families). Second, the taxonomic groups of greatest interest in conservation (generally plants, amphibians, reptiles, birds and mammals) should be correlated with one another in terms of richness.

Figure 13.6 The Kirtland's warbler (*Dendroica kirtlandii*), a species that exemplifies all three elements of rarity in having limited geographic distribution, high habitat specificity, and small populations. Photo by Ron Austing.

For example, areas with high richness in mammals should, ideally, be areas which are also rich in birds. Third, species richness and species endemism should be positively correlated. The greater the overall species richness of an area, the more endemic species should be present. Fourth, correlations in overall species richness and endemic species richness should also hold among rare and threatened species. In other words, preserving areas with high densities of rare and threatened bird species should also preserve high densities of rare and threatened species in other groups, like mammals or amphibians. If this is true, it would permit one species group to serve as a surrogate or 'umbrella' for another, and reserves within a hot spot that protect one group should protect all.

The first two of these assumptions have proved true, or at least true enough to be feasible in the work of identifying and protecting critical areas within hot spots. In plants, for example, tests of correspondence in diversity at different taxonomic levels have often demonstrated the expected correlation. In one study in Mexico, for example, area-specific variation in genera richness in plants explained 85% of plant species variation, and 95% if the correlations were made between plants of the same vegetation types (Villaseñor et al. 2005). If such assumptions proved dependable, it would mean that diversity at higher taxonomic levels, like families, which can be identified in the field more rapidly than species, should be an indicator of high diversity of species and other taxons as well. This correlation has been shown to hold in different kinds of organisms, supporting the second assumption. For example, variation in bird species richness has been shown to explain over 90% of variations in richness in mammals and amphibians in the same areas, and mammals do almost as well, explaining 79–86% of variations in species richness in birds and amphibians (Grenyer et al. 2006). The third assumption, the correlation between richness and endemism, has proved less reliable. John Lamoreaux and his colleagues examined this problem by looking at different levels of species richness and species endemism in the four major terrestrial vertebrate classes in many parts of the world (Table 13.2). The first two rows of Table 13.2 indicate that, considered separately, richness or endemism

Table 13.2 Pearson correlation coefficients of terrestrial vertebrate diversity measures

	Amphibians	*Reptiles*	*Birds*	*Mammals*	*Four Classes*
Richness[1]	0.591**	0.380**	0.715**	0.668**	
Endemism[2]	0.503**	0.587**	0.612**	0.490**	
Richness × endemism[3]	0.096**	0.085**	−0.068	−0.099	−0.025

* $p < 0.05$; ** $p < 0.01$

Notes: For richness and endemism, values reflect the correlation of that variable in the given class with a counterpart index of richness or endemism in the three other classes. Values for richness X endemism indicate the correlation between endemism and richness within a class and of the four classes combined. After Lamoreux et al. 2006.

[1] Correlation between class richness and a richness index of the three remaining classes.

[2] Correlation between class endemism and an endemism index of the three remaining classes.

[3] Correlation between richness and endemism within each class, and of the four classes combined.

in one class is generally correlated with richness or endemism in other classes. However, when richness and endemism are considered together, the correlation does not hold. For example, areas of high species richness in amphibians are not necessarily associated with areas of high species endemism in amphibians. In fact, the correlation between richness and endemism here is less than 10%. When all four classes of terrestrial vertebrates are considered together, the correlation is less than 3% (Lamoreux et al. 2006). Thus, in a given hotspot, high concentrations of endemic species and high concentrations of species richness are unlikely to exist in the same area. To protect both would require different preserves. The fourth assumption, that correlations in overall species richness and endemic species richness should also hold among rare and threatened species, has proven the most unreliable of all. On this point, birds have been the best indicators (Table 13.3). Areas selected to preserve all local or regional species of rare and threatened bird species also protected 62–64% of rare and threatened amphibians and 60–69% of rare and threatened mammals. When other groups were used as surrogates, however, correlations were much lower. For example, areas that protected all rare and threatened amphibians protected only 22–31% of rare and threatened species of birds (Grenyer et al. 2006).

Assumptions of Species-Area Relationships – As different conservation priorities lead different organizations to define different areas as hotspots, so do different mathematical models. The best model of the species-area relationship is not always the simple power model described earlier. Recent analyses have shown that multivariate models of species-area relationships are often more accurate than simple linear power model relationships, but they define entirely different areas as hotspots. For example, Guilhaumon and his colleagues examined the power of various models to explain the species-area relationship in different taxonomic groups. They determined that there was no single 'best' general model, but that the choice of the best model was area-specific, and must be determined by actual analysis of species distributions in the area, not *a priori* assumptions (Guilhaumon et al. 2008). The species-area relationship in amphibians, for example, is explained by a negative exponential model, mammals by a positive exponential model, reptiles by several different models, and the species-area relationship in birds is not adequately explained by any model (Guilhaumon et al. 2008). Complexities of these sophisticated analyses aside, the upshot is that different models will select different hotspots for different species groups in different ecosystems, and choosing the best model is difficult work.

Assumptions of Management – If preserves within hotspots that protect species richness do not protect all species to the same degree, managers must choose between richness and rarity in prioritizing conservation targets, and this requires different decision-making algorithms in selecting sites for purchases or easements. Although most of the world's global biodiversity hotspots are large, the actual targets of protection or other management actions are small, and the relative importance of richness or rarity affects which sites one chooses to protect.

At a local scale, there are three common algorithms for site selection among potential conservation reserves, and they can be easily remembered by their

Table 13.3 Patterns of cross-taxon surrogacy across birds, mammals, and amphibians

Richness index	*Surrogate groups*	$N_{spp.}$	N_{cells}	*Target groups** (*% of species represented in set*)		
				Birds	*Mammals*	*Amphibians*
Total	Birds	9,626	421	–	79.4 ± 0.3	55.5 ± 0.7
	Mammals	4,104	509	91.7 ± 0.1	–	61.1 ± 0.5
	Amphibians	5,619	831	90.9 ± 0.2	86.2 ± 0.2	–
	Birds, mammals	13,730	714	–	–	69.4 ± 0.5
	Birds, amphibians	15,245	1,028	–	89.8 ± 0.2	–
	Mammals, amphibians	9,723	1,077	95.0 ± 0.1	–	–
	All three groups	19,349	1,223	–	–	–
Rarity	Birds	2,424	380	–	43.3 ± 1.1	22.5 ± 1.3
	Mammals	1,026	432	68.3 ± 0.4	–	27.0 ± 0.7
	Amphibians	1,405	560	63.7 ± 0.5	51.6 ± 0.6	–
	Birds, mammals	3,450	656	–	–	35.7 ± 0.9
	Birds, amphibians	3,829	808	–	63.1 ± 0.6	–
	Mammals, amphibians	2,431	858	77.9 ± 0.2	–	–
	All three groups	4,855	1,033	–	–	–
Threat	Birds	1,096	282	–	51.7 ± 0.9	31.2 ± 1.4
	Mammals	1,033	357	60.7 ± 0.6	–	39.7 ± 0.9
	Amphibians	1,856	454	62.7 ± 0.4	59.7 ± 0.4	–
	Birds, mammals	2,129	518	–	–	49.2 ± 0.6
	Birds, amphibians	2,952	627	–	67.2 ± 0.5	–
	Mammals, amphibians	2,889	690	72.4 ± 0.4	–	–
	All three groups	3,985	821	–	–	–

Notes: Values represent the percentage of species in the target group represented in complementary sets of grid cells designed to contain all members of the surrogate group (mean ± standard deviation). N_{spp} is the total number of species in the surrogate group. N_{cells} is the number of cells in the optimal complementarity set. After Grenyer et al. 2006.

colorful names. The so-called *Greedy Richness Algorithm* is one in which the manager aims to get the most species with the fewest sites, essentially a classic 'hotspot' approach to species conservation. In contrast, the *Greedy Rarity Algorithm* is directed at saving the rarest species first, and then adding more common species with additional sites after the rarest species are protected. Finally, the *Connectivity Algorithm* assumes that the spatial distribution of protected populations is as important as the number of different species protected, and aims to maximize the likelihood of dispersal among protected sites by minimizing distances between sites or by providing stepping-stone habitats or dispersal corridors between sites. In this approach, the ultimate concern is long-term population viability, and no species is truly 'protected' unless its populations are sufficiently connected to place it above a critical threshold of extinction risk (Urban 2002).

No matter what maps reveal about site-specific species richness, conservation managers never have the power to arbitrarily designate a reserve simply because it will capture the greatest number of species, or the rarest ones. In almost every case, humans are living in or using potential preserves within and around hotspots. Politically, they must be involved in the identification and selection of such reserves. Economically, they must either be permitted some extractive use of the preserve if they are dependent on it for their livelihood, or they must be compensated for the loss, and be willing to accept the compensation offered as just and fair.

Historically, managers have often assumed that 'protection' of biodiversity in hotspots meant reduction or exclusion of human exploitation of physical and biological resources. There were some good reasons for this assumption. Not only do modern human populations create land use patterns that lower biodiversity, but even primitive and indigenous peoples, when equipped with modern technologies, show similar powers of negative effect. The conservation community has often lived under the assumption that satisfaction of human need in species-rich environments inevitably leads to ecosystem degradation and attendant loss of biodiversity, what ethicists and economists would refer to as a 'rivalry' view of human-nature interactions. But this assumption cannot be accepted uncritically, and there are ways to make human presence and biodiversity preservation operate as facilitative processes rather than rival ones, particularly in situations where the biodiversity of the hotspot has long-standing traditional and instrumental value to native residents, who in turn suffer serious material and instrumental loss from biodiversity decline. Describing how to foster such facilitative processes is beyond the scope of this article, but biodiversity does not exist independently of human social and cultural systems, and these systems are rooted in practices of land use and habitation. Increasingly conservation managers are exploring, and must continue to explore, strategies that integrate human presence into the sustainability of biodiversity. Human use is harder to manage, and more risky, than human exclusion, and precise restriction on some kinds of human use of hotspots is necessary to preserve biodiversity, but indiscriminant exclusion closes land use options to indigenous people and can exacerbate poverty. This is a tension not easily resolved, but one that is and will remain one of the critical issues of biodiversity conservation in the foreseeable future.

Assumptions of Environmental Stability – The hotspot approach is area specific and spatially explicit. It assumes that conditions in protected areas will remain relatively constant for the foreseeable future, and that species which are the targets of such protection will 'stay put' within the boundaries of the reserve. Historically this assumption has been considered so secure it was often not even recognized as an assumption. That can no longer be the case. The reality of climate change has powerful implications for achieving conservation goals through hotspots.

Climate change is creating worldwide effects in plant and animal distributions. In one of the most extensive and comprehensive meta-analyses of existing studies of changes in animal and plant distribution, as well as of biological timing-specific events (phenology), Parmesan and Yohe (2003) found, in an examination of over 1700 species examined in 271 published studies, that 87% of phenological changes (biologically timed events), such as flower emergence, onset of breeding behavior, or departure or arrival, dates in animal migrations changed in directions predicted by and consistent with global climate warming. Seventy-five percent of documented changes in distribution in lower latitude and equatorial regions changed in predicted directions, 81% in temperate and polar regions. Three specific kinds of climate risks are anticipated and even now occurring. The first, and most obvious, is that of local standardized climate change, which will shift existing climate regimes to new areas. The second will be the creation of entirely new climate regimes that have no current climate analog, and are therefore likely to create novel species associations. Finally, and of greatest concern, is the loss of current climate regimes which will not be replaced, leading to extinction of species with narrow geographic ranges and narrow range tolerances (Williams, Jackson and Kurtzbach 2007). The Monte Verde Golden Toad (*Bufo periglenes*) of Costa Rica (Figure 13.7), now believed to be

Figure 13.7 The Monte Verde Golden Toad (*Bufo periglenes*), a species unique to tropical cloud forests and their environments of high humidity. Under increasingly dry conditions in these forests, the Golden Toad has disappeared from the wild. Photo by Michael and Patricia Fogden. Copyright Michael and Patricia Fogden.

extinct in the wild, may be one of the first current examples of extinction of this third type.

In recent studies, extant species have shown remarkable resiliency and adaptability to climate change, primarily by tracking their 'climate niche' through migration to new areas. But precisely because most species show such ability, their distributions become a 'moving target' for the conservation manager, who must now treat species formerly assumed to have been protected by established reserves as a shifting distributional array, gradually (or, sometimes, not so gradually) moving out beyond reserve boundaries over time. For example, consider the current and projected ranges of the stiff sedge in Great Britain (Figure 13.8) (Pearson and Dawson 2003). Any current refuge in Wales designed to protect this sedge will, within 15 years, have no individuals left to protect. Conservationists, therefore, must develop means for moving refuge boundaries over time to protect targeted species, or must establish new refuges at a more rapid rate. Currently employed strategies for mitigating the effects of climate change include so-called 'assisted migration' (moving vulnerable species with low dispersal and migration abilities to new areas with more favorable climate conditions) and creating reserve 'networks' of sufficient scale that species can track their climate niche by moving from one refuge to another. Conservation agencies like Parks Canada are already developing Climate Change-Mitigated Conservation Strategies to meet conservation goals in a very different future world (Scott 2005). Yet even these kinds of tactics may not be enough. As John Williams and his colleagues at the University of Wisconsin, who studied and modeled the problem in detail, noted,

> There is a close correspondence between regions with globally disappearing climates and previously identified biodiversity hotspots; for these regions, standard conservation solutions ... may be insufficient to preserve biodiversity.
>
> (Williams, Jackson and Kurtzbach: 5738)

Assumptions of value in the hotspot concept: problems of scientific and moral integrity

It is impossible to fully address assumptions inherent in the hotspot concept without ultimately addressing assumptions of value, not only, or even mainly, of the hotspots themselves, but of the biodiversity they are intended to protect. Biodiversity, as already noted, is not only a measurable scientific parameter that provides information about the characteristics of ecological communities, it is also often treated, by conservation scientists and activists alike, as an intrinsic, value-laden quality of those communities to be preserved for its own sake. In this case, familiarity does not breed contempt, but affection. The more conservation biologists study hotspots and their biodiversity, the fonder they grow of them, and the more animated and passionate their appeals to protect both. When it comes to justifying the protection of biodiversity for ethical reasons, conservation biologists must solve three kinds of problems. First, they must address economic

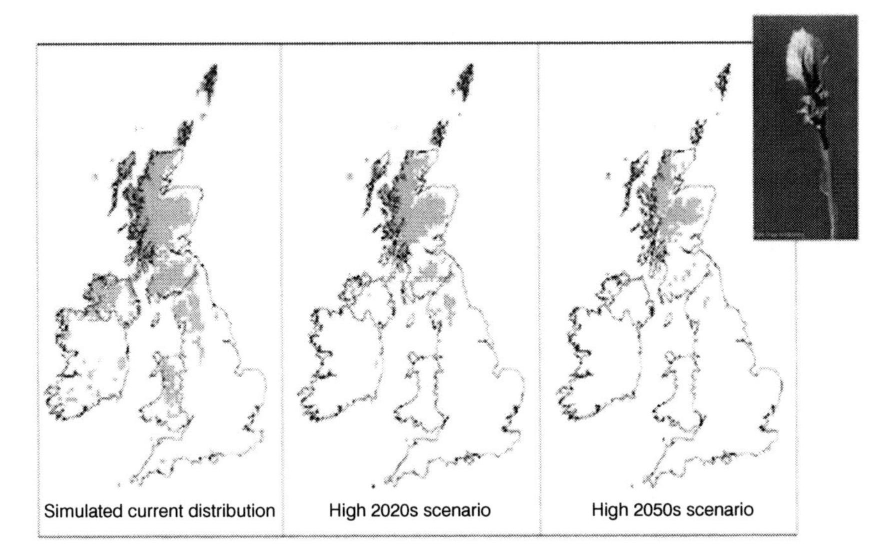

Figure 13.8 A simulated redistribution of suitable climate space for the stiff sedge (*Carex bigelowii*) in Great Britain and Ireland. As predicted by the SPECIES model (Pearson et al. 2002), a bioclimate envelope model used to track a species climate niche through future periods of predicted climate change. Note, in this species, the concurrent projection for northward migration and overall range reduction. After Pearson and Dawson 2003: 363.

problems of costs and benefits, weighing the advantages of protecting the hotspot, and the biodiversity it contains, with the disadvantage of lost resources, and presumably losses in aggregate human welfare, that come from restrictions on land and resource use needed to maintain the hotspot's biodiversity. The preservation of biodiversity is unlikely to generate popular or political support if it is perceived as an effort that violates principles of distributive justice to human beings. These kinds of problems are difficult, but they are definable, and therefore often the most amenable solution. The other kinds of problems are harder.

A second category of problem is that of conscience versus consumption, what could be called 'problems of design', because they are often rooted in the way humans design their interaction with nature (Sagoff 2000). Wood, for instance, can be a resource for human welfare and therefore can have an appropriate consumptive use, but design features associated with different kinds of logging cause more or less environmental harm. In this and similar examples, various combinations of appropriate technique and technology can reduce or prevent losses to biodiversity, especially if accompanied by reduction in human consumption of the resource.

The third category of ethical problems goes all the way down to how humans perceive the value of biodiversity, and nature in general, and what they perceive their responsibilities to nature to be. These are problems that might be called the 'needs of nature' or 'needs of species' problems. Non-human species have real physical and biological requirements for survival. Such needs cannot and will not be 'automatically' met simply by achieving optimal cost-benefit ratios of resource use or employing the latest or 'best' technology for every activity with actual or potential environmental impact. Problems in the third category point to the three most fundamental ethical questions that can be asked about the relationship between humans and their natural environment: (1) Do non-human species possess inherent or intrinsic value? (2) Are non-human species morally considerable? Is it possible to treat them rightly or wrongly with respect to interests or rights of their own? and (3) Are humans responsible for meeting the needs of other species? Although they often do not understand, possess or articulate a good reason for doing so, conservation biologists, in their own 'behavioral ethics', generally answer all three questions affirmatively. In fact, conservation biologists often hold strong positions about the intrinsic value of every species, about understanding and respecting the interests of other species, and about human responsibility to other species through protection and management that contributes to meeting the needs of such species. Unfortunately, conservation biologists often subsequently pretend that such values are not involved or important in determining the technical solution to the 'problem' of species' decline or potential extinction. 'Ethical' appeals to save species then default to instrumental utilitarianism and human welfare maximization (category 1 solutions) or technical and tactical proposals to reduce human 'impact' (category 2 solutions). Such approaches sometimes convince and succeed where values are not needed as conceptual categories for conflict resolution, but they are useless in solving problems that require an *explanatory basis* for the intrinsic value, moral significance, or human responsibility toward individual species or biodiversity in general, or toward the hotspots designed to protect both. Conservation organizations have historically aimed to avoid discussions, and consequently, solutions, to this third category of problems, usually by ignoring the question of intrinsic value altogether and presenting solutions in terms of enlightened human self-interest. In one of the main concourses of a major US airport, there was once displayed a large, full-color photograph of a giant tropical toad along several feet of one of the concourse walls. The toad looked out from the poster, face first, eye to eye with thousands of passing travelers. Below the picture, a caption read, 'My skin could save your life. Please save mine.' In smaller print below, the poster's authors explained that many wonder-working pharmaceuticals, all a blessing to human health, had been, were, or could be developed with chemicals from the toad's skin. But there could be no such benefit if there were no toads.

The irony of this poster was its sponsor. A pharmaceutical company, like Merck, Eli Lily or Pfizer, would have reason to advance an argument in favor of the toad's pharmaceutical value. They would derive *benefits* from the toad's skin that would accrue directly to them and to their stockholders. They could make millions from compounds derived from these, employ thousands of people, and treat, even cure, a multitude of diseases. But no drug company paid for this poster. The sponsor was not a 'company' at all, but an international conservation organization.

This appeal to save biodiversity, and acquire the hotspots to preserve it, is as common as it is duplicitous. As environmental ethicist Mark Sagoff noted:

> We environmentalists often appeal to instrumental arguments for instrumental reasons ... not because we believe them, but because we think that they work. (Sagoff 1991)

Scientists do not do any better than activists on this point. For example, the Amphibian Conservation Action Plan, the authoritative document for global amphibian conservation, stresses that the chief value of conserving amphibians is their potential use for medicines and biomedical models (Gascon et al. 2007), exactly the same message as the poster's. Alternatively, amphibian conservation is often reduced to that of the proverbial canary in the coal mine, 'reducing their worth', noted Douglas Woodhams, Senior Research Associate at the University of Zurich, 'to their utility as sentinels of an environmental crisis of which the public is already well aware' (Woodhams 2009). But there is a difference between asking people to save something on the basis of its *benefit* rather than its *value*. Benefits are things received directly by us for the improvement of our welfare, as in the 'health benefits' provided by our employer. Values, in contrast, represent moral principles or desired conditions we wish to uphold or maintain, even if doing so leads to a decline in our welfare, as when we give our money (the *benefit* of our work) to a charitable cause, to uphold the *value* of caring for the poor. The organization that put up the poster of the toad did not understand the difference. As a result, their poster, although perhaps effective, was disingenuous, and did nothing to move its readers to value the life of the toad for the toad's sake.

To avoid these messy ethical difficulties, conservation organizations often default to a strategy of simply ignoring ethical value, and the responsibilities and obligations this might require, and focusing solely on maximizing the number of species to be saved as an end in itself. Unfortunately, this turns conservation into something like a biological version of stamp collecting. Rare items are prized because they are rare, and each conservation organization becomes part of a race to be the one with the most rare items in its collection (preserves). Amaze your friends. Collect the whole set! If the analogy seems crass, it is nevertheless often an accurate depiction of reality. Under this approach, for example, island flora and fauna, because of their high levels of endemicity, may receive disproportionate attention and effort. But the public will ask, what's so important about saving the Mauritius kestrel, which occurs only on the island of Mauritius, when there are kestrels all over the world, most of them with stable or growing populations? I do not think conservationists have a good answer. Lack of transcendent value and meaning in conservation often translates into lack of hope for outcomes of conservation effort. Thus, left to themselves, conservationists often portray conservation as a hopeless cause. This 'hopelessness problem' has become so acute in the conservation community that two leading North American conservationists of the San Diego Zoo's Institute for Conservation Research, Ronald Swaisgood and James Sheppard, recently proposed in the pages of *BioScience* that conservation biologists should cultivate and establish 'rituals of hope' to better affirm and express positive prospects for conservation efforts. They warn that if conservation scientists continue to convey the message that nothing can be done, and that nothing that is done will matter, then nothing will be done (Swaisgood and Sheppard 2010).

Alternative approaches: conservation as part of value tradition

Such moral 'alerts' from scientists themselves should turn us, repentantly, to new approaches. To find traction with larger audiences, conservationists must be willing to explore, as well as respect, broader, more well-established traditions of ethical thought that address issues of species value at deeper levels. As the environmental ethicist J. Baird Callicott noted:

> Purely secular programs – bureaucratic, technological, legal, or educational – aimed at achieving environmental conservation may remain ineffective unless the environmental ethics latent in traditional worldviews animate and reinforce them.
>
> (Callicott 1994)

The best examples are those of major world religions and faith traditions, almost all of which speak directly to issues of the value of species, as well as to issues of human responsibility to them. It is not possible here to offer a global survey of the myriad of faith-based approaches to these questions, but we can learn many things from one example.

During the last 30 years, the coastline of Lebanon has experienced rapid development. Its shorelines and hills have been converted from natural areas to roads, homes and urban business centers. Yet it was these shorelines and hillsides that were the home of the famed 'cedars of Lebanon', forests of massive and ancient trees unique in the Mediterranean world. Most of these forests had been destroyed by the time the international conservation community realized the gravity of the situation, but the United Nations Environmental Programme (UNEP) and other conservation NGOs identified them as a conservation priority, ranking them among the 200 most important world ecosystems to be protected. But were any left? Upon investigation, the answer turned out to be 'Yes'. An ancient forest covering three hills, the Forest of Harissa, was located north of Beirut. The landowner turned out to be a church, the Maronite Church of Lebanon, which had held title to the land for centuries. To the Maronite Church, this forest was not 'the Forest of Harissa', but 'the Holy Forest of Our Lady of Lebanon'. In the center of the forest was the Cathedral of Our Lady, and an enormous outdoor statue of the Virgin Mary. UNEP and other conservation organizations prepared a 48-page proposal and delivered it to the church. The proposal demanded a promise from the church to abide by national and international laws to ensure protection of the forest. As the proposal put it, 'The area's custodians must have protection of biodiversity as a first order management objective. If other objectives take precedence over biodiversity protection, then the area as a whole, or those parts of the area where other objectives take precedence, should not be classified as a protected area.' Presented with the idea that protecting biodiversity was an ultimatum from outside interests who possessed neither sympathy nor understanding of the church's work and mission, the Maronite Church declined to enroll the forest as an area of protected status. Given the failure of this effort, two other conservation organizations, the Alliance for Religion and Conservation (ARC) and the World Wide Fund For Nature/World Wildlife Fund (WWF), tried a different tack. They met directly with the church's Patriarch, and almost immediately gained a commitment from him to protect the forest. The reasons for that protection were framed in a new statement, written mainly by the Patriarch himself, who expressed a different purpose for this conservation effort in different words. He wrote:

> For centuries the Church has defended the natural beauty and Godliness of the forests and hills of Harissa, as well as many other holy places in Lebanon. ... In so doing, we observe that the land and the flora and fauna on it, do not ultimately belong to us. We are simply the guardians of what belongs to God ... In protecting this area, the Church will continue to ensure that the diversity of plants, trees, animals and birds given by God, nurtured by the Church, will be maintained.
>
> (Palmer and Findlay 2003: 9)

In the original proposal, conservationists felt that the presence of unique biodiversity in the Forest of Harissa was sufficient motive for preserving it. The church did not. It had preserved the forest for centuries as a sacred trust to God, long before UNEP existed. Subsequent to making a formal declaration to protect the forest and its biodiversity, the Maronite Church created an ecology center for young people, formally protected two other woodland sites, and developed a program of environmental education and activism in 77 Lebanese villages and towns, making it one of the largest and most effective environmental organizations in Lebanon and one of the most influential advocates for environmental protection (Palmer and Findlay 2003). The view of the Maronite church is embedded in a larger, and older, Judeo-Christian tradition of theology and practice. Beginning with an understanding that the non-human world is good in itself and in the eyes of its Creator (Genesis 1), the Judeo-Christian tradition subsequently developed teaching, example, and illustration which treat plant, animals, and the land on which they reside as moral subjects. For example, in the Noachian Covenant that follows the Genesis story of the flood, God makes a covenant that protects not only human beings, but all living creatures, represented by the survivors in the ark, from future destruction (Genesis 9, 8–17). In Israelite law, the land is designated as the recipient of a Sabbath rest, provided every seventh year (Leviticus 25), and the failure to provide such rest is given as a prominent reason for the deportation of Israel and Judah from their homelands, and for the length of their exile. 'All the days of its [the land's] desolation it kept Sabbath until seventy years were complete.' Thomas Aquinas, the greatest Catholic scholar of the Medieval period, directly addressed the issue of the value of biodiversity (although he did not know or give his subject that name):

> God brought many things into being in order that his goodness might be communicated to creatures and represented in them; and because this goodness could not be adequately represented by one creature alone, God produced many and diverse creatures, so that what was wanting to one in the representation of divine goodness might be supplied by another. For goodness which in God is simple and uniform, in creatures is manifold and diverse. Hence the whole universe together participates in the divine goodness more perfectly, and represents it better than any single creature whatever.
>
> (Thomas Aquinas 1947: 248)

Teachings and traditions in faith communities worldwide address fundamental issues in the relationship between human beings and nature, and have grown into contemporary expression into an entirely new kind of conservation entity, the conservation faith-based organization (FBO), which has become one of the most significant developments in conservation activism during the past twenty years (Van Dyke 2008). Among the most effective of these have been A Rocha, a Christian conservation FBO instrumental in the preservation of the Aamique Wetland in Lebanon, the Islamic Foundation for Ecology and Environmental Sciences (IFEES), which has, among other initiatives, gained government support in Indonesia for a land and forest management system in West Sumatra based on Islamic principles of stewardship (McKay 2010), and the Buddhist Perception of Nature Project, supported by the WWF, which has, among other efforts, built environmental teachings based on Buddhist principles of sustainability into conservation education curricular throughout southeast Asia (Kabilsingh 1990; Van Dyke 2008).

Traditional conservation organizations are increasingly recognizing the importance, contribution and value of faith communities in biodiversity preservation and hotspot management, recently illustrated in the joint effort in Papua New Guinea by Conservation International, Papua New Guinea's Institute for Biological Research, and A Rocha International, in which scientific teams composed of members of all three organizations discovered over 200 new species in only two months of rapid assessment surveys (A Rocha 2011). Many hotspots, particularly in the developing world, are in areas where faith communities are active, or even dominant. Cooperative efforts like that in Papua reflect the potential, and necessity, to integrate protection of biodiversity into a larger fabric of broader ethical value systems that address multiple dimensions of human interaction with nature.

Imagining the future: prognosis and prescription

Hotspots will continue to be centers of conservation concern for the foreseeable future, and the procedures that identify high biodiversity areas at even smaller spatial scales will increase in technological sophistication and efficiency. Technical prowess in identifying new hotspots, skill at managing their flora and fauna, and effectiveness in mitigating threats to their habitats will increase. But despite increasing scientific and professional acumen, hotspots will be under stress in accomplishing their fundamental objective, the preservation of biodiversity, because of the political instability, human population growth, climate change, and economic development that surround them.

To preserve biodiversity, the scientific and conservation community must move away from traditional preserve approaches that succeeded through the exclusion of humans and their influence on 'natural' areas, recognize existing deficiencies in their own ability to make normative prescriptions in management, and increasingly communicate and cooperate with other communities with longer, more ethically grounded traditions of valuing biodiversity while attempting to reconcile human presence in and needs associated with hotspot areas in ways that do not degrade biodiversity. Without such cooperation, scientists and conservationists will be frustrated with and even excluded from decision-making processes that determine current and future management of hotspots. To that end I offer five prescriptions for those who are now or will someday be charged with responsibility to protect existing hotspots and associated preserves within them, or establish new ones.

1. Strategize globally. Manage locally. Management and protection, not designation and description, are the real means of effective conservation. Scale and place preserves appropriate to the creatures they are to protect, and put no trust in surrogate species.
2. Clearly determine the conservation objective, whether to protect diversity, endemism or rarity, and whether to protect what is most unique or what is most vulnerable.
3. Integrate these objectives when possible. Separate and clarify them when not.
4. Make no assumption of environmental stability, but work skillfully and persistently to develop mitigation strategies for changing the locations and boundaries of protected areas, or acquiring new ones, in anticipation of climate change affecting distribution of protected species.
5. Be ready to give a good answer about the value of biodiversity and the significance of protecting it, and be prepared to work with and learn from communities who enjoy continuing traditions of conservation grounded in ethical and moral principles intrinsic to those traditions.

References

A Rocha (2011) *Incredible array of new species found in remote Papua New Guinea*. Online. Available HTTP: <http://www.arocha.org/int-en/news/media/9444-DSY.html> (accessed 13 June 2013).

Aquinas, Thomas (1947) *Summa Theologica*, vol. I., trans. Fathers of the English Dominican Province, New York: Benzinger Brothers.

Brooks, T.M., Mittermeier, R.A., da Fonseca, G.A., Gerlach, J., Hoffmann, M., Lamoreux, J.F., Mittermeier, C.G., Pilgrim, J.D. and Rodrigues, A.S. (2006) 'Global biodiversity conservation priorities', *Science* 313: 58–61.

Callicott, J.B. (1994) *Earth's insights: a survey of ecological ethics from the Mediterranean Basin to the Australian Outback*, Berkeley, CA: University of California Press.

Conservation International (2011) *Hotspots defined*. Online. Available HTTP: <http://www.conservation.org/where/priority_areas/hotspots/Pages/hotspots_defined.aspx> (accessed 13 June 2013).

Darlington, J.P. (1957) *Zoogeography: the geographical distribution of animals*, New York: Wiley.

Darwin, C. (1859) *On the origin of species by means of natural selection or the preservation of favored races in the struggle for existence*, 5th edn, London: Appleton and Company.

Gascon, C., Collins, J.P., Moore, R.D., Church, D.R., McKay, J.E. and Mendelson III, J.R. (eds) (2007) *Amphibian Conservation Action Plan. IUCN/SSC Amphibian Specialist Group*. Online. Available HTTP: <http://www.amphibians.org/wp-content/uploads/2011/09/ACAP.pdf> (accessed 13 June 2013).

Grenyer, R., Orme, C.D., Jackson, S.F., Thomas, G.H., Davies, R.G., Davies, T.J., Jones, K.E., Olson, V.A., Ridgely, R.S., Rasmussen, P.C., Ding, T.S., Bennett, P.M., Blackburn, T.M., Gaston, K.J., Gittleman, J.L. and Owens, I.P. (2006) 'Global distribution and conservation of rare and threatened vertebrates', *Nature* 444: 93–6.

Guilhaumon, F., Gimenez, O., Gaston, K.J. and Mouillot, D. (2008) 'Taxonomic and regional uncertainty in species-area relationships and the identification of richness hotspots', *Proceedings of the National Academy of Sciences USA* 105: 15458–63.

Kabilsingh, C. (1990) 'Early Buddhist views on nature', in A. Hunt Badiner (ed.) *Dharma gaia*, Berkeley, CA: Parallax, 34–53.

Keller, C. and Schradin, C. (2008) 'Plant and small mammal richness correlate positively in a biodiversity hotspot', *Biodiversity and Conservation* 17: 911–23.

Lamoreux, J.F., Morrison, J.C., Rickets, T.H., Olson, D.M., Dinerstein, E., McKnight, M.W. and Shugart, H.H. (2006) 'Global tests of biodiversity concordance and the importance of endemism', *Nature* 440: 212–4.

MacArthur, R.H. and Wilson, E.O. (1967) *The theory of island biogeography*, Princeton, NJ: Princeton University Press.

Mayer, P. (2006) 'Biodiversity – the appreciation of different thought styles and values helps to clarify the term', *Restoration Ecology* 14: 105–11.

McKay, J.E. (2010) 'Islamic beliefs and Sumatran forest management', EcoIslam 7. April: 1–2. Online. Available HTTP: <http://www.ifees.org.uk/pdf/newsletter_EcoIslam7.pdf> (accessed 13 June 2013).

Myers, N. (1988) 'Threatened biotas: hotspots in tropical forests', *Environmentalist* 8: 187–208.

Palmer, M. and Findlay, V. (2003) *Faith in conservation: new approaches to religion and the environment*, Washington, D.C.: World Bank.

Parmesan, C. and Yohe, G. (2003) 'A globally coherent fingerprint of climate change impacts across natural systems', *Nature* 421: 37–42.

Pearson, R.G. and Dawson, T.P. (2003) 'Predicting the impacts of climate change on the distribution of species: are bioclimate envelope models useful?', *Global Ecology and Biogeography* 12: 361–71.

Pearson, R.G., Dawson, T.P., Berry, P.M. and Harrison, P.A. (2002) 'SPECIES: a spatial evaluation of climate impact on the envelope of species', *Ecological Modelling* 154: 289–300.

Pimm, S.L. (1998) 'Extinction', in W.J. Sutherland (ed.) *Conservation science and action*, Oxford: Blackwell Science, 20–38.

Pimm, S.L. (2001) *The world according to Pimm: a scientist audits the Earth*, New York: McGraw Hill.

Rabinowitz, D., Cairns, S. and Dillon, T. (1986) 'Seven forms of rarity and their frequency in the flora of the British Isles', in M.E. Soulé (ed.) *Conservation biology: the science of scarcity and diversity*, Sunderland, MA: Sinauer, 182–204.

Sagoff, M. (1991) 'Zuckerman's dilemma: a plea for environmental ethics', *Hastings Center Report* September–October: 32–40.

Sagoff, M. (2000) 'Environmental economics and the conflation of value and benefit', *Environmental Science and Technology* 34:1426–32.

Scott, D. (2005) 'Integrating climate change into Canada's national park system', in T.E. Lovejoy and L. Hannah (eds) *Climate change and biodiversity*, New Haven, CT: Yale University Press, 342–5.

Shannon, C.E. and Weaver, W. (1949) *The mathematical theory of communication*, Urbana, IL: University of Illinois Press.

Swaisgood, R.R. and Sheppard, J.K. (2010) 'The culture of conservation biologists: show me the hope!', *BioScience* 60: 626–30.

Thompson, R. and Starzomski, B.M. (2006) 'What does biodiversity actually do? A review for managers and policy makers', *Biodiversity and Conservation* 16: 1359–68.

Urban, D.L. (2002) 'Prioritizing reserves for conservation', in S.E. Gergel and M.L. Turner (eds) *Learning landscape ecology: a practical guide to concepts and techniques*, Berlin: Springer, 293–305.

Van Dyke, F. (2008) *Conservation biology: foundations, concepts, applications*, 2nd edn, Dordrecht: Springer.

Van Dyke, F., Schmeling, J.D., Starkenburg, S., Yoo, S.H. and Stewart, P.W. (2007) 'Responses of plant and bird communities to prescribed burning in tallgrass prairies', *Biodiversity and Conservation* 16: 827–39.

Villaseñor, J., Ibarra-Manríquez, G., Meavem, J.A. and Ortiz, E. (2005) 'Higher taxa as surrogates of plant diversity in a megadiverse country', *Conservation Biology* 19: 232–8.

Williams, J.W., Jackson, S.T. and Kurtzbach, J.E. (2007) 'Projected distributions of novel and disappearing climates by 2100 AD', *Proceedings of the National Academy of Sciences USA* 104: 5738–42.

Wilson, E.O. (1988) *Biodiversity*, Washington, D.C.: National Academy Press.

Woodhams, D.C. (2009) 'Converting the religious: putting amphibian conservation in context', *BioScience* 59: 462f.

14 Whose biodiversity is it?

Challenges in managing migratory species

Aline Kühl-Stenzel[1]

Introduction: a shared resource

Who owns the White storks arriving in Germany in spring and in the Sahel in autumn? Who owns the High Seas, its sharks, marine turtles and other highly migratory species in the world's oceans? The management of animal populations across national jurisdictional boundaries is notoriously challenging and has received growing attention, not least since the 1970s when the Convention on the Conservation of Migratory Species of Wild Animals (CMS) was first drafted in response to increasing threats to this particular group of species. The reasons for the inherent vulnerability of animals on the move are as multi-fold as this group is taxonomically diverse. Across the animal kingdom we find migrants amongst the insects and other invertebrates, fish, reptiles, birds and mammals, both on land and in the sea. Even the largest vertebrates to ever walk on earth, herbivorous sauropod dinosaurs, used to migrate to upland regions during the hotter summer season to sustain their high need for vegetation and water (Fricke, Hencecroth and Hoerner 2011: 513).

One overarching factor that threatens migratory species is their qualification as a 'common-pool resource', as defined by Elinor Ostrom et al. (e.g. 1999: 278). Migrants travel long distances, often across many countries, and thus their survival depends on the quality and management of many habitats. In essence, they inhabit the global 'commons'. Unlike a pond or a small forest, migratory species are shared by huge numbers of stakeholders, sometimes spread across different continents. Value systems are likely to vary strongly between different stakeholder groups, as well as people's culture, language and other factors. These are far from ideal conditions for sustainable management.

There are, however, many shades of grey and certain factors make the effectiveness of conservation action for migratory species more likely. This chapter will discuss and contrast some of these factors, including biological and socio-economic ones. Emphasis will be placed on migratory species which cross national borders and the resultant challenge of international biodiversity conservation. The chapter will further highlight how international legal instruments can create an additional level of ownership and thereby act as a catalyst to conserve wide-ranging migratory species across many different countries. Firstly, however, the chapter will give an overview of migratory species biology and why they are in urgent need of conservation action, not least for economic reasons. Case studies from terrestrial and marine environments will be compared and recommendations for management and policy made.

What is a migratory species?

Our planet is covered with invisible pathways used by billions of migratory animals each year. Across all continents one can find animals which cyclically move in large numbers from one seasonal range to another. Many of these mass migrations, such as those of the wildebeests (*Connochaetes taurinus*) in the Serengeti-Mara ecosystem or the Monarch butterfly (*Danaus plexippus*) in the Americas, are well known across the globe. For the majority of migratory species, however, especially in the world's oceans, we only have a poor understanding of their migratory route and ecology. Animals migrate for many different reasons, often taking advantage of seasonally overabundant food sources. Many herbivores follow the rains and the greenest vegetation, but also often migrate to avoid predators, disease or unsuitable climatic conditions. Some animals migrate to access optimal breeding grounds or to be in a safe place when they are vulnerable (e.g. when moulting). Migratory species move along latitudinal (e.g. Siberian crane *Leucogeranus leucogeranus*) or altitudinal gradients (e.g. moose *Alces alces*) or follow different patterns altogether (e.g. Mongolian gazelle *Procapra gutturosa*). Some species have a high level of site fidelity, meaning they return to the same sites again and again, such as female marine turtles coming to lay their eggs on the beaches where they themselves hatched, or Red knots (*Calidris canutus*) and many other waterbirds overwintering, stopping and breeding at the same sites every year (Figure 14.1).

Other migratory species have a low level of site fidelity and follow an overall pattern but use different paths year after year, such as many songbirds. In some species only one sex migrates, such as in Tibetan antelope (*Pantholops hodgsonii*) where only females move about 200 km to specific sites every year to give birth (Buho et al. 2011: 43). Because of this huge variation in migratory behaviour no single definition of animal migration exists which suits both ecological and policy purposes (Hoare 2009: 12). Opportunism is, however, one feature which does unite all migratory species (Wilcove 2010: 23). Animals on the move manage to take advantage of beneficial circumstances even in the remotest and harshest of habitats, such as baleen whales migrating to polar waters to feast on overabundant krill (see pp. 274–5). Depending on the definition, at least 8,000 animal species are considered to be transboundary migratory today (Johnson and Vagg 2010, 7). Interestingly, it has been shown that migratory blackcaps (*Sylvia atricapilla*) can stop migrating and become resident in just a few generations in response to intense selection pressure mimicking climate change (Pulido and Berthold 2010: 7341). It is, however, not yet known how common such genetic adaptation is.

The distance that animals cover on their annual journeys, generally between breeding and non-breeding sites, varies from only a few hundred metres to

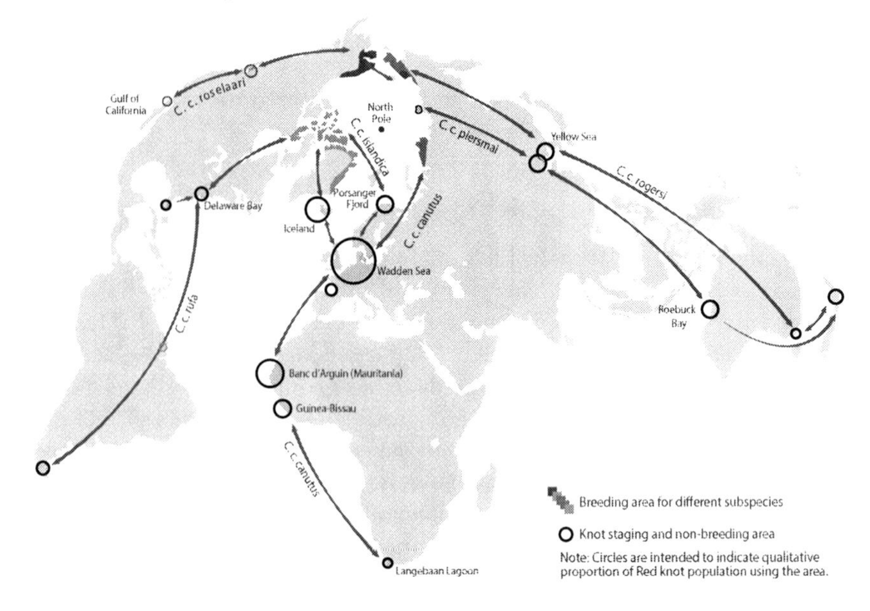

Figure 14.1 Global flyways of the six subspecies of Red knot
Source: Kurvits et al. 2011: 59.

trans-continental movements. The longest migration distances have been recorded for birds, specifically for Arctic terns (*Sterna paradisaea*) and Sooty shearwater (*Puffinus griseus*), which both travel across the oceans between northern and southern hemispheres, covering up to 80,000 and 65,000 kilometres per year respectively (Egevang et al. 2010: 2078; Shaffer et al. 2006: 12799). Given that Arctic terns can live to more than 30 years, the oldest individuals might fly the equivalent of three return trips to the moon in their lifetime. Out of the approximately 10,000 bird species identified worldwide, about 1,800 undertake transboundary migrations (Sekercioglu 2007: 284). Runners-up in terms of distance records *per annum* include Blue whales (*Balaenoptera musculus*) for marine mammals with 17,000 kilometres, Leatherback turtles (*Dermochelys coriacea*) for reptiles with 16,000 kilometres (Figure 14.2.), 'Globe skimmer' dragonflies (*Pantala flavescens*) for insects with 14–18,000 kilometres, and caribou (*Rangifer tarandus*) for terrestrial mammals with 5,000 kilometres (Hoare 2009; Hobson et al. 2012). Some experts even consider daily movements a migration, such as the passage of jellyfish up and down the oceans' water column following the light and plankton (Hoare 2009: 49).

Migratory species are threatened by the same anthropogenic pressures as sedentary (non-migratory) species, with the primary ones being habitat destruction/fragmentation, overexploitation, pollution, the introduction of invasive species and climate change. Almost one third of the global land area has been converted into agricultural land and another third has been fragmented, mostly through roads, fences, wind turbines and other barriers to migration

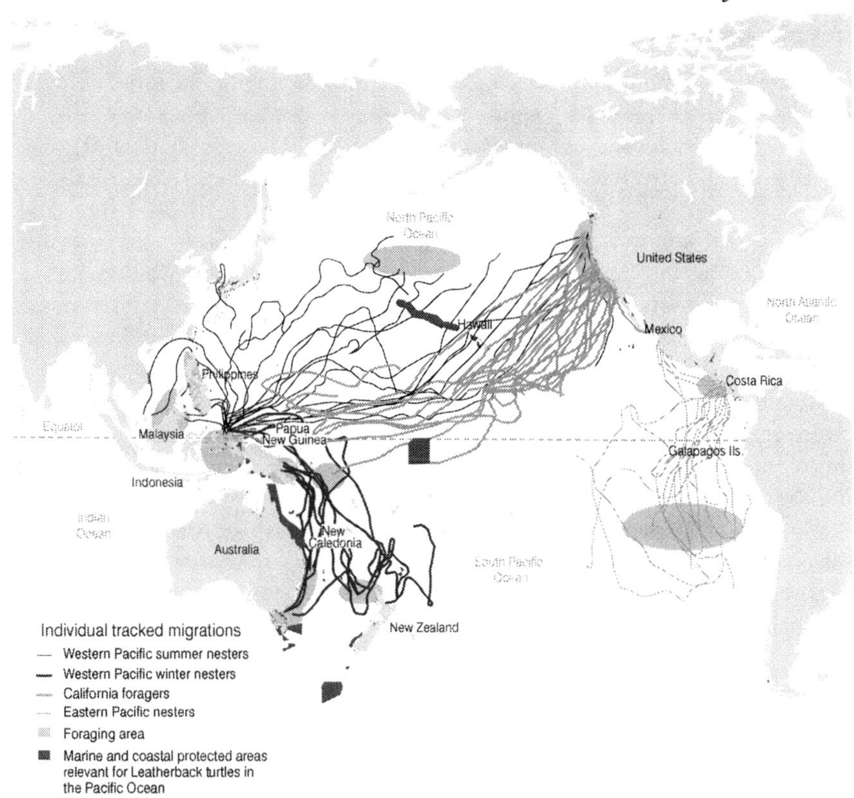

Figure 14.2 Leatherback turtle migration in the Pacific illustrating variability in migration strategies in a single ocean basin.
Source: Kurvits et al. 2011: 51.

(Alkemade et al. 2009: 374; Pereira et al. 2010: 1496). The area covered by wetlands worldwide has declined by 50% in the last century (UNEP 2010a). The reason why long-distance migrants are particularly vulnerable compared to sedentary species is that the removal of one single critical site on the migratory route can jeopardise the entire migration, not unlike the removal of a single card from a house of cards (Bolger et al. 2008). The feeding strategy of an animal is carefully adjusted to the migration ahead and thus even minor declines in habitat quality and resultant declines in animal fitness can have a detrimental impact on the survival of a species. The multitude of threats already impacting migratory species today have brought many populations to the edge of their physical resilience with little ability to buffer against any further detrimental impacts. This is why climate change, for example, is predicted to become a number one threat for migratory species despite their ability to successfully adjust their annual journeys to climatic changes in the past (Kühl and Mrema 2011: 176).

People and animal migration

The physical and navigational challenges of animal migration have fascinated and challenged experts for centuries. It is maybe not surprising that Aristotle thought that swallows overwinter in the mud below lakes (Aristotle 1991: 49b, 632b14–633a28). In the sixteenth century in Sweden this belief was still upheld (Magnus 1998: 980). To hypothesise that these small birds travel thousands of kilometres south to their wintering sites would no doubt have seemed unbelievable at the time. Today experts are regularly amazed when analysing the results from satellite transmitters showing that yet again a migratory animal has broken a record in terms of distance, depth, height or speed. Humpback whales (*Megaptera novaeangliae*), for example, are able to swim across 200 km in a straight line with remarkable precision, losing only 1 degree from their target (Horton et al. 2011: 1) (Figure 14.3). It is not known how they achieve such outstanding accuracy. Equipped with only a compass, such a feat would be impossible to accomplish for humans on a boat.

Rock carvings and ancient corral hunting structures pay tribute to the close relationship between people and migratory species dating at least back to the Bronze Age (e.g. Bull and Esipov 2013: 18). Many of the larger mammals, especially ungulates, were already hunted in prehistoric times and frequently feature at archaeological sites today (e.g. Lister 1992: 329). The tendency for migratory species to aggregate in large numbers at predictable places and times throughout the year contributes to the attention given to these by hunters and fishermen over thousands of years. Today many communities continue to have a close relationship with migratory species, valuing these animals either as a resource or for cultural, religious and other reasons. In Bhutan, for example, Black-necked cranes (*Grus nigricollis*) are considered holy and feature throughout the literature, in folksongs, dances and paintings. In the Autonomous Republic of Kalmykia in the Russian Federation illustrations, songs and stories of the Saiga antelope (*Saiga tatarica*) are common, even on bus stops out in the steppe. Despite this high cultural value, poaching continues to threaten this migratory species across its range in Central Asia and Russia (Kühl et al. 2009: 1442; Box 14.1).

Migration: a challenge for sustainable management

The Saiga antelope presents a typical example of a migratory species that roams over such large distances that many human settlements can potentially exploit the same saiga population, even across different countries (Figure 14.4). Managing such a 'common-pool resource' is challenging in many ways. Access to a migratory population tends to be open since it is usually too difficult and costly to enforce the law over such vast areas. In the Ustiurt region between the Caspian and Aral Seas in the west of Kazakhstan, for example, on average less than two ranger cars patrolled a territory of approximately 600 × 300 km^2 in 2004–2007. Finding saiga poachers (as well as the saiga antelopes themselves) in such an area is like finding a needle in a haystack.

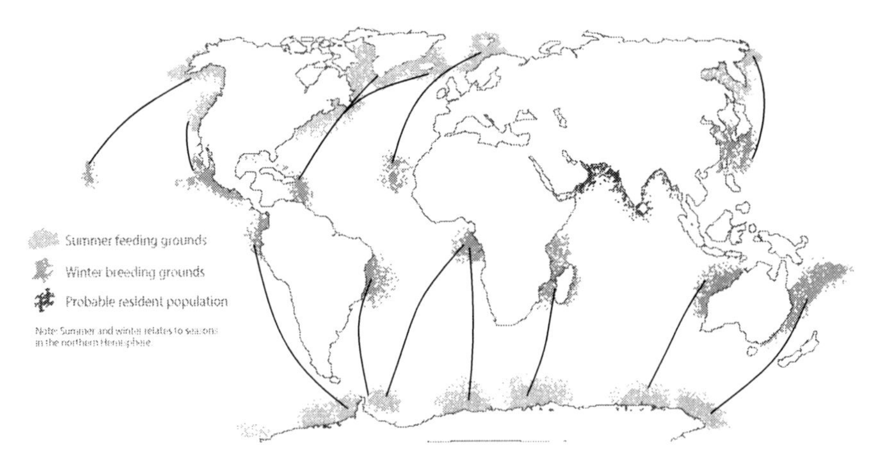

Figure 14.3 The long migration of the humpback whale
Source: Kurvits et al. 2011: 49.

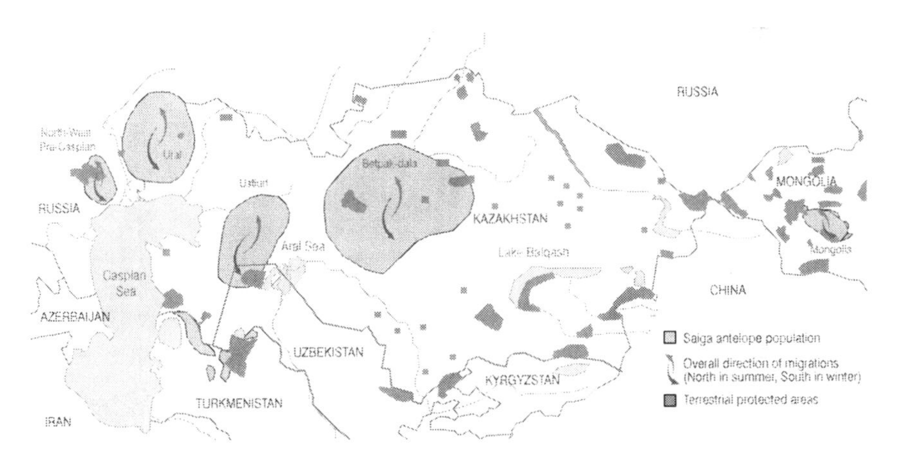

Figure 14.4 Saiga antelope populations
Source: Kurvits et al. 2011: 36.

When access is open, populations are more likely to be harvested unsustainably than when access is restricted (Hardin 1968: 1243; Sutherland and Gill 2001). Exploited migratory species often fulfil all the conditions for a situation known as a 'Tragedy of the Commons' to evolve: there are many users with free access to a finite resource and exclusion is difficult (Hardin 1968: 1243). Under such circumstances individuals increasing their use of the resource retain the full private benefit, but the costs (such as ecological degradation) are borne by all. There is therefore little incentive for an individual to manage the resource sustainably and open-access behaviour often results. International

Box 14.1 The case of the Saiga antelope

In the early 1990s there used to be around one million saiga antelopes (*Saiga* spp.), today only ~150,000 saigas remain in Kazakhstan, Mongolia, Uzbekistan, Russia and Turkmenistan (Milner-Guland et al. 2001; Figure 14.4). The decline of the saiga is closely tied to the collapse of the Soviet Union in 1991. Collective farms fell apart across the saiga's range, leading to widespread unemployment and poverty. Saiga meat and especially the species' horn, which is used in traditional Chinese medicine, provided a much-needed alternative source of income. There was little to stop the poachers since the law enforcement system had also fallen apart. The opening of international borders to China and other neighbouring countries connected the supply of saiga horn in range states with strong demand across South East Asia, leading to the fastest species decline documented for a large mammal in recent decades: a 95% crash in total population size in 10 years (Milner-Gulland et al. 2001: 340; CMS 2006: 5). The speed of this decline is partly caused by a reproductive collapse of populations in response to a severe lack of adult males, since only these bear the precious horn and are therefore targeted by poachers. In 2001 less than 1% of adult males were recorded in the Kalmyk steppes, which led to a steep decline in fecundity levels (Milner-Gulland et al. 2003: 135).

Following their sharp decline, saigas have become subject to considerable international policy attention because of the species' transboundary migrations and the international trade in saiga horn. Already in 1995 the saiga was listed on Appendix II of the Convention on International Trade in Endangered Species of Wild Fauna and Flora (CITES) to try and reduce the international horn trade. In 2002 CMS listed *Saiga tatarica tatarica* on its respective Appendix II and IUCN uplisted this subspecies by two levels from 'Vulnerable' to 'Critically Endangered'. In 2006 a special agreement for the international conservation of the saiga antelope came into force under the auspices of CMS to facilitate transboundary conservation action in all range states and to halt the trade in saiga horn in close dialogue with CITES and consumer states. This agreement has since then been administered by the UNEP/CMS Secretariat in Bonn, Germany, in close collaboration with the CITES Secretariat in Geneva, Switzerland. This 'Memorandum of Understanding concerning the Conservation, Restoration and Sustainable Use of the Saiga Antelope (*Saiga* spp.)' has been signed by all range states and also includes the Mongolian subspecies (*Saiga tatarica mongolica*) since 2009. Technical coordination to the Memorandum is being provided by two Non-Governmental Organisations (NGOs): the Association for Conservation of Biodiversity of Kazakhstan (ACBK) and the Saiga Conservation Alliance (SCA). The Medium-Term International Work Programme under the Memorandum of Understanding (MoU) benefits from biological and socio-economic research, which helped to draft policy action with a close understanding of what drives saiga poaching, the attitudes of local people and the conservation solutions supported by communities within the saiga range (e.g. Bekenov et al. 1998: 1; Kühl et al. 2009: 1442). The recognition of the pivotal role that communities play in saiga conservation and the explicit inclusion of local people, scientists and other stakeholders have to some extent empowered these to implement the MoU. Within the framework of the MoU a considerable amount of conservation action has been implemented ranging from law enforcement action to community-based conservation and research (for more information see CMS 2013a; Saiga News 2013). A sizable network of range state and donor governments, as well as NGOs, research bodies and civil society, has made this agreement one of the most successful international policy instruments for biodiversity in terms of impact on overall population size. Overall population numbers have increased from an all-time low in the mid-2000s of less than 50,000 to ~150,000 saigas today. However, stronger incentives and financial means are urgently needed to drive saiga conservation action from the bottom.

One saiga population in central Kazakhstan (Betpak-dala) has stabilised and is starting to increase in numbers, possibly in response to improved socio-economic conditions in rural villages, strengthened law enforcement, awareness raising and improved research to inform decision-making. One population in Russia (Precaspian) and another in Mongolia are starting to stabilise, but at low population numbers. The remaining two *Saiga tatarica tatarica* populations are transboundary and unfortunately continue to be in a precarious state. In Ustiurt (Kazakhstan, Uzbekistan, Turkmenistan), aerial survey data suggest that the population size has declined to less than 10,000 animals since 2009 which is a very low density given that saigas migrate up to 1000 km between summer and winter ranges (until the late 1990s this population numbered > 200,000). The other transboundary population, Ural (Kazakhstan, Russian Federation), has suffered from several mass mortality events and was in 2012 estimated to number ~20,000 animals.

regulation provides one avenue to counteract such open-access behaviour, as explained further below.

There are a number of factors that are positively correlated with the development of a 'Tragedy of the Commons'. The longer the migration route and the more users there are along the way, the more likely it is that the species in question will be a 'common-pool resource' and access will be difficult to restrict. Similarly, the more countries an animal crosses on their annual migrations, the more challenging sustainable management can become. Cultural, geographical, linguistic, legal and other barriers between communities may further exacerbate this.

The habitats which migratory species use on their annual journeys to stopover, refuel, breed or feed are thankfully not necessarily 'common-pool resources'. Individual wetlands might, for example, be under national protection or might be managed by a single community. Even larger sites such as the Wadden Sea shared between Denmark, Germany and the Netherlands can be managed effectively, while carefully balancing biodiversity, fisheries, shipping, tourism and other interests. International agreements can play an important role to facilitate such cooperation, especially for common-pool resources.

The most challenging type of common-pool resource to manage worldwide is probably the High Seas, the open ocean area outside the Exclusive Economic Zones (EEZ) and other national waters. In the High Seas the restriction of access, such as the establishment of a Marine Protected Area (MPA), depends on the agreement of a large number of countries. Depending on the legal instrument under which such an MPA or other protection measure is being negotiated, there might be more than 100 Parties that have to find consensus (e.g. under the Memorandum of Understanding on the Conservation of Migratory Sharks under UNEP/CMS). However, even when it has been agreed to protect a certain area of the High Seas or to limit the fishing of a shark species, for example, this is notoriously difficult to enforce due to the vast size of the territory, limited resources and the challenges involved with shared sovereignty.

For specific sites along the migration route it is theoretically possible to restrict access and to maintain the biodiversity value of the site. However, the entire range of a migratory species is generally too large to put entirely under protection. It is therefore important to determine which sites are critical for the survival of the species and at what times of the year the species is present. These might be important breeding, feeding or nesting sites, such as Delaware Bay on the East coast of the United States in spring when thousands of migratory shorebirds arrive specifically to feed on the energy-rich eggs of Horseshoe crabs (*Limulus polyphemus*). The higher the level of site fidelity in a species, i.e. if a species returns to the same places year after year, the easier it becomes to determine the network of critical sites which a particular species needs. The Red knot (*Calidris canutus*) is one of many shorebirds with high site fidelity making it relatively straightforward to determine where protection is needed once the migration route is fully understood (Figure 14.1). The development of the 'Critical Site Network Tool' under one of the CMS instruments, the Agreement on the Conservation of African-Eurasian Migratory Waterbirds (AEWA), has benefited from overall strong site fidelity for this group of species. Different geo-referenced population data are overlaid, including census results, and specific sites can then be identified which are critical to the species' survival. Such information is very helpful for spatial planning of conservation measures at the national and regional level.

When site fidelity is low it is much more difficult to spatially plan protected areas and other conservation measures. Sometimes, such as in the case of the Saiga antelope, site fidelity is generally low, but high for certain parts of the life cycle in some populations. In the Precaspian in Russia, for example, saigas tend to give birth in the same area year after year and it has been possible to protect this broad area through two adjacent reserves, the Stepnoi Reserve and the Chernye Zemli State Biosphere Reserve. Jointly the reserves cover 208,901 hectares, as well as a 90,000-hectare buffer zone. Even though the total reserve area makes up less than 5% of the saiga range in the Precaspian, the protection of the calving grounds makes a significant contribution to overall saiga conservation. Nowhere else in the Precaspian is the protection of 200 hectares likely to have a bigger impact on saiga conservation. Regular monitoring and research is therefore important to determine the migration and the level of site fidelity throughout the life cycle for individual populations.

While it is rare to find examples where migratory species are still successfully sustainably harvested, these do exist. In Costa Rica, for example, community cooperatives collect approximately four million eggs per year from Olive ridley turtles (*Lepidochelys olivacea*) as a valued delicacy (Campbell 2002: 1229). The government has given the community the right to self-regulate the collection of eggs on the nesting beaches of the turtles. Thanks to strict social enforcement within the community this system appears to have been sustainable for decades (Campbell 2002: 1229). This is, however, an unusual case since only one community has access to the resource (the eggs in this instance). Usually a great number of communities have access to a migratory species population or its habitat, making such ownership and successful self-enforcement very challenging. Assuming that saiga numbers will recover, trophy hunting has been proposed as a novel future source of alternative income for communities (Brown et al. 2010). During the Soviet Union saiga antelopes were hunted sustainably, thus from a biological perspective this is possible. From an economic point of view, however, the sustainable use of saiga is not straightforward. Currently saiga horn may not be sufficiently valuable to create a sizable incentive for saiga conservation across so many stakeholders. In 2013 in Kalmykia a kilogram of horn sold for 12,000 Rubels (€ 295; containing 4–6 horns), which is equivalent to an average monthly income in rural poaching villages in the region (S. Aitkulov, pers.comm.). This is a very low value compared to other successful community-based trophy hunting schemes, such as Suleiman Markhor (*Capra falconeri jerdoni*) in Pakistan, for example, where a single trophy can fetch US$ 40,000 (Baldus, Damm and Wollscheid 2008: 29). However, it is worth noting that if the possession of saiga horn was legalised and marketed (once populations have recovered to allow for sustainable use), prices would most likely rise. Furthermore, the sale of saiga meat can contribute to up to 80% of the income of saiga poachers (Kühl et al. 2009: 1448), thus the integration of this second and often neglected saiga product needs to be considered in any potential future scenario of the sustainable use of saiga antelopes. However, the greatest challenge above all others is likely to be the joint ownership, decision-making and fair sharing of income between the many communities across the various countries. Such coordination can to some extent be facilitated through international legal instruments.

International policy action for the conservation of transboundary animal migration

Since 1972, when the first global environmental conference took place in Stockholm, policy makers have been busy 'weaving a web of environmental law' to facilitate the management of migratory species and specific habitats such as wetlands, as well as the international trade in wildlife products and many other aspects of environmental governance (for a historical account cf. Lausche 2008). This web is becoming tighter on an annual basis with new environmental treaties being signed and existing ones evolving. Out of more than 500 environmental agreements drawn up in the last 50 years under the aegis of the United Nations of one of its agencies, an estimated 155 are biodiversity-related (Najam, Papa and Taiyab 2006). After trade matters in first place, the governance of environmental resources has become a priority area for international law when measured by the number of treaties.

The management of transboundary migratory species is naturally a subject of international concern. In addition to CMS there is a multitude of other relevant global instruments that contribute to the conservation of migratory species and their habitats; however, a full description of these is beyond the scope of this chapter. An exhaustive list and analyses of biodiversity-related treaties can be found elsewhere (e.g. Najam, Papa and Taiyab 2006; Chambers 2008; Lausche 2008; Johnson 2012).

The most relevant treaties for migratory species are found amongst the Biodiversity-Liaison Group (BLG): Convention on Biological Diversity (CBD), Convention on the Conservation of Migratory Species of Wild Animals (CMS), Convention on International Trade in Endangered Species of Wild Fauna and Flora (CITES), International Treaty on Plant Genetic Resources for Food and Agriculture, Ramsar Convention on Wetlands and World Heritage Convention. CMS is one of the oldest treaties amongst this group, being first available for signature in 1979 and coming into force in November 1983. CMS is also known as the 'Bonn Convention' since this is where the treaty was finalised and first signed. It is the only global treaty aimed at conserving animal migration across all taxonomic groups. Approximately 150 countries have signed one or more of the CMS instruments today (Figure 14.5).

However, since the well-being of migratory species is closely linked to overall biodiversity well-being, wetland and other habitat status, as well as international trade and other matters, CMS works closely with the members of the BLG, as well as a whole suite of other treaties and organisations (for an overview cf. UNEP 2011).

The primary drafters of CMS from the IUCN Environmental Law Centre had long-distance migrants, such as Lesser white-fronted geese (*Anser erythropus*; Figure 14.6) and other waterfowl, in mind when they prepared a first outline of the treaty (F. Burhenne-Guilmin, pers.comm.). They were aiming to create one global structure, which would not only lead to broader conservation measures within range states but would also place considerable emphasis on restricting the exploitation of species. Migratory species are defined as follows within CMS: 'the

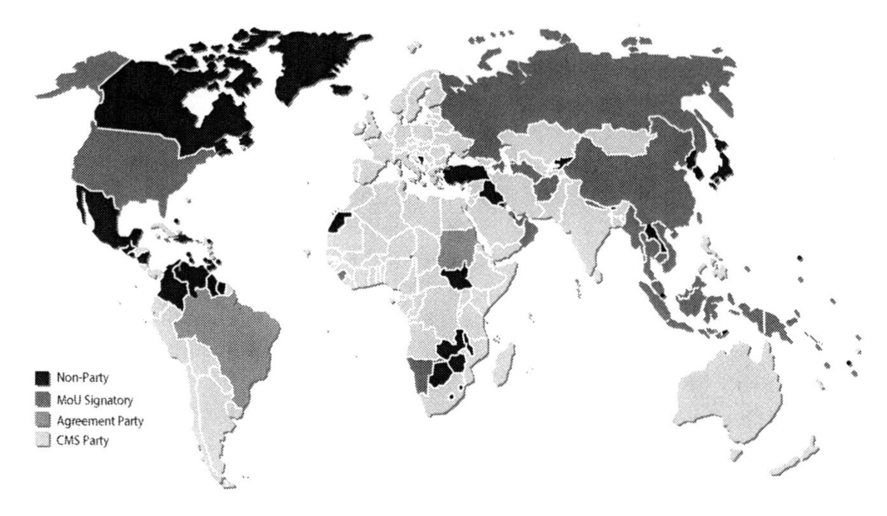

Figure 14.5 Who protects them? Convention on the Conservation of Migratory Species of Wild Animals
Source: Kurvits et al. 2011: 61

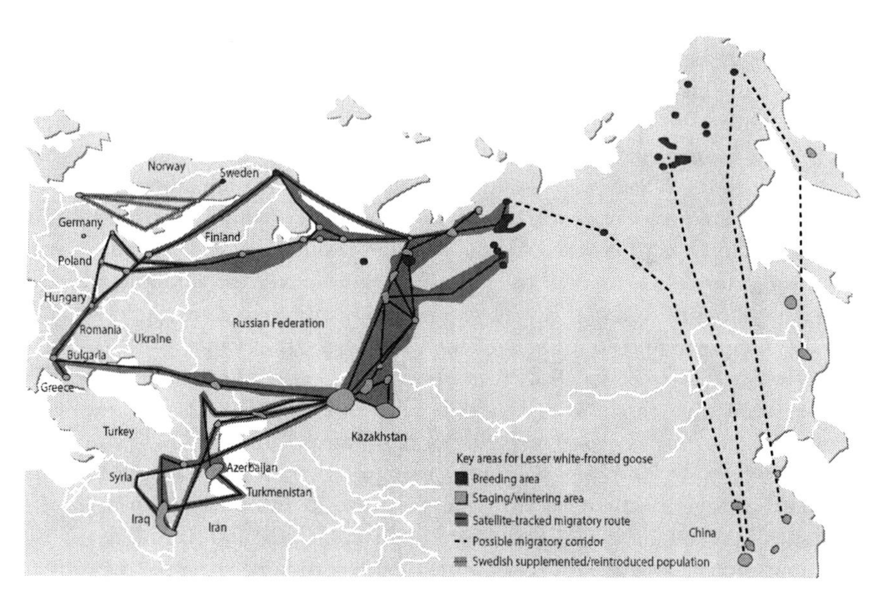

Figure 14.6 Lesser white-fronted goose migratory routes in Eurasia
Source: Adopted from the AEWA International Single Species Action Plan for the Conservation of the Palearctic population of the Lesser White-fronted Goose, 2008. Original data from Birdlife Norway

entire population or any geographically separate part of the population of any species or lower taxon of wild animals, a significant proportion of whose members cyclically and predictably cross one or more national jurisdictional boundaries' (CMS: Art. I). Relatively liberal interpretations of the terms 'cyclically' and 'predictably' in the 1980s and 1990s have meant that CMS has become a much broader tool, not only targeting biologically migratory species, but also many other transboundary animal populations. The new Strategic Plan 2015–2023 of CMS is likely to reflect the full breadth and impact of the Convention once adopted (UNEP 2013a).

After the final negotiations in Bonn (Germany) in 1979 it was agreed that no species listed on Appendix I must be harvested, albeit with similar exceptions as under the International Whaling Commission that species can be taken for scientific, captive breeding, traditional use and other exceptional reasons (cf. CMS: Art. III). Appendix I also calls for habitat conservation and threat reduction for listed species, although the language is less strong compared to the clear prohibition of any harvesting. Unlike under CITES, Appendix II is very different to Appendix I. The establishment of new agreements is mandated for those species listed on CMS Appendix II, which has essentially turned CMS into an umbrella convention with seven legally binding agreements and another 19 Memoranda of Understanding (such as the one for Saiga antelopes) attached today. This group of instruments is known as the 'CMS Family' and has spread across the globe with different Secretariats for a number of instruments. To some extent this large number of smaller treaties makes up the strength and applied nature of CMS. The case of the Saiga MoU illustrates how policy action can be designed specifically for the needs of different populations across five range states. Clear prioritised objectives for specific populations are being agreed at regular meetings with the range states, after technical preparation by scientists, NGOs and other stakeholders. The resultant internationally agreed work programme has assisted to empower people across the steppes and semi-deserts of Central Asia and Russia, as well as in consumer states and elsewhere, to take integrated and coordinated action for saiga conservation.

However, CMS is far from an optimal tool for transboundary wildlife conservation and much work is still required to strengthen this Convention, assuming that this is what the 120 Parties (status 2013) will support. Most fundamentally, the Convention requires an enforcement and sustainable financial mechanism to implement its vast mandate (for an overall assessment of the strengths and weaknesses of the Convention cf. UNEP 2013b). Many of the decisions under CMS are excellent from a conservation perspective, but without sticks and carrots they may not be implemented. This is an inherent problem that applies to environmental governance in general, including all of the biodiversity-related treaties.

In order to counterbalance this natural risk it is vital to create incentives for all stakeholders, including communities, to use existing international law for the conservation action needed right outside their own front door. Stronger emphasis should therefore be placed on the design of suitable incentives to implement existing policy rather than creating new policy. A shift, maybe even a revolution, is needed in international policy to make the design of incentives for implementation the number one priority. It is not sufficient to find the scientifically optimal solution. Just as much attention needs to be given to the implementation process and its long-term management. Otherwise policy may never reach the ground. Furthermore, there is much to be said for governments 'risking' to give local people control and ownership over the biodiversity where they live. The well-known Campfire project in Zimbabwe is inspiring in the sense that while it showed that in some communities this experiment can go wrong, the vast majority of communities sustainably used the biodiversity on their land (e.g. Taylor 2009). Modern communication tools have opened up novel avenues for connecting communities and maybe it is worth experimenting with more self-regulated management, even for long-distance migratory species. Participatory monitoring of migratory species has already illustrated the effectiveness of parts of such an experiment (Whitebread 2008; Howe, Medzhidov and Milner-Gulland 2011). The feasibility of such self-application is, however, likely to be highly dependent on socio-economic, political, legal and other factors. Thus in the medium term the locations for experimental Community-Based Natural Resource Management (CBNRM) for migratory species will have to be carefully selected. Given the mandate, biodiversity-related treaties could no doubt facilitate such initiatives.

Conclusion

Transboundary migratory species are owned by everyone and nobody. These jetsetters of the animal kingdom are part of the global commons, but ultimately do not belong to any one government or community. It can be remarkably challenging for a group of countries, sometimes numbering more than 100 nations in the case of certain marine species, to jointly manage this important component of global biodiversity. International law can facilitate such coordination and is indeed essential in the High Seas, which are shared by the international community. While the majority of migratory species find themselves at this extreme end of the spectrum where international regulation is vital, it is important to recognise that often the effectiveness of conservation action can be strengthened if stakeholders other than governments are given direct access and ownership. Emphasis of policy makers needs to urgently shift from drafting instruments for governments to developing and trialling incentives for different levels of stakeholders, specifically communities.

Note

1 The views expressed in this chapter are those of the author and do not necessarily reflect the views of the United Nations Environment Programme or the Convention on Migratory Species.

References

Alkemade, R., van Oorschot, M., Miles, L., Nellemann, C., Bakkenes, M. and ten Brink, B. (2009) 'GLOBIO3: a framework to investigate options for reducing global terrestrial biodiversity loss', *Ecosystems* 12: 374–90.

Aristotle (1991) *History of Animals: Books VII–X*, trans. D.M. Balme, Cambridge, MA: Harvard University Press.

Baldus, R.D., Damm, G.R. and Wollscheid, K.-U. (2008) *Best Practices in Sustainable Hunting – A Guide to Best Practices From Around the World*, Budapest: CIC – International Council for Game and Wildlife Conservation. Online. Available HTTP: <ftp://ftp.fao.org/docrep/fao/010/aj114e/aj114e.pdf> (accessed 13 June 2013).

Bekenov, A.B., Grachev, I.A. and Milner-Gulland, E.J. (1998) 'The ecology and management of the Saiga antelope in Kazakhstan', *Mammal Review* 28: 1–52.

Bolger, D.T., Newmark, W.P., Morrsion, T.A. and Doak, D.F. (2008) 'The need for integrative approaches to understand and conserve migratory ungulates', *Ecology Letters* 11: 63–77.

Brown, M. (2010) 'Exploring a sustainable trophy hunting model for saiga antelope of the Bepak-dala population of Kazakhstan', MSc thesis, Imperial College London.

Buho, H., Jiang, Z., Liu, C., Yoshida, T., Mahamut, H., Kaneko, M., Asakawa, M., Motokawa, M., Kaji, K., Wu, X., Otaishi, N., Ganzorig, S. and Masuda, R. (2011) 'Preliminary study on migration pattern of the Tibetan antelope (*Pantholops hodgsonii*) based on satellite tracking', *Advances in Space Research* 48 (1): 43–8.

Bull, J.W. and Esipov, A. (2013) 'Ancient techniques for hunting saigas in Ustyurt: the remains of arrans', *Saiga News* 16: 18f.

Campbell, L.M. (2002) 'Science and sustainable use: views of marine turtle experts', *Ecological Applications* 12: 1229–46.

Chambers, B. (2008) *Interlinkages and the Effectiveness of Multilateral Environmental Agreements*, Tokyo: United Nations University Press. Online. Available HTTP: <http://archive.unu.edu/unupress/sample-chapters/1149-InterlinkagesAndEffectivenessOfMultilateralEnvironmentalAgreements.pdf> (accessed 13 June 2013).

CMS – United Nations Environment Programme (UNEP) (1979) *Convention on the Conservation of Migratory Species of Wild Animals. Bonn, 23 June 1979, in force 1 November 1983, 1651 United Nations Treaty Series 28395*. Online. Available HTTP: <http://www.cms.int/documents/convtxt/cms_convtxt_english.pdf> (accessed 13 June 2013).

Egevang, C., Stenhouse, I.J., Phillips, R.A., Petersen, A., Fox, J.W. and Silk, J.R.D. (2010) 'Tracking of Arctic terns *Sterna paradisaea* reveals longest animal migration', *Proceedings of the National Academy of Sciences (USA)* 107: 2078–81.

Fricke, H.C., Hencecroth, J. and Hoerner, M.E. (2011) 'Lowland-upland migration of sauropod dinosaurs during the Late Jurassic epoch', *Nature* 480: 513–5.

Hardin, G. (1968) 'The tragedy of the commons', *Science* 162: 1243–7.

Hoare, B. (2009) *Animal Migration: Remarkable Journeys in the Wild*, Berkeley, CA: University of California Press.

Hobson, K.A., Anderson, R.C., Soto, D.X. and Wassenaar, L.I. (2012) 'Isotopic evidence that dragonflies (*Pantala flavescens*) migrating through the Maldives come from the Northern Indian Subcontinent', *PLoS ONE* 7 (12). Online. Available HTTP: <http://www.plosone.org/article/info:doi/10.1371/journal.pone.0052594> (accessed 13 June 2013).

Horton, T.W., Holdaway, R.N., Zerbini, A.N., Hauser, N., Garrigue, C., Andriolo, A. and Clapham, P.J. (2011) 'Straight as an arrow: humpback whales swim constant course tracks during long-distance migration', *Biology Letters*. Online. Available HTTP:

<http://rsbl.royalsocietypublishing.org/content/early/2011/04/14/rsbl.2011.0279.full> (accessed 13 June 2013).

Howe, C., Medzhidov, R. and Milner-Gulland, E.J. (2011) 'Evaluating the relative effectiveness of alternative conservation interventions in influencing stated behavioural intentions: a case study of the saiga antelope in Kalmykia', *Environmental Conservation* 38: 37–44.

Johnson, S. (2012) *UNEP – The First 40 Years. A Narrative*, Nairobi: United Nations Environment Programme.

Johnson, S. and Vagg, R. (2010) *Survival: Saving Endangered Migratory Species*, Northampton, Mass.: Interlink Pub Group.

Kühl, A. and Mrema, E. (2011) 'Impacts of climate change on biodiversity, with a focus on migratory species' in E. Couzens and T. Honkonen (eds) *International Environmental Law-making and Diplomacy Review 2010, University of Eastern Finland – UNEP Course Series 10*, Joensuu: University of Eastern Finland, 176–85. Online. Available HTTP: <http://www.uef.fi/documents/1508025/1508072/Review+2010+final+%282%29.pdf/c8c35cc7-e404-4867-8805-ea5de7bf68c9> (accessed 13 June 2013).

Kühl, A., Balinova, N., Bykova, E., Arylov, I.N., Esipov, A., Lushchekina, A.A. and Milner-Gulland, E.J. (2009) 'The role of saiga poaching in rural communities: linkages between attitudes, socio-economic circumstances and behaviour', *Biological Conservation* 142: 1442–9.

Kurvits, T., Nellemann, C., Alfthan, B., Kühl, A., Prokosch, P., Virtue, M., Skaalvik, J.F. (eds) (2011) *Living Planet: Connected Planet – Preventing the End of the World's Wildlife Migrations through Ecological Networks. A Rapid Response Assessment*, United Nations Environment Programme/GRID-Arendal. Online. Available HTTP: <http://www.grida.no/publications/rr/living-planet/> (accessed 13 June 2013).

Lausche, B. (2008) *Weaving a Web of Environmental Law*, Berlin: Schmidt Erich Verlag.

Lister, A.M. (1992) 'Mammalian fossils and quaternary biostratigraphy', *Quaternary Science Reviews* 11: 329–44.

Magnus, O. (1998) *Description of the Northern Peoples*, vol. 3, trans. P. Fisher and H. Higgens, London: Hakluyt Society.

Milner-Gulland, E.J., Kholodova, M.V., Bekenov, A., Bukreeva, O.M., Grachev, I.A., Amgalan, L. and Lushchekina, A.A. (2001) 'Dramatic declines in saiga antelope populations', *Oryx* 35: 340–5.

Milner-Gulland, E.J., Bukreevea, O.M., Coulson, T., Lushchekina, A.A., Kholodova, M.V., Bekenov, A.B. and Grachev, I.A. (2003) 'Conservation – reproductive collapse in saiga antelope harems', *Nature* 422: 135.

Najam, A., Papa, M. and Taiyab, N. (2006) *Global Environmental Governance: A Reform Agenda*, Winnipeg: International Institute for Sustainable Development (IISD). Online. Available HTTP: <http://www.iisd.org/pdf/2006/geg.pdf> (accessed 13 June 2013).

Ostrom, E., Burger, J., Field, C.B., Norgaard, R.B., Policansky, D. (1999) 'Revisiting the commons: local lessons, global challenges', *Science* 284: 278–82.

Pereira, H.M., Leadley, P.W., Proença, V., Alkemade, R., Scharlemann, J.P.W., Fernandez-Manjarrés, J.F., Araújo M.B., Balvanera, P., Biggs, R., Cheung, W.W.L., Chini, L., Cooper, H.D., Gilman, E.L., Guénette, S., Hurtt, G.C., Huntington, H.P., Mace, G.M., Oberdorff, T., Revenga, C., Rodrigues, P., Scholes, R.J., Sumaila, U.R. and Walpole, M. (2010) 'Scenarios for global biodiversity in the 21st century', Science 330: 1496–501.

Pulido, F. and Berthold, P. (2010) 'Current selection for lower migratory activity will drive the evolution of residency in a migratory bird population', *Proceedings of the National Academy of Sciences (USA)* 107: 7341–6.

Sekercioglu, C.H. (2007) 'Conservation ecology: area trumps mobility in fragment bird extinctions', *Current Biology* 17: 283–6.

Shaffer, S.A., Tremblay, Y., Weimerskirch, H., Scott, D., Thompson, D.R., Sagar, P.M., Moller, H., Taylor, G.A., Foley, D.G., Block, B.A. and Costa, D.P. (2006) 'Migratory shearwaters integrate oceanic resources across the Pacific Ocean in an endless summer', *PNAS* 103: 12799–802.

Sutherland, W.J. and Gill, J.A. (2001) 'The role of behaviour in studying sustainable exploitation', in J.D. Reynolds, G.M. Mace, K.H. Redford and J.G. Robinson (eds) *Conservation of Exploited Species*, Cambridge: Cambridge University Press.

Taylor, R. (2009) 'Community based natural resource management in Zimbabwe: the experience of CAMPFIRE', *Biodiversity and Conservation* 18: 2563–83.

UNEP (United Nations Environment Programme) (2010a), *Medium-term International Work Programme for the Saiga Antelope 2011–2015, UNEP/CMS/SA-2/Report Annex 5.* Online. Available HTTP: <http://www.cms.int/species/saiga/2ndMtg_Mongolia/Mtg_Rpt/Annex_5_MTIWP_2011_2015_E.pdf> (accessed 13 June 2013).

UNEP (United Nations Environment Programme) (2010b) *Memorandum of Understanding on the Conservation of Migratory Sharks under UNEP/CMS.* Available HTTP: <http://www.cms.int/species/sharks/MoU/Migratory_Shark_MoU_Eng.pdf> (accessed 13 June 2013).

UNEP (United Nations Environment Programme) (2011) *Report on Synergies and Partnerships, UNEP/CMS/Conf.10.28.* Online. Available HTTP: <http://www.cms.int/bodies/COP/cop10/docs_and_inf_docs/doc_28_synergies_partnerships_e.pdf> (accessed 13 June 2013).

UNEP (United Nations Environment Programme) (2013a) *Draft Strategic Plan 2015–2023.* Available HTTP: <http://www.cms.int/bodies/StC/strategic_plan_2015_2023_wg/strpln_wg_drafts.html> (accessed 13 June 2013).

UNEP (United Nations Environment Programme) (2013b) *Inter-sessional Working Group on the Future Shape of CMS.* Available HTTP: <http://www.cms.int/bodies/future_shape/future_shape_mainpage.htm > (accessed 13 June 2013).

Whitebread, E. (2008) 'Evaluating the potential for participatory monitoring of Saiga antelope by local villagers in Kalmykia, Russia', MSc thesis, Imperial College London. Online. Available HTTP: <http://www.iccs.org.uk/wp-content/thesis/consci/2008/Whitebread.pdf> (accessed 13 June 2013).

Wilcove, D.S. (2010) *No Way Home: The Decline of the World's Great Animal Migrations*, Washington, D.C.: Island Press.

15 Facing the biodiversity challenge

Plant endemism as an appropriate biodiversity indicator

Ines Bruchmann

Introduction

The loss of biodiversity is probably the most critical global environmental threat alongside climate change. With the Convention on Biological Diversity during the World Summit in 1992, the United Nations accepted the challenge to prevent species loss, and declared the ambitious 2010 target which reads as 'to achieve by 2010 a significant reduction in the current rate of biodiversity loss'. The importance of the target was underlined by its incorporation in the Millennium Development Goals (MDG 7(2)) and by the proclamation of an International Year of Biodiversity in 2010 (United Nations General Assembly 2000).

In 2010, however, it was recognised that this target had not been met. Many extinction events show that the general decline of biodiversity continues (Secretary of the Convention on Biological Diversity 2008, 2010). Although the decline of a few threatened target species has been prevented and despite mounting conservation efforts, more and more species are being classified as 'endangered' or even 'extinct', as the general pressure on ecosystems continues to accelerate (e.g. habitat fragmentation and destruction, land-use intensification, pollution; cf. European Environment Agency 2010; Secretary of the Convention on Biological Diversity 2010; IUCN 2011). The vision of halting the loss of biodiversity features had to be extended to 2020 (European Commission 2011).

In view of this failure of politics, we now need to re-consider the issue of how to reach this new target. What are the major values, the leading ideas and the central aspects of taking conservation action: biological and ecological data, the needs of civil society, economic dictates and trends, or normative, ethical reflection (Lanzerath et al. 2008)?

In order to stop the global loss of biodiversity it is imperative to establish maximally effective and at the same time cost-efficient actions which will protect and conserve the different features of biodiversity. This has to be done at different spatial scales (local to global) and in different time frames (sustainability). Should conservation start at the species or at the ecosystems level?

What regions or species should be prioritised in the short-, mid-, or long-term? Should conservation action be focused on species of economic value (medicinal plants, crop species), on species that have cultural or aesthetic values (landscape-forming species, ornamental plants), or is there a greater need to save those species and ecosystems that provide regulating services, such as regulation of air quality by trees and woods? How can species that do not as yet appear to provide any anthropocentric value but that are rare or threatened with extinction be protected?

The plant kingdom – as the primary producer and a major ecosystem component – forms the basis on which the rest of the world's biological diversity depends, and it is for this reason that the conservation of plant diversity should be of vital interest in preventing species loss (UNEP 2002; Reid et al. 2005). The focus of the present study is on plants – and more precisely on so-called endemic plants. It is assumed that, due to their limited range of occurrence, endemic plants are generally more in danger of extinction than widespread species. Because of this assumed vulnerability, endemic species are treated today as an important indicator of the 'quality' or the 'uniqueness' of biodiversity in a given area, and have often been used to delineate regions of rich biodiversity (e.g. the well-established concept of Biodiversity Hotspots).

The present chapter is written from an ecological perspective regarding the conservation of biodiversity and is mainly based on a biocentric approach valuing features of biodiversity. On the basis of a comprehensive investigation of Europe's endemic plants (Bruchmann 2011), the applicability of plant endemism as an indicator of biodiversity at regional scales was evaluated. I aim to give a general overview of the phenomenon of endemism, the evolution of terms and concepts, and trace the career of the term in conservation and politics. Some of the biases and problems involved in utilising the concept as an (in-situ) conservation tool are shown, as are the striking qualities of endemic plants as effective indicators identifying regions with high biodiversity values.

The term 'endemism'

Despite the apparently great importance of endemic species and despite the frequent use of the term 'endemism', its definition remains diffuse. The term 'endemic' comes from the Greek *endemos* which means 'native to a place'. Today, two different scientific disciplines use and define this term. In the medical context, 'endemic' denotes a disease that is typically found among the inhabitants of a particular region and is prevalent only in this area (i.e. malaria diseases in tropical regions; Haubrich 2003). The concept of endemism in biogeography dates back to the Swiss botanist De Candolle in 1820 who described botanical phenomena in his 'Essai élémentaire de géographie botanique' (De Candolle 1820). In the biogeographical sense the term endemic refers to any taxonomic entity (species, genus, family), the occurrence of which is restricted entirely to a defined area. It should be noted that the concept of endemism, as it is presently conceived, is strongly scale dependent. The reference area of an endemic taxon can vary greatly in size. It might be confined by either 1) natural or 2) artificial boundaries. 1) Natural boundaries can be geomorphological structures such as mountain ranges or islands but can also be given by edaphic conditions, e.g. serpentine soils which are characterised at the same time by extremely low nutrient content and high heavy metal concentrations (Werger, Wild and Drummond 1978). 2) Artificial boundaries are most often national or administrative boundaries. Even though artificial boundaries seldom follow naturally given delineations such as biogeographical zones, it seems reasonable to define endemism in terms of artificial boundaries when evaluating national efforts towards species conservation (Heywood 1996; Bruchmann 2011).

Over the last few decades the number of definitions, concepts, theories and hypotheses on the phenomenon of endemism has increased continuously, focusing on different dimensions of endemism. Today there are a large number of terms associated with the idea of endemism, e.g. local endemics, microendemics, subendemics, near endemics, palaeoendemics, neoendemics, and many others (Kruckeberg and Rabinowitz 1985; Heywood 1996). Despite the very concrete nature of some of these definitions there is still no consistent usage of the term endemism. McDonald and Cowling (1995) noted that the interpretation of endemism data strongly depends on the mode in which the assessed data is quantified and reported. Current reviews of data on worldwide reported numbers of regional plant endemism suggest that scientists apply quite different categories of endemism in their studies (Hobohm 2013, especially chapters 3–5). This is not only true for the application of the concept at very different geographical scales (continental, regional, local scales) but also for the different stringency in usage of the terms (e.g. taxonomic interpretation and taxonomic rank of the endemic entities). Thus, endemism is quantified in a number of very different evaluation modes, hence making the data partly incommensurable, e.g. in comparative studies.

The ambiguities and problems involved in defining endemic taxa according to political or natural divisions at different scales will be shown by looking at the Alpine rock-jasmine (*Androsace alpina*, *Primulaceae*), which is a showcase for many other taxa. This plant species inhabits high altitude rock and scree habitats (2,000–2,400 m a.s.l.) in France, Switzerland, Austria and Italy. In none of these countries is this plant listed as an endemic species because its natural range is not confined to any of the national territories. Thus, none of the countries has special responsibility for the protection of this plant. However, if the reference area is changed and the Alps as a natural site or ecoregion is the focus of consideration it becomes obvious that *A. alpina* is an endemic of the Alps. The Alpine rock-jasmine is most likely also a species that is endemic to the habitat type 'rocks and scree' – a naturally confined division. If we enlarge the scale again to the dimension of the continent, Europe, *A. alpina* becomes endemic (European endemic plant). However, if classification is based on the political division European Union (EU) then *A. alpina* loses its status as an endemic plant because Switzerland is not a member of the EU.[1]

Endemic plants as indicator species and conservation efforts

The idea of seeking out regions with high numbers of plant species has quite a long history and goes back to the beginnings of biogeography (e.g. von Humboldt 1805; Grisebach 1866; Drude 1890). To identify regions of high biodiversity the pure quantitative counting of species occurring in a distinctive area has over the years been progressively combined with qualitative aspects, e.g. with aspects of the uniqueness of a habitat (e.g. the Wadden Sea) or the incidence of rare species (Heywood 1996; Hobohm 2000; also Lanzerath et al. 2008). Recently, different concepts of nature conservation (e.g. the Centres of Plant Diversity or the Global 200: Priority ecoregions for global conservation) have used the number of endemic species as an indicator for quantifying the diversity and the uniqueness of a certain region (Davis, Heywood and Hamilton 1994, 1995; Davis et al. 1997; Olson and Dinerstein 2002). The concept of *Biodiversity Hotspots*, in particular, attracted much media attention as this claimed that protection of the 34 identified Hotspot regions, which cover just a small percentage of the Earth's surface, would ensure the survival of about 75% of the planet's most threatened mammals, birds and amphibians (Mittermeier et al. 2005). To identify these areas of high biodiversity value the Biodiversity Hotspot concept utilises two strict criteria: 1) an indicator for endemism and 2) an indicator for threat of habitats.[2]

However, this combining of different qualitative indices at different scales to measure biodiversity values remains controversial. Orme et al. (2005) studied Hotspot areas for birds using the three indices 1) species richness, 2) endemism or 3) threats, and found that there is only little geographical congruence across the three types of Hotspot areas. This knowledge has explicit implications for the use of the Hotspot concept as a conservation tool. As the different Hotspot or diversity indices do not work as a surrogate, the question arises which of these diversity indices should be used to guide conservation policy. It was shown that endemism was the most suitable indicator for identifying Hotspot regions (also capturing a substantial proportion of both threatened and overall species richness; Orme et al. 2005).

In fact, endemism is often used as a powerful political argument, as endemics are said to be more in danger of extinction than widespread species, due to their limited distribution range (Fontaine et al. 2007; European Communities 2008).

Global strategies such as the Convention on Biological Diversity (CBD) underline the prioritisation of biotic-rich regions (e.g. megadiverse regions) but also emphasise the particular importance of endemic species in the course of in-situ conservation. This is concretised in strategies at (trans-)regional and local scales. The European Plant Conservation Strategy, for example, which adapts the targets of the Global Strategy for Plant Conservation (GSPC; UNEP 2002) to the European level, emphasises that all national endemic plants should be included in the European Red List (Planta Europa 2002: Objective 1.02). Almost all national strategies on biological diversity mention the great importance of endemic species and utilise endemism figures in their national reports.[3] The National Strategy on

Biological Diversity of the Federal Republic of Germany, for example, states that 'Germany has a particular responsibility to conserve ... species which are endemic in Germany or Central Europe', i.e. which only occur within the national and administrative boundaries of the Republic or in Middle Europe (Federal Ministry for the Environment 2007).

However, if a species is endemic this does not necessarily also mean it is rare or in danger of extinction.[4] Some endemic taxa show very wide geographical distribution ranges (e.g. *Narthecium ossifragum* – endemic to large areas of Europe with an Atlantic-influenced climate), while others have extremely narrow ranges of occurrence (e.g. *Oenanthe conioides* – endemic to a small area of about 100 km^2 close to Hamburg, Germany). Fontaine et al. (2007) stressed that local endemic species 'are by far the most at risk of extinction' but also underlined the aspect of demographic rarity, which means that small isolated populations that are distributed over a large geographic range are endangered because of local extinction events. Extreme habitat specialist species may be endangered as well because they are not able to buffer habitat changes or to adapt within adequate time periods.

The plant *Atractylis preauxiana* (Asteraceae), for example, which is found exclusively in stony habitats at the southeastern coastal fringes of the islands of Tenerife and Gran Canaria, is an extremely local endemic species and is also extremely endangered. It has a very narrow distribution range, small local population sizes, strict ecological requirements, and seems to be unable to shift from its original habitats. Almost all subpopulations of *A. preauxiana* are now declining in size and some have already become extinct (Caujapé-Castells et al. 2008) because of strong human pressure on the remaining habitat fragments.

The critical assessment of endemics may, thus, also help to evaluate the efficiency of implemented conservation action (in-situ) and give a good indication of the conservation status of the respective habitats (favourable/unfavourable conservation status). In the context of in-situ conservation it is not only important to ask where the endemic species live but – and this should be even more important – which habitat provides the necessary prerequisites for the species' survival. In order to effectively pursue in-situ conservation (CBD: Art. 8), data on endemism related to natural and ecological (e.g. habitats) divisions is needed. As patterns of endemism do not generally conform to political territories, it is evident that the currently available data is not what is needed to take conservation action.

Interestingly, to date, there has been no overall assessment of Europe's endemic inventory either on a spatial level or showing the ecological distribution of endemic populations (Bilz et al. 2011; Bruchmann and Hobohm 2011).

Material and methods

For the purpose of evaluating the endemic flora in Europe, a comprehensive database EvaplantE (Endemic vascular plants in Europe) was established which currently comprises about 6,200 endemic plant taxa (Hobohm 2008; Bruchmann 2011). The study area covered by EvaplantE (Figure 15.1) comprises Europe as defined in Fontaine et al. (2007).

To obtain the data on Europe's endemic taxa, their distribution and ecological affinities or habitat preferences, a large number of European floras and floristic monographs were evaluated. Being well aware of the difficulties and biases that result either from the different habitat terminologies in the various European languages or from different national regulations and standards of classification, the decision was made to use eight habitat categories that correspond well with those defined by the European Habitats Directive (European Commission 2007). As far as possible, the endemic taxa were assigned to the predefined habitat categories: rocky habitats and screes; grassland ecosystems; scrubs and heaths; forests; coastal and saline habitats; urban and other man-made habitats; inland water bodies; and mires (including bogs, fens, swamps). The endemism data obtained was combined with geographical datasets and visualised in digital maps using GIS applications. Spatial distribution, floristic aspects of endemism and also the ecological affinities of the endemics were analysed by querying the numbers of the most endemic-rich regions, plant families, and habitat types.

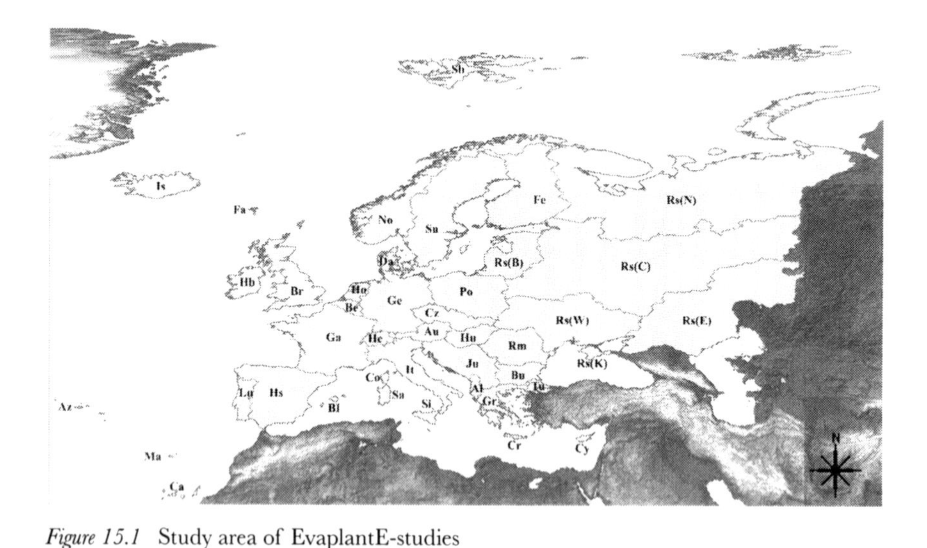

Figure 15.1 Study area of EvaplantE-studies
Abbreviations: Al – Albania; Au – Austria; Az – Azores Archipelago; Be – Belgium with Luxembourg; BI – Balearic Islands; Br – Great Britain; Bu – Bulgaria; Ca – Canary Islands; Co – Corsica; Cr – Crete; Cy – Cyprus; Cz – Czech Republic with Slovakia; Da – Denmark; Fa – Faero; Fe – Finland; Ga – France (mainland); Ge – Germany; Gr – Greece; Hb – Ireland; He – Switzerland; Ho – The Netherlands; Hs – Spain (mainland); Hu – Hungary; Is – Iceland; It – Italy (mainland); Ju – former Yugoslavia; Lu – Portugal (mainland); Ma – Madeira Archipelago; No – Norway; Po – Poland; Rm – Romania; Rs (B) – Russia Baltic division; Rs (C) – Russia central division; Rs (E) – Russia Southeastern division; Rs (K) – Russia Crimean division; Rs (N) – Russia Northern division; Rs (W) – Russia Western division; Sa – Sardinia; Sb – Svalbard; Si – Sicily; Su – Sweden; Tu – European Turkey

Results

The study on Europe's plant endemism revealed some important spatial relationships, such as a general north–south gradient in endemism with extraordinarily high numbers of endemic plants in the southern, Mediterranean regions, and underlines the distinctiveness of isolated islands and mountainous areas (Bruchmann 2011). The results suggest that the geographical patterns of small-scale endemics follow, to an extent, different rules from those of large-scale endemics. Spatial patterns of large-scale endemics are mostly influenced by the impact of the glacial periods in Europe (mediated by the richness of the regional species pools) and the diversity of habitats, while the spatial patterns of small-scale endemics are largely explained by the isolation degree (e.g. islands), the regional species pools (species richness), and the diversity of habitats (Figure 15.2).

Floristic aspects of endemism show that Europe's endemic vascular plants belong to 110 plant families and 719 genera. The generally most species-rich plant families are also richest in endemics.

The evaluation of distribution patterns of European endemics according to habitat categories shows that the large majority of endemics inhabit rocky habitats (2,792), followed by grassland (1,336), shrub and heath habitats (1,150), and forests (733). Lower rates were found for coastal and saline habitats (449), man-made habitats (446), inland waterbodies (275), and finally mires, bogs and fens, which are inhabited by only about 100 endemics (Figure 15.3; Bruchmann 2011).

When habitat specificity is the focus of an analysis, coastal and saline habitats and rocks and scree habitats have the highest proportions of habitat-specific endemics, while shrub- and heathland habitats and bogs, mires and fens have the lowest proportions: about 60% of the assigned endemics are habitat specific and strictly bound to one of the eight habitat categories. Most of the habitat-specific

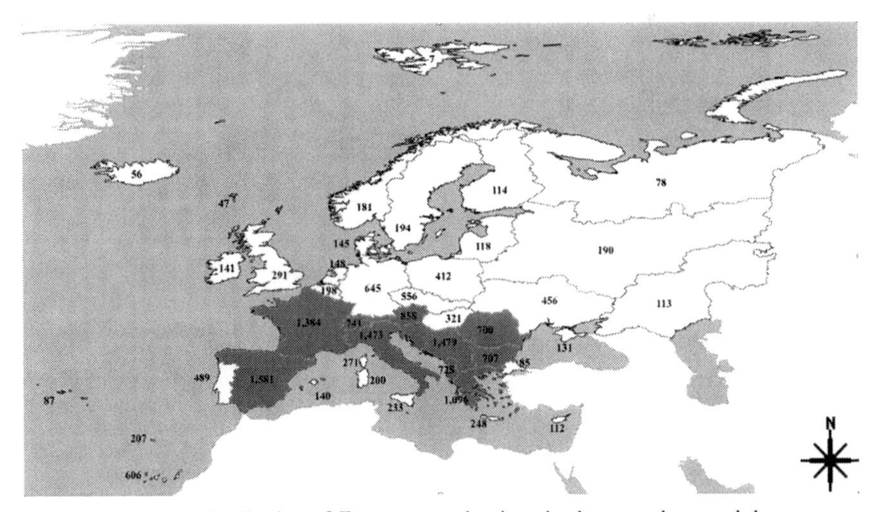

Figure 15.2 Spatial distribution of European endemics: absolute numbers and the ten most endemic-rich regions (dark grey coloured)

endemics of rocky habitats occur in Europe's mountainous regions: the Alps, the Apennines, the Pyrenean mountains, and the Balkan region with Yugoslavia; they also occur in Greece and the volcanic-origin Canary Archipelago. Depending on the habitat category under consideration, some temperate or even northern regions gain importance. The Atlantic islands of Great Britain and Ireland, for example, reach very high scores in endemic plants specific for coastal and saline habitats. High numbers of habitat-specific grassland endemics are found in Central- and Southeastern Europe. Many of the habitat-specific European endemics in the generally endemic-poor habitats of bogs, mires and fens are reported for central and northern parts of Europe (e.g. Germany, Austria, Switzerland, as well as for Britain, Ireland and even for the northern European country Norway).

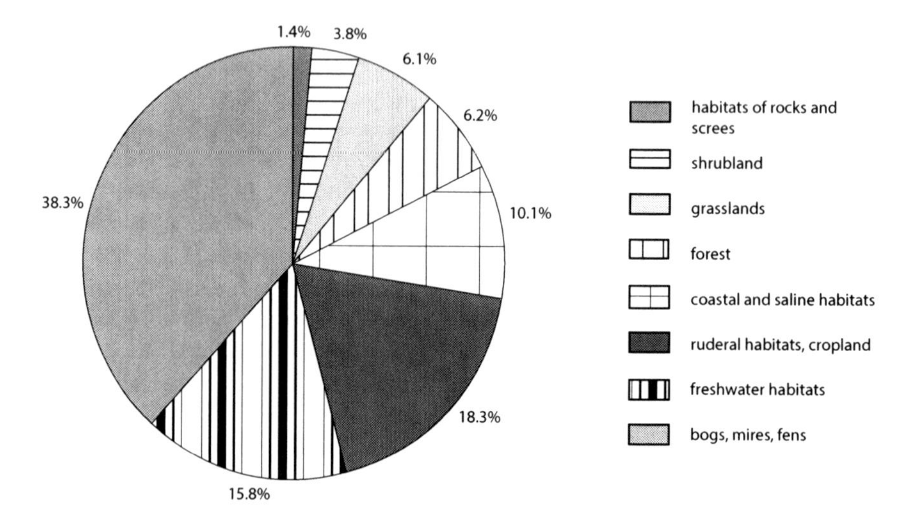

Figure 15.3 Distribution of European endemics per habitat category (percentage values)

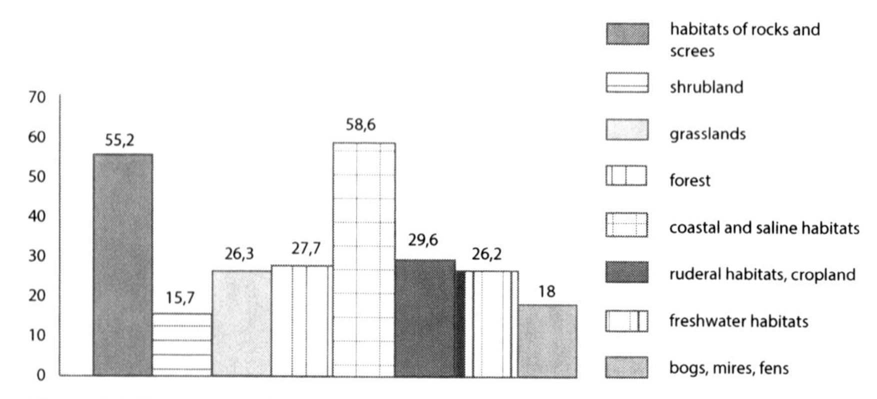

Figure 15.4 Percentage values of habitat-restricted European endemic taxa per habitat category

Discussion

Relating endemic taxa to habitat categories, and thus to ecological features, in order to find habitat dependencies of endemic plants is a relatively new approach recently advanced by Ricketts et al. (1999) for North America, Wikramanayake et al. (2002) for the terrestrial ecoregions of the Indo-Pacific,[5] by Burgess et al. (2004) for Africa and Madagascar, by Mucina and Rutherford (2006; at smaller scales), and by Hobohm and Bruchmann (2009) for the European continent. This approach is certainly vulnerable to changes in taxonomical interpretation, gaps in ecological knowledge and the biases which result from different international classification standards in habitat terminology (cf. also comprehensive discussion in Bruchmann 2011); nevertheless, the present investigation was able to detect missing data (floristic, geographic and ecologic), and reveals unsolved questions in conservation sciences, e.g. the problems resulting from delineating endemism according to political divisions.

Before assessing whether plant endemism could function as an appropriate indicator system for evaluating biodiversity, several biases of the concept and contradictions in the way it is applied have to be discussed.

Endemism in Europe is mostly defined according to artificial boundaries (borders of the national states), and only few data on endemism are related to natural (e.g. mountain ranges or islands) or to ecological (e.g. biomes or habitats) divisions. The study confirms that about 52% of Europe's endemics occur across national borders. Of these, 20% are endemic to two study regions and a further 15% of endemic taxa are confined to three (9%) and four (6%) study regions. Because the EvaplantE still comprises only binary data on the presence or absence of endemics in geographical regions and habitats, it is not possible to determine how many of the cross-border endemics may be vulnerable to extinction. The wider range of occurrence, however, does not necessarily imply that these endemics are abundantly distributed or not vulnerable towards extinction.

For many species the status of rarity or vulnerability is not yet known: only about three quarters of all endemic taxa are assigned to habitats. This lack of data makes it difficult to decide about the potential rarity of the species given by ecological restrictions (e.g. habitat specialists cf. also Kruckeberg and Rabinowitz 1985).

Generally, data availability at population level is even worse. Unfortunately, only some endangered species which already have critical population sizes receive the frequent attention of nature conservation, e.g. monitoring of plants listed in the Red Lists (national or IUCN Red List) or in Annex 1 of the European Habitats Directive (good or bad conservation status; Bundesamt für Naturschutz; Council of the European Communities 1992; Ozinga et al. 2005; European Communities 2008). Thus, it is to be feared that many endemic plants fall through the conservation net simply because their range extends across the border of individual countries and their respective administrative responsibilities, or because the respective species are as yet not so rare that they have found their way into the Red Lists of endangered species.

In the course of in-situ conservation the habitat affinities of endemic plants should be of much greater interest than the pure confinement of plants to political borders. Here, the approach of conservation responsibility as suggested by Ludwig et al. (2007) is promising.

The evaluation of habitat affinities shows that more than half of Europe's endemics (that were assigned to habitats) do have a narrow ecological amplitude and can only be found in one habitat type. Interestingly, the ruderal and human-influenced habitats (including semi-natural grasslands) have equally high or even higher proportions of habitat-specific endemics than Europe's forest ecosystems. The special role of Europe's grassland habitats in endemic and also in general species richness should be highlighted at this point (EDGG 2010; Jones 2010).

When focusing on the habitat specificity of the endemic plants, it became evident that it is not only the endemic-rich Mediterranean and the isolated island regions which have particular responsibility for the protection of the habitats in which endemics live, but also the temperate or even northern regions, such as Britain and Ireland for coastal endemics, or Germany or Norway for the habitats of bogs, mires and fens.

The latter results clearly underline the value of Europe's open cultural landscapes as species-rich, and thus high-value grassland, coastal and wetland habitats. The first assessment of habitats listed in Annex I of the Habitats Directive confirms that 'grassland, wetland and coastal habitats are under the most pressure' because of today's trend towards land abandonment and intensification of land-use (Pignatti 1978, 1983; European Communities 2008; Commission of the European Communities 2009; Bruchmann and Hobohm 2010).

The value of endemism as an indicator of priority regions for nature or, more precisely, habitat conservation is stated throughout all political strategies from global to local scales. The fact that several concepts of nature conservation such as the Biodiversity Hotspots use endemism as a baseline indicator for delineating regions of high biodiversity value appears logical because of the presumption that endemic species suffer higher risk of extinction than widely distributed species.

However, the perception of endemism has to be concretised and the term used with more precision than is the case in the current political debate. If endemism is to gain more significance in conservation as an indicator of biodiversity, the classification of the factual rarity or vulnerability status of endemic plants must be tackled. Otherwise, it is to be feared that the population sizes of more abundant endemic plants will decrease unobserved due to habitat changes and ongoing trends in intensification of land-use and that this will occur without attracting conservationists' attention. The categorisation of endemics according to their rarity and vulnerability status should consider their geographical range, habitat specificity and abundance.

Considerable work remains to be done in evaluating endemics' factual distribution ranges and in bringing together data on habitats preferences and monitoring data on population trends. A better knowledge of Europe's endemic plants will help to span a systematic and tight net of conservation to contain the loss of Europe's biodiversity.

Summary

Conservation policy must face up to this challenge very soon if the loss of biodiversity in Europe is to be seriously taken in hand. Plant endemism can function as an appropriate indicator of biodiversity. To gain significance in conservation reality, endemism should be confined according to natural boundaries. Conservation action must be taken across borders transcending administrative boundaries. Considerable work remains to be done in collecting missing data so as to be able to classify endemic plants according to their rarity.

Notes

1 Fortunately, *A. alpina* is protected in Switzerland by legal ordinance (Appendix 2 of the NHV; cf. also Moser et al. 2002).
2 To qualify as a Hotspot, a region must meet two strict criteria: it must contain at least 1,500 species of endemic vascular plants, and it has to have lost at least 70% of its original habitat (Myers et al. 2000).
3 <www.cbd.int/reports> (accessed 13 June 2013).
4 Comprehensive summary on aspects of rarity in Kruckeberg & Rabinowitz (1985).
5 However, a criticism of the works of Wikramanayake et al. and Burgess et al. is that they are based upon bad quality data, as most endemism data was roughly measured in only four endemic-richness categories (coarse interval scale!).

References

Bilz, M., Kell, S.P., Maxted, N. and Lansdown, R.V. (2011) *European Red List of vascular plants*, Luxembourg: Publication office of the European Union.

Bruchmann, I. (2011) 'Plant endemism in Europe: Spatial distribution and habitat affinities of endemic vascular plants', thesis, University of Flensburg.

Bruchmann, I. and Hobohm, C. (2010) 'Halting the loss of biodiversity: Endemic vascular plants in grasslands of Europe', *Grassland Science in Europe* 15: 776–8.

Bruchmann, I. and Hobohm, C. (2011) 'Über Grenzen hinweg: Schutz endemischer Gefäßpflanzen in Europa?', *Treffpunkt Biologische Vielfalt* 10: 119–25.

Bundesamt für Naturschutz (BfN) *FloraWeb – Daten und Informationen zu Wildpflanzen und zur Vegetation Deutschlands*. Online. Available HTTP: <http://www.floraweb.de> (accessed 13 June 2013).

Burgess, N., D'amico Hales, J., Underwood, E., Dinerstein, E., Olson, D., Itoua, I., Schipper, J., Ricketts, T. and Newmann, K. (2004) *Terrestrial ecoregions of Africa and Madagascar*, Washington, D.C.: Island Press.

Caujapé-Castells, J., Naranjo-Suarez, J., Santana, I., Baccarani-Rosas, M., Cabrera-Garcia, N., Marrero, M., Carqué, E. and Mesa, R. (2008) 'Population genetic suggestions to offset the extinction ratchet in the endangered Canarian endemic *Atractylis preauxiana (Asteraceae)*', *Plant Systematics and Evolution* 273: 191–9.

Commission of the European Communities (2009) *Report from the commission of the Council and the European Parliament: Composite report on the conservation status of habitat types and species as required under article 17 of the habitats directive*, Brussels: Commission of the European Communities.

Council of the European Communities (1992) *Directive 92/43/EEC on the conservation of natural habitats and of wild fauna and flora*.

Davis, S.D., Heywood, V.H. and Hamilton, A.C. (1994) *Centres of plant diversity Vol. 1: Europe, Africa, South West Asia and the Middle East*, Cambridge: IUCN Publications.

Davis, S.D., Heywood, V.H. and Hamilton, A.C. (1995) *Centres of plant diversity Vol. 2: Asia, Australasia and the Pacific*, Cambridge: IUCN Publications.

Davis, S.D., Heywood, V.H., Herrera-Macbryde, O., Villa-Lobos and Hamilton, A.C. (1997) *Centres of plant diversity Vol. 3: The Americas*, Cambridge: IUCN Publications.

De Candolle, A.B. (1820) *Essai élémentaire de geographie botanique. Dictionnaire des sciences naturelles*, Strasbourg: Flevrault.

Drude, O. (1890) *Handbuch der pflanzengeographie*, Stuttgart: Engelmann.

EDGG (European Dry Grassland Group) (2010) *Smolenice grassland declaration. Bulletin of the European Dry Grassland Group*, 7, 7.

EEA (European Environment Agency) (2010) *The European environment: State and outlook 2010 – biodiversity*, Copenhagen.

European Commission (2007) *Interpretation manual of European Union habitats – eur 27. In, p. 144*, Brussels.

European Commission (2011) *Communication from the Commission to the European Parliament, the economic and social committee and the committee of the regions: Our life insurance, our natural capital: An EU biodiversity strategy to 2020*, Brussels.

European Communities (2008) *LIFE and endangered plants: Conserving Europe's threatened flora. Office for Official Publications of the European Communities*, Luxembourg.

Federal Ministry for the Environment (2007) *National strategy on biological diversity*, Paderborn: Bonifatius GmbH.

Fontaine, B.E. et al. (2007) 'The European Union's 2010 target: Putting rare species in focus', *Biological Conservation* 139: 167–85.

Grisebach, A. (1866) 'Die Vegetationsgebiete der Erde, übersichtlich zusammengestellt', *Petermann's Mitteilungen* 12: 45–53.

Haubrich, W.S. (2003) *Medical meanings – a glossary of word origins*, 2nd edn, American College of Physicians, US.

Heywood, V.H. (1996) *Global biodiversity assessment*, Cambridge: United Nations Environment Programme.

Hobohm, C. (2000) *Biodiversität*, Wiebelsheim: Quelle and Meyer.

Hobohm, C. (2008) 'Endemische Gefäßpflanzen in Europa', *Tuexenia* 28: 10–22.

Hobohm, C. (2013) (ed.) *Endemism in vascular plants*, Dordrecht: Springer.

Hobohm, C. and Bruchmann, I. (2009) 'Endemic vascular plants in European grasslands', in M. Partzsch and U. Jandt (eds) *6th Meeting of the European Dry Grassland Working Group (EDGG) – Dry grasslands – species interaction and distribution*, Halle: Martin-Luther-Universität Halle, 44.

IUCN (2011) *IUCN Red List of Threatened Species. Version 2011.1.* Online. Available HTTP: <http://www.iucnredlist.org> (accessed 16 June 2011).

Jones, G. (2010) 'Endemics in european grasslands', *La Cañada* 25: 4. Online. Available HTTP: <http://www.efncp.org/download/la-canada25.pdf> (accessed 25 November 2013).

Kruckeberg, A.R. and Rabinowitz, D. (1985) 'Biological aspects of endemism in higher plants', *Annual Review of Ecology and Systematics* 16: 447–9.

Lanzerath, D., Mutke, J., Barthlott, W., Baumgärtner, S., Becker, C. and Spranger, T.M. (2008) *Biodiversität*, Freiburg: Alber.

Ludwig, G., May, R. and Otto C. (2007) *Verantwortlichkeit Deutschlands für die weltweite Erhaltung der Farn- und Blütenpflanzen – vorläufige Liste. BfN-Skripten 220*, Bonn: Bundesamt für Naturschutz.

Mcdonald, D.J. and Cowling, R.M. (1995) 'Towards a profile of an endemic mountain fynbos flora: Implications for conservation', *Biological Conservation* 72: 1–12.

Mittermeier, R.A., Gil, P.R., Hoffman, M., Pilgrim, J., Brooks, T., Mittermeier, C.G., Lamoreux, J. and da Fonseca, G.A.B. (2005) *Hotspots revisited: Earth's biologically richest and most endangered terrestrial ecoregions*, Mexico City: Cemex.

Moser, D.M., Gygax, A., Bäumler, B., Wyler, N. and Palese, R. (2002*) Rote Liste der gefährdeten Farn- und Blüten-Pflanzen der Schweiz*, Bern: BUWAL.

Mucina, L. and Rutherford, M.C. (2006) *The vegetation of South Africa, Lesotho and Swaziland*, Pretoria South African National Biodiversity: Institute.

Myers, N., Mittermeier, R.A., Mittermeier, C.G., da Fonseca, G.A.B. and Kent, J. (2000) 'Biodiversity hotspots for conservation priorities', *Nature* 403: 853–8.

Olson, D.M. and Dinerstein, E. (2002) 'The global 200: Priority ecoregions for global conservation', *Annals of the Missouri Botanical Garden* 89: 199–224.

Orme, C.D.L., Davies, R.G., Burgess, M., Eigenbrod, F., Pickup, N., Olson, V.A., Webster, A.J., Ding, T.-S., Rasmussen, P.C., Ridgely, R.S., Stattersfield, A.J., Bennett, P.M., Blackburn, T.M., Gaston, K.J. and Owens, I.P.F. (2005) 'Global hotspots of species richness are not congruent with endemism or threat', *Nature* 436: 1016–9.

Ozinga, W.A., Schaminée, J.H.J., Heer, M., De Hennekens, S.M., van Opstal, A.J.F.M., Sierdsma, H., Smits, N.A.C., Stumpel, A.H.P. and van Swaay, C. (2005) *Target species – species of European concern: A database driven selection of plant and animal species for the implementation of the pan-European ecological network*, Wageningen: Alterra.

Pignatti, S. (1978) 'Evolutionary trends in Mediterranean flora and vegetation', *Vegetatio* 37: 175–85.

Pignatti, S. (1983) 'Human impact in the vegetation of the Mediterranean basin', in W. Holzner, M.J.A. Werger and I. Ikusima (eds) *Man's impact on vegetation*, The Hague: Dr. W. Junk Publishers, 151–61.

Planta Europa (2002) *European plant conservation strategy: Saving the plants of Europe*, London.

Reid, W.V., Mooney, H.A., Cropper, A., Capistrano, D., Carpenter, S.R., Chopra, K., Dasgupta, P., Dietz, T., Duraiappah, A.K., Hassan, R., Kasperson, R., Leemans, R., May, R.M., Mcmichael, T., Pingali, P., Samper, C.N., Scholes, R., Watson, R.T., Zakri, A.H., Shidong, Z., Ash, N.J., Bennett, E., Kumar, P., Lee, M.J., Raudsepp-Hearne, C., Simons, H., Thonell, J. and Zurek, M.B. (2005) *Millennium ecosystem assessment – ecosystems and human well-being: Synthesis*, Washington, D.C.: Island Press.

Ricketts, T.H., Dinerstein, E., Olson, D.M., Louks, C.J., Eichbaum, W., Dellasala, D., Kavanagh, K., Hedao, P., Hurley, P.T., Carney, K.M., Abell, R. and Walters, S. (1999) *Terrestrial ecoregions of North America: A conservation assessment*, Washington D.C.: Island Press.

Secretary of the Convention on Biological Diversity (2008) *Plant conservation report: A review of progress in implementing the global strategy of plant conservation*, Montreal.

Secretary of the Convention on Biological Diversity (2010) *Global biodiversity outlook 3*, Montreal.

United Nations Environment Programme (UNEP) (2002) *Global strategy for plant conservation*, Montreal.

United Nations General Assembly (2000) *United Nations millennium declaration*, New York.

von Humboldt, A. (1805) *Essai sur la géographie des plantes; accompagné d'un tableau physique des régions équinoxiales*, Paris: Schoell.

Werger, M.J.A., Wild, H. and Drummond, B.R. (1978) 'Vegetations structure and substrate of the northern part of the Great Dyke, Rhodesia: Gradient analysis and dominance-diversity relationships', *Plant Ecology* 37: 151–61.

Wikramanayake, E., Dinerstein, E., Loucks, C.J., Olson, D.M., Morrison, J., Lamoreux, J., Mcknight, M. and Hedao, P. (2002) *Terrestrial ecoregions of the Indo-pacific. A conservation assessment*, Washington, D.C.: Island Press.

16 Biodiversity and technical innovations: bionics

Wilhelm Barthlott, Walter R. Erdelen and M. Daud Rafiqpoor

Biodiversity and its dimensions

The tremendous diversity of life on Earth is the only specific feature of our planet. Hence, it is all the more surprising that our knowledge of this biological diversity is distressingly limited. About 1.7 million different living organisms have been described and scientifically documented. According to all estimates, however, planet Earth is home to at least 8.6 million (probably even far more than 10 million) different species (Mora et al. 2011).

Based on a rather conservative estimate of 10 million species, Figure 16.1 shows that the major groups of organisms account for very different proportions of global biodiversity. Arthropods (depicted in grey) make up the largest category, with roughly 5 million species. Our level of knowledge, i.e. the number of documented species, extends to some 1.7 million species; in other words, more than 80% of all species are unknown. Within the arthropods, the most speciose group are insects (e.g. beetles, hymenoptera, butterflies). Only about 350,000 species of beetles, representing 179 families, have been scientifically described. What a striking contrast to our knowledge of the more conspicuous vertebrates (e.g. mammals, birds, reptiles): they only comprise about 62,000 species, but are relatively well known (level of knowledge above 83%) by virtue of their size alone.

With an estimated 320,000 species, the terrestrial plants (higher plants or vascular plants, i.e. flowering plants, gymnosperms and ferns) form a group that is relatively poor in species. Compared to insects, however, they are quite large and eye-catching organisms. Besides, they are 'sessile'; in other words, they do not run away. This is the simple reason for the very high level of knowledge (about 90%) (Figure 16.1).

So does this mean that when dealing with global biodiversity in a changing environment we should mainly focus on arthropods, as they account for 50% of global biodiversity? From an ecosystems perspective, this would be a dangerous fallacy. Arthropods and other animals are only the consumers in the system. They all depend on the producers – the giant global powerhouse covering the planet as a global green solar collector: the higher plants. Figure 16.2 illustrates this relationship in the form of pyramids based on species numbers (left: estimate 10 million) and estimated biomass (right). The consumers with their high diversity comprise about 69% of the species; plants, on the other hand, make up only some 5%. Hence, they are the most important structural elements of every terrestrial biocoenosis. To a large extent, our food, clothing and medical care is also based on plants.

The most important normative instrument dealing with biodiversity is the Convention on Biological Diversity (CBD) which emerged from the UN Conference on Environment and Development, held in Rio de Janeiro in 1992. Currently 192 countries and the European Union are Parties to the Convention. All UN member states except for Andorra, the United States of America, and South Sudan have ratified the Convention which is an international legally binding treaty. The Convention refers to three major goals, i.e. conservation of biodiversity, the sustainable use of its components, and the fair and equitable sharing of benefits arising from genetic resources. Biodiversity continues to decline worldwide despite all efforts, in particular the global commitment in 2002

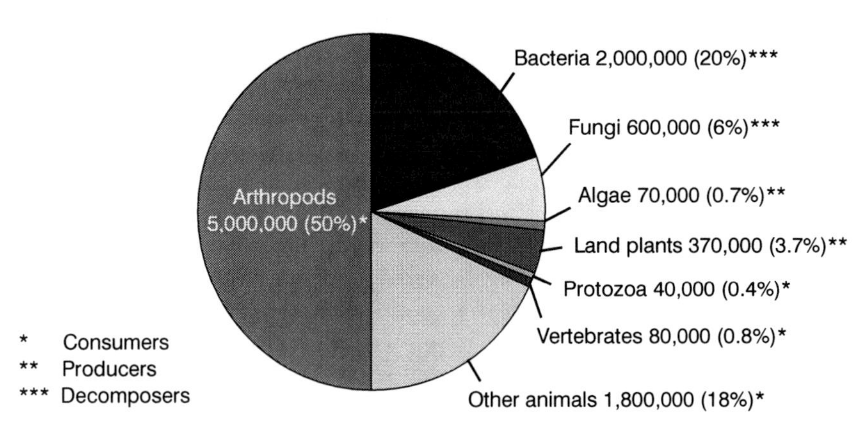

Figure 16.1 Estimated numbers of species. Biodiversity: 10 million species estimated, 1.7 million species are known, more than 80% of all species unknown.
Sources: Chapmann 2009; Guiry 2012; LeCointre and Guyader 2001; Mora et al. 2011; Sweetlove 2011; IUCN 2010

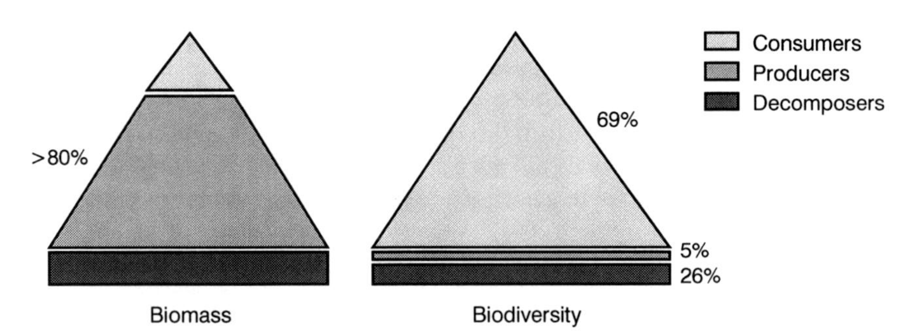

Figure 16.2 Biomass versus biodiversity.
Sources: Chapmann 2009; Guiry 2012; LeCointre and Guyader 2001; Mora et al. 2011; Sweetlove 2011; IUCN 2010

to significantly reduce biodiversity loss by 2010 (the '2010 Target') which was not met. Future commitments were made at the Nagoya summit of the Convention in 2010, including a strategic plan (2011–2020) and the 20 Aichi Targets (for further details cf. Erdelen, this volume). New hope rests now with the Intergovernmental Science-Policy Platform on Biodiversity and Ecosystem Services (IPBES), established in 2012. In addition, a paradigm shift in economics towards a 'green global agenda' may foster our efforts towards a significant reduction of biodiversity loss. Bionics could play a crucial role in this context as discussed below.

The uneven distribution of global biodiversity

The numbers of species per area vary considerably in different geographical regions. For instance, on a surface of one hectare in the Siberian Tundra, the forest may well be composed by only one tree species, whereas we may find 450 different tree species on an area of the same size in Amazonian Ecuador. By way of comparison, it should be noted that the Federal Republic of Germany (surface: 375,121 km^2) is home to only some 40 tree species. In the following, our considerations and the maps will focus on terrestrial plants, which are relatively well known. However, we would come up with similar results if we looked at animals or non-marine ecosystems. For instance, the comparison of a rocky reef in the North Atlantic and a tropical coral reef reveals similar relations. Maps of global terrestrial diversity can only be based on plants. With regard to animals, our knowledge is too limited (e.g. arthropods), and only a few non-representative groups (such as birds or butterflies) have been sufficiently studied so far.

Our world map of biodiversity (Figure 16.3) is based on the analysis of several thousands of flora inventories, checklists and databases (to learn more about the method used to generate the map, cf. Barthlott, Lauer and Placke 1996; Barthlott et al. 1999; Barthlott, Mutke and Kier 1999). Since its first publication (Barthlott, Lauer and Placke 1996), the map has been continuously refined (e.g. Barthlott et al. 2005, 2007) and is presented here in its current version, which has been enriched with new data. The map displays the geographical distribution of global biodiversity with the highest resolution that is currently possible.

When investigating the causal dependencies of these diversity patterns, we can identify a number of underlying principles. As one would expect, both the polar and subpolar regions as well as the extreme deserts of planet Earth are relatively poor in species, whereas the highest diversity occurs in the subtropics and especially in the tropics. This latitudinal gradient has been known for a long time; its reason is the increasingly favourable influence of hygrothermal parameters in the area of the equator (Kreft and Jetz 2007). A sea surface temperature of above 26 °C correlates surprisingly well with highly diverse tropical areas (cf. Figure 16.3). However, in very arid zones, such as the Sahara, or regions with unfavourable pedological conditions (lack of nutrients in the Gran Sabana in Venezuela), even subtropical or tropical areas may be poor in species.

A high level of biodiversity is by no means only characteristic of tropical regions. The Caucasus, for instance, is richer in species than certain parts of the lowland

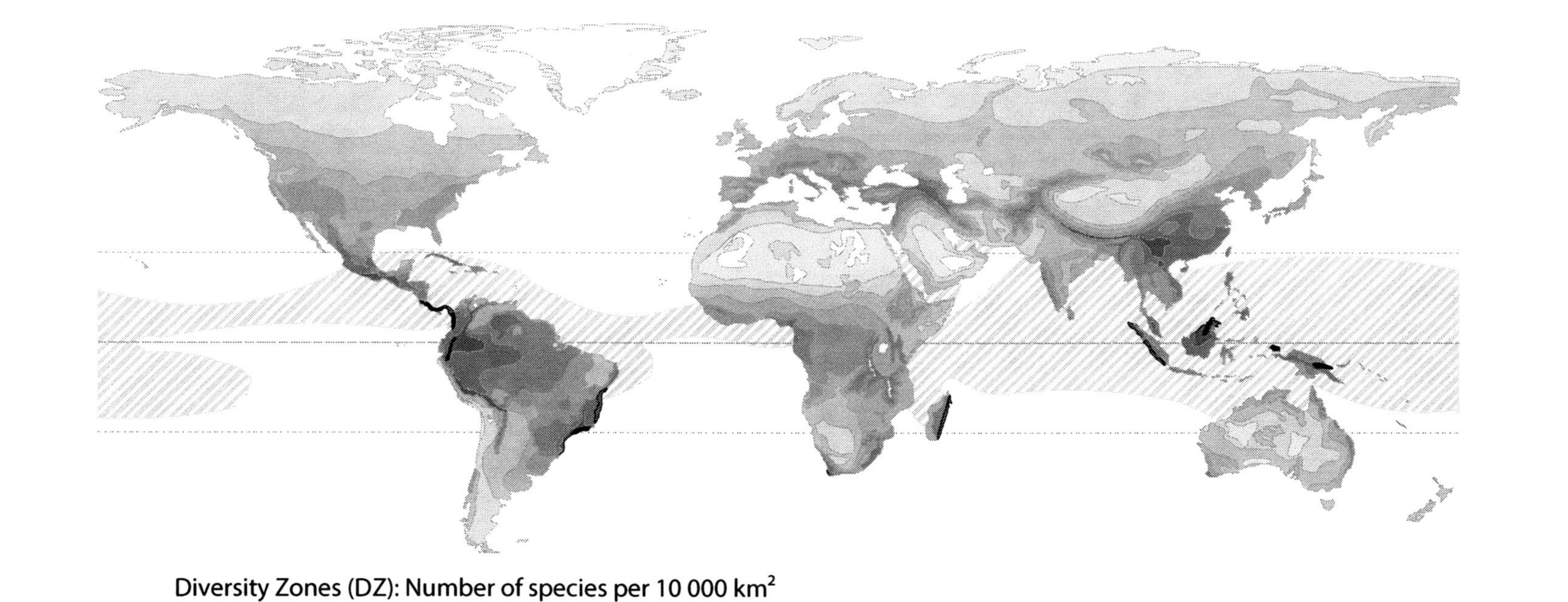

Figure 16.3 Uneven distribution of global biodiversity: species numbers of plants per 10,000 km^2.
Source: W. Barthlott, M.D. Rafiqpoor, J. Mutke (2013) Nees Institute for Biodiversity of Plants, University of Bonn ©Barthlott
Based on: Barthlott, Kier, Kreft, Küper, Rafiqpoor, Mutke (2005)

rainforests in the Congo basin. Here, a second important factor comes into play. The biodiversity of a given area very much depends on 'habitat heterogeneity', i.e. the diversity of abiotic factors (such as climate, geomorphology, geology, soils) within this space, which we subsume under the term geodiversity (Barthlott et al. 1999) – in contrast to Gray (2004), who defines geodiversity as the sum of all geological and geomorphological structures and processes. Mountain areas nearly always exhibit very high levels of diversity.

When we pool areas with more than 3,000 species per 10,000 km^2 in a map, we obtain 20 centres of biodiversity on planet Earth (Figure 16.4). Those centres clearly correlate with the mountain areas of the tropics and subtropics and represent the so-called mega-diverse countries, which are *in most cases* developing and emerging countries. It has been demonstrated elsewhere (Kier et al. 2009) that not only quantitative but also qualitative aspects play a most significant role in this respect, such as the level of endemism (in other words, the 'specific quality' of an area). To illustrate this, we may use the example of Hawaii and Thuringia. With roughly the same surface (16,000 km^2) and similar numbers of species (Hawaii 1,140 spp., Thuringia 1,570 spp.), these two regions show a fundamental difference: while 86% of Hawaii's flora is endemic, Thuringia has no endemic plant species. This also reflects the particularity of island systems, which is not depicted in our map. In other papers (Kier et al. 2009; Kreft et al. 2008) we have demonstrated that the oceanic islands only comprise 3% of the Earth's land surface, but are home to 25% of all known plant species. Among the 10 areas with the highest levels of endemism, six are islands: New Caledonia, Polynesia-Micronesia, the Caribbean islands, other Atlantic islands, the Eastern Melanesian islands and Taiwan.

Another interesting – but not at all surprising – aspect is the contrast between mega-diverse centres in the tropics and subtropics on the one hand, and 'mega-research centres' in the industrial countries in regions with a predominantly temperate climate on the other. In Figure 16.5 – which is by no means exhaustive – major research institutes that deal with biodiversity issues have been projected onto the biodiversity map. This North–South gradient between research intensity and biodiversity makes it obvious that the industrialised world must assume a major responsibility in terms of 'capacity building', in order to raise greater awareness for the protection of biodiversity in mega-diverse countries and to protect natural resources for future generations in a sustainable manner.

The value of endangered diversity

The loss of biodiversity, which began with the industrial era, but has only become blatantly apparent since the 1960s, is alarming (Hammond 1995; Perrings et al. 1997; Duffy 2003). Much like the growth of the world population, this decline seems to progress exponentially. Despite all international conventions adopted since the Rio de Janeiro summit in 1992 and all political declarations of intent, there have been no fundamental improvements. The destruction of habitats, including the deforestation of tropical rainforests, has continued at the same or

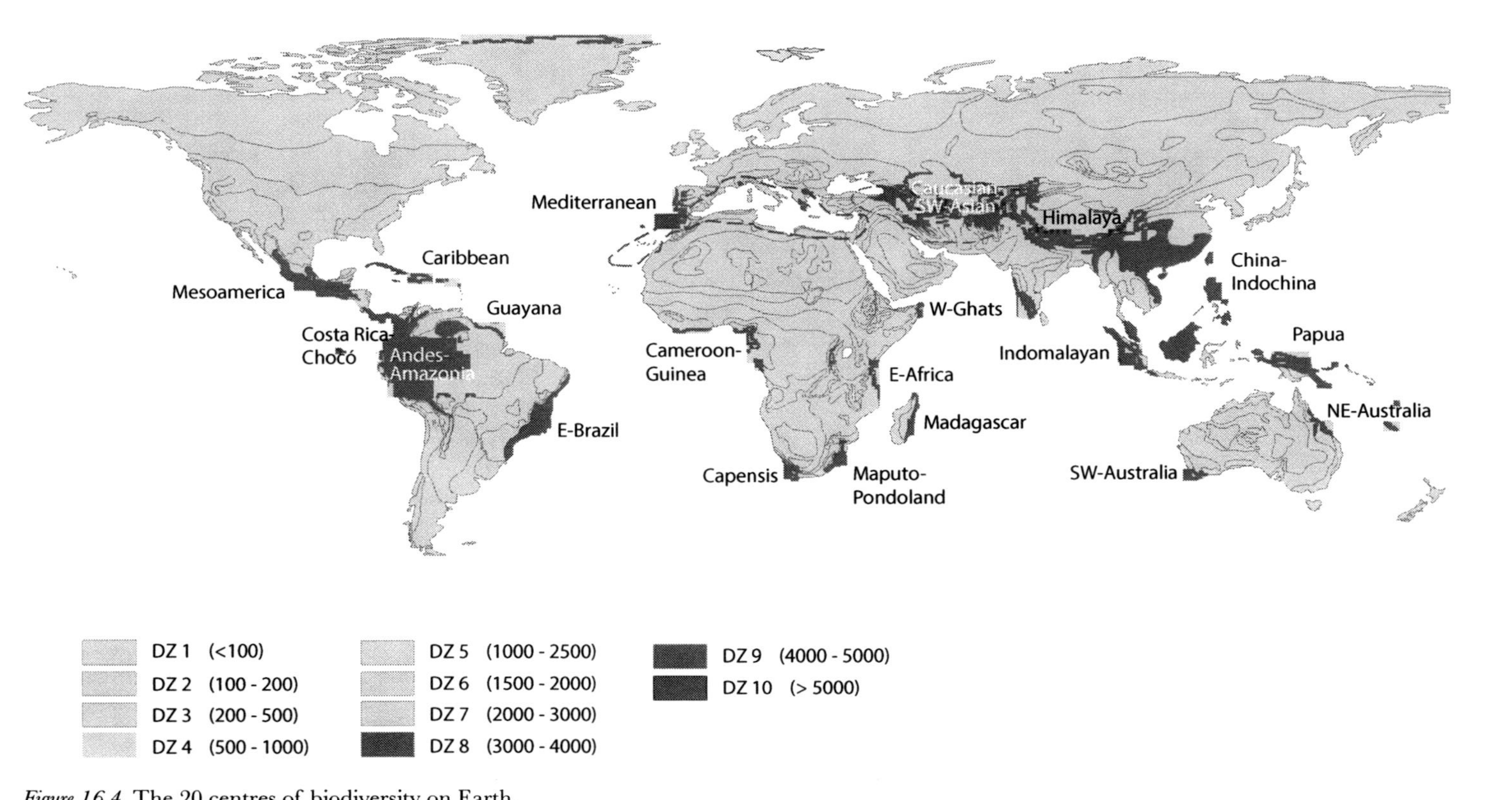

Figure 16.4 The 20 centres of biodiversity on Earth.
Source: W. Barthlott, M.D. Rafiqpoor, J. Mutke (2013) Nees Institute for Biodiversity of Plants, University of Bonn © W. Barthlott
Based on: W. Barthlott, W. Lauer, A. Placke (1996)

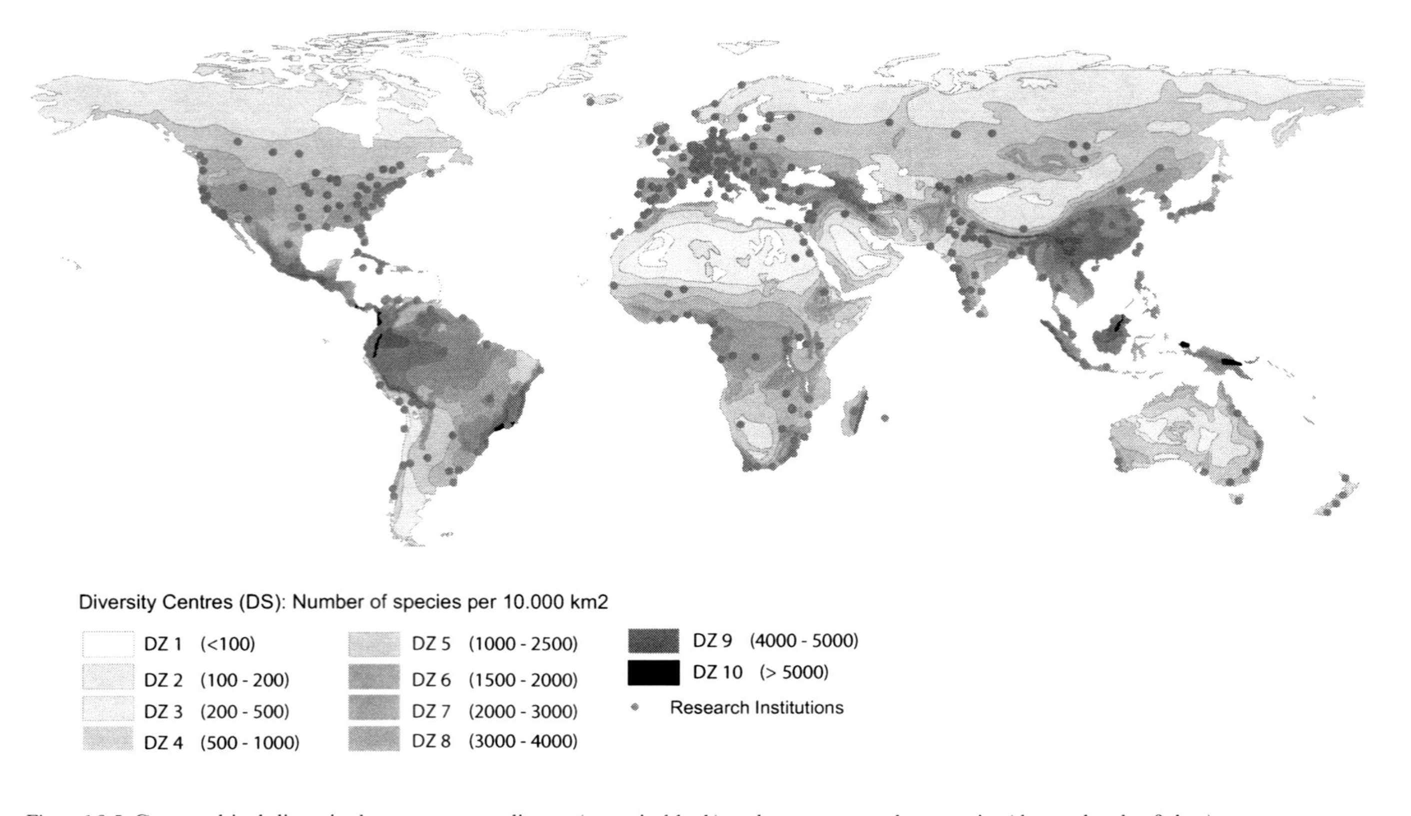

Figure 16.5 Geographical disparity between mega-diverse (areas in black) and mega-research countries (dense clouds of dots).
Source: W. Barthlott, M.D. Rafiqpoor, J. Mutke (2013) Nees Institute for Biodiversity of Plants, University of Bonn © W. Barthlott

Figure 16.6 Example of endangered biodiversity on islands: a fossil egg of elephant bird (Aepyornis maximus), who was native to Madagascar until the seventeenth century. These birds were more than 3 m (10 ft) tall and weighed close to 400 kg (880 lb). The birds became extinct after humans settled in Madagascar. The birds survived as the enormous legendary bird 'roc' or 'rukh' in *One Thousand and One Nights* and in many other Arabian myths and fairy-tales.

even an increased rate. At the same time, species numbers increase locally as a phenomenon of globalisation (for instance, Germany and the USA now exhibit a higher diversity than ever before due to invasive species), but decrease on a global scale. Bioglobalisation has become a core challenge. In the context of research on species extinction, invasive species with their superior competitive abilities play a central role (Essl et al. 2008; Klingenstein and Otto 2008; Nehring et al. 2010).

The conservation of global biodiversity depends, among other things, on our willingness to change our economic systems – an idea which is unacceptable to industrial countries in particular, but also to many emerging countries. And it is extremely difficult to impart the reasons why we should preserve high levels of diversity. While ethical aspects are important, they will not be considered here – we will use mainly utilitarian arguments to sustain the importance of biodiversity conservation. In this context, two aspects are of particular concern (overviews in TEEB 2010): first, the ecosystemic function of biodiversity (stability of ecosystems, climate etc.), which can indirectly be translated into numbers and economic benefit; and second, the direct value of plants and animals, as foodstuff (e.g. rice), medicine or building material (e.g. wood) or source of fibre (e.g. cotton).

More than 20,000 plant species are directly used by man. The number of animal species is even higher; for instance, 100,000 different marine organisms are used to produce 200,000 extracts which are examined for their pharmaceutical utility.[1] And still, food supply for more than 6 billion people strongly depends (>50%) on four grass species: wheat, rice, maize and millet. In this context, diversity of species and varieties is an important livelihood for humankind. Only one in 6,200 examined rice varieties was resistant to a virus that threatened the entire rice crop of South East Asia in the 1970s.

New active ingredients for medicine are discovered in plants and animals every year. For instance, the extracts of the Madagascar periwinkle (*Catharanthus roseus*) are used in the treatment of leukaemia and testicular cancer. Nearly 80% of antibacterial agents that are currently being introduced are biological products or have been extracted from them. The same is true for approximately 60% of new ingredients that have been introduced in the area of cancer therapeutics.

2010 was the UN international Year of Biodiversity. On this occasion, countless research institutes that are active in the field of biodiversity distributed information about the conservation and sustainable use of biodiversity. The UN Secretariat of the Convention on Biological Diversity (UNCBD) encouraged many organisations, institutions, companies and individuals to get engaged in this important agenda, in order to raise public awareness for the ongoing loss of global biodiversity.

Nature as a model: biodiversity and bionics

It is extremely surprising that all these considerations do not take into account one aspect of biodiversity that is gaining more and more importance: the value of plants and animals as sources of ideas for technical construction, i.e. bionics or biomimetics (surveys e.g. in Benyus 1997; Forbes 2005; Cerman, Barthlott and Nieder 2011; Barthlott and Koch 2011). Even the Convention on Biological Diversity does not mention it.

The idea of using 'nature as a model' is not new. Daedalus and Icarus in Greek mythology already copied (with little success) biological models, and there is an unbroken tradition from them to Leonardo da Vinci and to the construction of the first airplanes. But a systematic analysis of biological systems under the aspects of technical applicability started astonishingly late. *Bionics* or *Biomimetics* are (like *nanotechnology* and *biodiversity*) terms created after 1960 and the discipline only really became influential after 1990. Bionics meant to quite some extent robotics and mechanics. Bio-inspired materials and biomimetic surfaces came into focus only after 1995.

Technical products are developed in a comparatively extremely short time, by trial and error processes or modelling. Species evolved in some 500 million years of continuous undirected (mostly negative) mutation and selection. Plant and animal species are optimised, but they are *not* necessarily optimal, sustainable or environmentally friendly.

Each species is an optimised solution to a particular environmental condition. We know of some 1.7 million solutions – this is the number of animals, plants and micro-organisms described in the literature. However, all extrapolations indicate a total number of more than 10 million species. This means that more than 80%

of all living organisms and their biomimetic potentials inhabiting our planet are still unknown.

Engineers and biological species have a common interest: to construct (e.g. airplanes) or to evolve (e.g. wind-dispersed seeds in *Zanonia*) highly functional 'products'. However, the processes of development und evolution are basically different – and biologists and engineers or material scientists speak different languages.

Classical examples of biomimetic applications include the first glider aircraft (Etrich-Rumpler Taube 1910), a popular plane up to 1914 that was modelled after a dove, and in particular the famous wind-dispersed seeds of the tropical cucumber plant *Zanonia (Alsomitra) macrocarpa*. But the most widely known example of biomimicry is Velcro, invented in 1941 by the Swiss engineer George de Mestral after he removed burrs (fruits of *Arctium*) from the fur of his dog.

Surfaces only began to play a role in bionics relatively late (surveys in Gorb 2009; Koch, Bhushan and Barthlott 2010). This is all the more surprising given that the most interesting processes in both animate and inanimate nature occur on surfaces – every interaction between a solid body (airplane or submarine) or a living being (*Zanonia* seed or shark) and its gaseous (air) or liquid (water) environment is determined by this interface. Today, biomimetic materials play a significant role (Barthlott and Koch 2011). Their development was initiated by the palaeontologist W.-E. Reif (University of Tübingen), who studied the skin of shark fossils in the 1970s. He noticed delicate rib structures (riblets), which the engineer D.W. Bechert at the Technical University of Berlin successfully tested for their drag-reducing properties for the first time in 1982. The first artificial riblet foils manufactured by the company 3M were glued to an Airbus A320 in 1996. Although the aircraft was only coated in parts, its fuel consumption was reduced by 1.5% – a rather significant quantity. Unexpected applications were discovered in the world of sports. It all started with Michael Phelps' gold medals at the 2008 Summer Olympics – and in the following years, all swimming world records were broken. However, this was not attributable to the athletes' performance, but to their biomimetic swimsuits – modelled after shark skin.

Nearly at the same time, a second biological discovery and its biomimetic technical implementation took place which brought about a paradigm shift in surface technology: the lotus effect.[2] When performing scanning electron microscopic analysis on plant surfaces, the lead author (WB) observed for the first time in 1972 a peculiar phenomenon. While he worked with plants from the Botanical Garden of the University of Heidelberg, he noticed that water completely beaded off the surface of certain leaves, taking dirt particles with it without leaving any residues (Barthlott and Ehler 1977). Later, it became obvious (Barthlott 1990) that this system of physical interaction was optimised in the case of the leaves of the lotus flower (*Nelumbo*) due to a complicated hierarchically structured roughness. We researched the phenomenon in detail and created the abbreviation and trademark Lotus-Effect® to describe it (surveys in Koch, Bhushan and Barthlott 2010; Yan, Gao and Barthlott 2011). Cooperation with the industry started in 1995 and the first product, the facade paint 'Lotusan', was rolled out in 1999. Today an internet search on major search engines delivers more than 50,000 hits for 'lotus effect'.

This was a biological discovery, which resulted in the description of a new physical effect (v. Bayer 2000; Forbes 2005) that is now used all over the world for innovative high-tech products to create superhydrophobic anti-adhesive surfaces.

Based on work with the lotus plant, current research in the field of biomimetic surfaces focuses on the air-retaining surfaces of the floating fern *Salvinia*. Due to a surprising physical trick (Barthlott et al. 2010), its highly complicated surfaces permanently keep an air layer submerged under water (Figure 16.8). Ships with a Salvinia-effect surface that would glide through the ocean on an air layer would reduce their friction by more than 10%, meaning a reduction of 1% of the fuel consumption. As the mass transportation of goods is carried out by big container vessels, the Salvinia®-Effect could help to reduce global oil consumption by about 1%.

Figure 16.7 Technical implementation of the self-cleaning surface properties of the lotus plant, using the example of the facade paint Lotusan

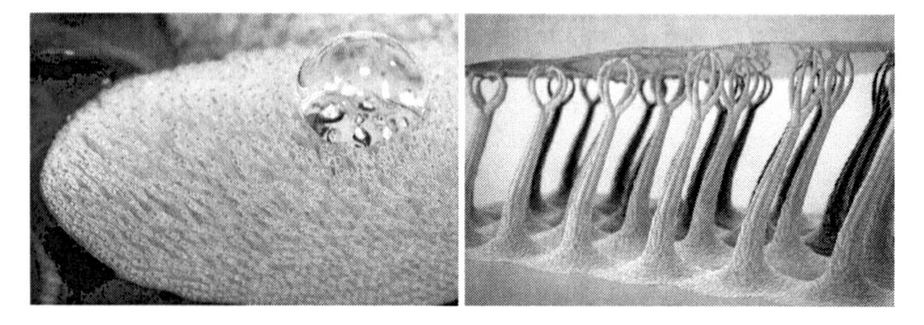

Figure 16.8 The floating fern *Salvinia* with its complex egg-beater shaped superhydrophobic elastic hairs and a water-droplet (left) © W. Barthlott. An underwater permanent air layer is maintained (right)

The Lotus effect, i.e. hierarchically structured, superhydrophobic self-cleaning surfaces, has brought about a paradigm shift in surface technology. Before the discovery of this physical effect, companies around the globe had optimised their products to be ultra-smooth in their effort to obtain anti-adhesive surfaces. But while the discovery and the principle may be new, the underlying knowledge is ancient. *Nelumbo* is a religious symbol; among other things, the *Holy* Lotus Flower is a symbol of purity in Buddhism and Hinduism (Barthlott and Neinhuis 1997). Its leaves rise from the water and are impeccably dry; mud and dirt simply glide off them. The knowledge of the anti-adhesive and self-cleaning properties of Lotus is as old as mankind and subject to two-thousand-year-old Chinese poems. In Hinduism it represents beauty and non-attachment: 'A man who reliquishes attachment … is not stained by evil, like a lotus leaf unstained by water' (Bhagavad Gita, 5th teaching, verse 10; cf. Stoler Miller 1986: 60). However, modern research in science is untouched by texts almost as old as mankind. We had ignored it until 1990.

Lotus and Salvinia are only two examples of probably more than 10 million species that inhabit our planet. Every single one of them has been optimised in terms of function in a process of millions of years of evolution. Thus, physicists and engineers have millions of years of development work (trial and error without computers or modelling) on hand – free of charge. However, we are still ignorant of some 80% of global biodiversity – and the rate of extinction progresses much faster than our taxonomic expertise in the field. Conservation of biodiversity will therefore remain an enormous challenge.

Bionics and green economy

As outlined above, bionics aims at developing and applying new technologies based on the examination of how nature has addressed issues faced in the course of evolution. Essentially bionic products 'emulate' natural products and processes. A thorough understanding of these therefore is a necessary prerequisite for the development of such products.

Bionic products reflect systemic characteristics found in nature such as dependency on solar energy, complete recycling of waste, structural adaptations to particular environments and so forth (cf. examples in this paper). These characteristics are commonly summarised under 'biomimicry principles' not further discussed here (from Edwards 2006, based on Benyus 1997):

- Nature runs on sunlight.
- Nature uses only the energy it needs.
- Nature fits form to function.
- Nature recycles everything.
- Nature rewards cooperation.
- Nature banks on diversity.
- Nature demands local expertise.
- Nature curbs excesses from within.
- Nature taps the power of limits.

Basically, bionics may contribute to using natural resources not only more efficiently by using – qualitatively and quantitatively – fewer resources but in a more environmentally friendly and sustainable manner. Secondly, the process of developing new bionics-based technologies may optimise efficiency in the production process and may, in contrast to 'classical' production processes, already from the beginning be an integral part of a cyclic production and decomposition process (from cradle to cradle). Thirdly, the innovative principle underpinning the process from insight into nature and its 'mechanisms' may lead to generations of new products we need in our time of global resource depletion and major environmental problems such as global climate change and loss of biodiversity.

This close nexus between bionic approaches and nature facilitates developing new bionics-based technologies with little or no negative impact on the environment. In other words, they may be more environmentally friendly than many of the technologies currently in use, present solutions to pressing problems the global community is currently facing, and finally increasingly contribute to realising sustainable development at all scales. Examples are minimising energy needs for vessels with a bionics-based coating, self-cleaning surfaces which save costs for cleaning and maintenance and so forth (see above).

As for the current global discussion of sustainable development, the latest big gathering took place in 2012, twenty years after the famous Earth Summit of 1992 (commonly referred to as Rio+20). The 2012 Summit focused on the process towards a green economy as a vital ingredient for a sustainable future for humankind. This move towards a global green economy, a process triggered by the UN under UNEP's leadership, was a major objective of the meeting in Rio in 2012. As indicated in a speech by Achim Steiner, the Executive Director of the United Nations Environment Programme (UNEP), about Rio+20: '*Heads of State and more than 190 nations did embrace the Green Economy in the context of sustainable development and poverty eradication. Nations agreed that such a transition could be an "important tool" when supported by policies that encourage decent employment, social welfare and inclusion and the maintenance of the Earth's ecosystems, from forests to freshwaters*' (Steiner 2013).

These most recent international developments underscore not only the potential of bionics towards a new generation of technological and therefore economic development, but also the importance bionics may have for sustainable development and the new economic paradigms underpinning this process.

Moreover, the continued effort to investigate solutions nature has already produced in its treasure trove of inventions and innovations will be a growing pool of knowledge adaptable to a totally changed modern world which addresses the major problems of humankind in a more nature-friendly manner, in fact from 'a literally natural angle'. As so eloquently expressed by Janine Benyus: '*doing it nature's way has the potential to change the way we grow food, make materials, harness energy, heal ourselves, store information, and conduct business*' (Benyus 1997).

Bionics-based products may address virtually every component of this syndrome of issues humankind is currently facing, including for instance the demand on renewable energy and the phasing out of traditional resources of energy as well as global climate change and the problem of stabilising greenhouse gas emissions.

To date, however, bionics does not (yet) feature as prominently as it should despite its potential to contribute to sustaining the Earth system and its processes. For instance, in UNEP's green economy report (UNEP 2011) none of the terms bionics, biomimicry or biomimetics appears, not even in the chapter on enabling conditions for a green economy. This also applies to the report of the United Nations Environment Management Group on green economy (UNEMG 2011).

Better awareness of the usefulness of nature's products and processes in the current development debate is urgently required. In addition, more thorough examination is needed of how bionics might play a more prominent role in the context of what is so desperately needed: a paradigm change towards a new green, i.e. nature-based, economy.

To realise such a change a number of initiatives should be taken as suggested in a recent position paper on the future of bionics, published by the Association of German Engineers (Verband Deutscher Ingenieure) and *BIO*KON, the German Bionics Competence Network. These include the fostering of applied research and development, the implementation of a bionics-related education initiative, participation of small and medium-sized companies, the inclusion of bionics into regional research and innovation programmes, and the linkage of bionics to national biodiversity strategies as an opportunity to develop sustainable and environmentally friendly products, to create better environmental awareness and to be a leitmotiv for the transformation to a Green Economy (VDI and BIOKON 2012).

This chapter concludes with a quotation from the first chapter of Janine Benyus's book on biomimicry (Benyus 1997): '*At this point in history, as we contemplate the very real possibility of losing a quarter of all species in the next thirty years, biomimicry becomes more than just a new way of looking at nature. It becomes a race and a rescue.*'

Notes

1 Cf. <www.pharmamar.com/partnering.aspx> (accessed 2 July 2013).
2 Cf. < www.lotus-salvinia.de> (accessed 2 July 2013).

References

Barthlott, W. (1990) 'Scanning electron microscopy of the epidermal surface in plants', in D. Claugher (ed.) *Application of the Scanning EM in Taxonomy and Functional Morphology. Systematics Association's Special Volume*, Oxford: Clarendon Press, 69–94.

Barthlott, W. and Ehler, N. (1977) *Raster-Elektronenmikroskopie der Epidermis-Oberflächen von Spermatophyten. Trop. subtrop. Pflanzenwelt 19*, Stuttgart: Steiner.

Barthlott, W. and Koch, K. (2011) 'Biomimetic materials', *Beilstein J. Nanotechnol.* 2: 135f.

Barthlott, W. and Neinhuis, C. (1997) 'Purity of the sacred lotus, or escape from contamination in biological surfaces', *Planta* 202: 1–8.

Barthlott, W., Lauer, W. and Placke, A. (1996) 'Global distribution of species diversity in vascular plants: towards a world map of phytodiversity', *Erdkunde* 50: 317–28.

Barthlott, W., Mutke, J. and Kier, G. (1999) 'Biodiversity – the uneven distribution of a treasure', *NNA-Repots* 12, *Special Issue* 2: 18–31.

Barthlott, W., Mutke, J., Rafiqpoor, M.D., Kier, G. and Kreft, H. (2005) Global centres of vascular plant diversity, *Nova Acta Leopoldina* 92: 61–83.

Barthlott, W., Biedinger, N., Braun, G., Feig, F., Kier, G. and Mutke, J. (1999) 'Terminological and methodological aspects of the mapping and analysis of the global biodiversity', *Acta Bot. Finnica vol.* 162: 103–10.

Barthlott, W., Hostert, A., Kier, G., Küper, W., Kreft, H., Mutke, J., Rafiqpoor, M.D. and Sommer, J.H. (2007) 'Geographic patterns of vascular plant diversity at continental to global scales', *Erdkunde* 61: 305–15.

Barthlott, W., Schimmel, T., Wiersch, S., Koch, K., Brede, M., Barczewski, M., Walheim, S., Weis, A., Kaltenmaier, A., Leder, A. and Bohn, H.F. (2010) 'The Salvinia paradox: superhydrophobic surfaces with hydrophilic pins for air-retention under water', *Advanced Materials* 22: 1–4.

Benyus, J.M. (1997) *Biomimicry*, New York: William Morrow.

CBD – United Nations (1992) *Convention on Biological Diversity*. Online. Available HTTP: <http://www.cbd.int/doc/legal/cbd-en.pdf> (accessed 12 June 2013).

Cerman, Z., Barthlott, W. and Nieder, J. (2011) *Erfindungen der Natur. Bionik – Was wir von Pflanzen und Tieren lernen können*, 3rd edn, Reinbeck: Rowohlt.

Chapman, A.D. (2009) *Numbers of Living Species in Australia and the World*, 2nd edn, Canberra: Australian Government, Department of the Environment, Water, Heritage and the Arts. Online. Available HTTP: <http://is.gd/k8ljSQ> (accessed 15 October 2013).

Duffy, J.E. (2003) 'Biodiversity loss, tropic skew and ecosystem functioning', *Ecology Letters* 6: 680–7.

Edwards, A.R. (2006) *The Sustainability Revolution. Portrait of a Paradigm Shift*, Gabriola, Canada: New Society Publishers.

Essl, F., Klingenstein, F., Nehring, S., Otto, C., Rabitsch, W. and Stöhr, O. (2008) 'Schwarze Listen invasiver Arten – ein Instrument zur Risikobewrtung für die Naturschutzpraxis', *Natur und Landschaft* 83: 418–24.

Forbes, P. (2005) *The Gecko's Foot. Bio-inspiration Engineered from Nature*, London: Fourth Estate.

Gorb, S.N. (ed.) (2009) *Functional Surfaces in Biology*, Berlin: Springer.

Gray, M. (2004) *Geodiversiy – Valuing and Conserving Abiotic Nature*, Chichester: Wiley.

Guiry, M.D. (2012) 'How many species of algae are there?', *Journal of Phycology* 48: 1057–63.

Hammond, P.M. (1995) 'The current magnitude of biodiversity', in V.H. Heywood and R.T. Watson (eds) *Global Biodiversity Assessment*, Cambridge: Cambridge University Press, 113–38.

IUCN (2010) *IUCN Red List of Threatened Species. Table 1: Numbers of Threatened Species by Major Groups of Organisms (1996–2010)*. Online. Available HTTP: <http://www.iucnredlist.org/documents/summarystatistics/2010_1RL_Stats_Table_1.pdf> (accessed 15 October 2013).

Kier, G., Kreft, H., Leeb, T.M., Jetz, W., Ibisch, P.L., Nowicki, C., Mutke, J. and Barthlott, W. (2009) 'A global assessment of endemism and species richness across island and mainland regions', *PNAS* 106: 9322–7.

Klingenstein, F. and Otto, C. (2008) 'Zwischen Aktionismus und Laisser-faire: Stand und Perspektiven eines differenzierten Umgangs mit invasien Arten in Deutschland', *Natur und Landschaft* 83: 407–11.

Koch, K., Bhushan, B. and Barthlott, W. (2010) 'Functional plant surfaces, smart materials', in B. Bhushan (ed.) *Handbook of Nanotechnology*, 3rd edn, Heidelberg: Springer, 1399–436.

Kreft, H. and Jetz, W. (2007) 'Global patterns and determinants of vascular plant diversity', *PNAS* 104: 5925–30.

Kreft, H., Jetz, W., Mutke, J., Kier, G. and Barthlott, W. (2008) 'Global diversity of island floras from a macroecological perspective', *Ecology Letters* 11: 116–27.

LeCointre, G. and Le Guyader, H. (2001) *Classification phylogenetique du vivant*, Paris: Belin.

Mora, C., Tittensor, D.P., Adl, S., Simpson, A.G.B. and Worm, B. (2011) 'How many species are there on earth and in the ocean?', *PLoS Biology* 9 (8). Online. Available HTTP: <http://www.plosbiology.org/article/info:doi/10.1371/journal.pbio.1001127> (accessed 2 July 2013).

Mutke, J. and Barthlott, W. (2005) 'Patterns of vascular plant diversity at continental to global scales', *Biol. Skr.* 55: 521–31.

Nehring, S., Essl, F., Klingenstein, F., Nowack, C., Rabitsch, W., Stöhr, O., Wiesner, C. and Wolter, C. (2010) *Schwarze Liste invasiver Arten: Kriteriensystem und Schwarze Listen invasiver Fische für Deutschland und für Österreich*, Bonn: Bundesamt für Naturschutz.

Perrings, C., Mähler, K.-G., Folke, C., Holling, C.S. and Jansson, B.-O. (1997) *Biodiversity Loss – Economic and Ecologica Issues*, Cambridge: Cambridge University Press.

Steiner, A. (2013) *Has Rio+20 Made a Difference to International Cooperation on the Environment? Speech at Natur 2103 by Achim Steiner, UN Under-Secretary and Executive Director, UN Environment Programme*. Online. Available HTTP: <http://www.unep.org/NewsCentre/default.aspx?DocumentID=2708&ArticleID=9430> (accessed 2 July 2013).

Stoler Miller, B. (1986) *The Bhagavad-Gita*, New York: Bantam Classics.

Sweetlove, L. (2011) 'Number of species on Earth tagged at 8.7 million', *Nature News* 23 August 2011. Online. Available HTTP: <http://www.nature.com/news/2011/110823/full/news.2011.498.html> (accessed 15 October 2013).

TEEB (2010) *The Economics of Ecosystems and Biodiversity: Mainstreaming the Economics of Nature: A Synthesis of the Approach, Conclusions and Recommendations of TEEB*. Online. Available HTTP: <http://www.teebweb.org/wp-content/uploads/Study%20and%20Reports/Reports/Synthesis%20report/TEEB%20Synthesis%20Report%202010.pdf> (accessed 2 July 2013).

UNEMG (United Nations Environment Management Group) (2011) *Working Towards a Balanced and Inclusive Green Economy: A United Nations System-wide Perspective*, New York: United Nations.

UNEP (United Nations Environment Programme) (2011) *Towards a Green Economy: Pathways to Sustainable Development and Poverty Eradication*. Online. Available HTTP: <http://www.unep.org/greeneconomy/Portals/88/documents/ger/GER_synthesis_en.pdf> (accessed 2 July 2013).

VDI (Verband Deutscher Ingenieure) and BIOKON (2012) *Positionspapier Zukunft der Bionik. Interdisziplinäre Forschung stärken und Innovationspotenziale nutzen*. Online. Available HTTP: <http://www.vdi.de/uploads/media/Broschuere_TLS_Positionspapier_Zukunft_der_Bionik_06.pdf> (accessed 2 July 2013).

von Bayer, H.C. (2000) 'The Lotus Effect. – the sciences', *J. New York Academy of Sciences* 40: 12–5.

Yan, Y.Y., Gao, N. and Barthlott, W. (2011) 'Mimicking natural superhydrophobic surfaces and grasping the wetting process: a review on recent progress in preparing superhydrophobic surfaces', *Advances in Colloid and Interface Science*: 80–105.

17 Nature's chemical treasure trove

Biodiversity and pharmaceuticals

Tobias A. M. Gulder and René Richarz

Introduction

The use of natural remedies for the treatment of human diseases has a long history. A prominent example from former times is the application of willow extracts for the treatment of pain. While ancient users did not know about the biologically active ingredients in such preparations, it is now common knowledge that the curative power came from salicin, a small organic molecule produced by the plant that has inspired the development of acetylsalicylic acid, one of the most successful drugs to date. The application of purified and chemically defined natural substances in modern medicine was most likely initiated by the isolation of morphine from *Papaver* sp. (poppy) by Friedrich Sertürner in 1804. Since then, the development of modern analytical equipment as well as chemical and biomolecular methodology has rendered natural products one of the most important sources of new drugs in a broad variety of medical applications, in particular in the fields of antibiotics, anti-infectives, pain relief and cancer chemotherapy. Of utmost importance for the continued success of nature-derived drugs is the access to and increasing understanding of diverse organisms that produce such metabolites, e.g. for communication or protection. This chapter briefly showcases selected success stories of natural products in modern medicine and outlines current and future potential developments in this exciting field of (bio-)chemical and medical research, emphasising the importance of the preservation of biodiversity also from a medical and economic perspective.

Natural products from plants – the traditional source of biomedical agents

With the evolution of the human race, terrestrial plants and animals have started to not only be exploited as sources of food, but also to increase the quality of life by, e.g., providing material to craft clothes, housing, weapons or tools of all kinds. With modern societies developing, mankind started to use extracts made from living beings for diverse purposes, not knowing about the actual chemical constituents exhibiting the properties they were striving to use. Examples include the use of venomous preparations to facilitate hunting and the application of fragrant or colourful plants and animals to create perfumes or dyed clothes, respectively. The importance of nature-derived chemicals, so-called natural products, is even evident in our modern life. In particular, the ancient use of herbal remedies for the treatment of diverse maladies has translated into highly important biomedical agents developed by pharmaceutical companies that are of utmost importance to date (for selected reviews on the importance of natural products for the development of pharmaceuticals, cf. Cragg, Newman and Snader 1997; Newman, Cragg and Snader 2003; Butler 2004; Newman and Cragg 2007, 2012).

Plants produce various types of natural products, including phenols (e.g. salicin, flavonoids) and polyphenols (e.g. lignins, tannins), phytosterols, terpenoids, and alkaloids, many of which strongly interfere with biochemical targets of pathogens attacking human health or directly with processes in the human body (Dewick 2009). Two everyday examples of plant-derived compounds inducing such direct effects are caffeine and nicotine that both have stimulating effects by binding to certain receptors inside our body. Both compounds belong to the chemical class of alkaloids that comprises a large set of natural products with in many cases very strong biological activities. Two other well-known examples illustrating this phenomenon are the tropane alkaloid cocaine (**1**) isolated from coca plants, a very strong serotonin-norepinephrine-dopamine reuptake inhibitor and thus a powerful stimulant of the nervous system (Ruetsch, Böni and Borgeat 2001) that is unfortunately mostly abused as an illegal drug, and morphine (**2**), a constituent of poppy used as a potent opiate analgesic drug to relieve severe pain conditions (cf. Figure 17.1; Iversen 1996 and references cited therein). The alkaloid quinine (**3**), isolated from cinchona trees, likewise has analgesic properties, combined with antipyretic (fever-reducing), anti-inflammatory, and, most importantly, antimalarial activities (Kaufman and

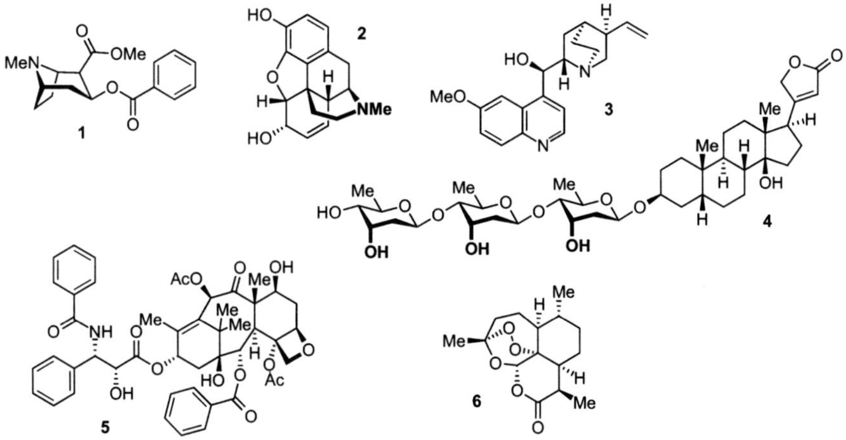

Figure 17.1 Structures of selected biomedical drugs of plant origin: cocaine (**1**), morphine (**2**), quinine (**3**), digitoxin (**4**), taxol (**5**), and artemisinin (**6**)

Rúveda 2005). As such, **3** was the first available effective treatment for malaria caused by *Plasmodium falciparum*. The molecular structure of this plant metabolite additionally served as a blueprint for the development of even more efficient antimalarial drugs, such as chloroquine or mefloquine. The mechanism of action of quinines against malarial parasites is highly remarkable. The parasite needs to degrade human haemoglobin to produce sufficient essential amino acids for its own metabolism, thereby producing heme as a side product (Tilley, Dixon and Kirk 2011). As heme is toxic to plasmodia, the parasite induces biocrystallisation of heme to give the nontoxic hemozoin (Egan et al. 2002). The quinines disturb this biocrystallisation process by coordinating to the heme molecules, thus leading to sufficiently high concentrations of free heme resulting in self-intoxication of the parasite by its own metabolic product (Combrinck et al. 2013 and references cited therein). Despite this elaborate mechanism of action, quinines unfortunately lost most of their therapeutic value due to exploding resistance of the malaria parasite, a result of the mass use of these molecules in malaria treatment and prevention.

Interesting examples of medically important phytosterols are the so-called cardiac glycosides, such as digitoxin (**4**) from the foxglove *Digitalis lanata* (Warren 1986). The molecular structure of these plant metabolites consists of a steroid backbone equipped with a sugar portion consisting of one to four sugar building blocks. Cardiac glycosides are medically applied in the treatment of a series of heart conditions, including irregular heart beat (artrial fibrillation, artrial flutter) and heart failure (Gheorghiade, Adams and Colucci 2004). The compounds increase cardiac output by increasing intracellular calcium levels and thus contraction by calcium-induced calcium release (CICR).

Another large group of plant-derived molecules are terpenoids (Dewick 2009). This class of natural products comprises a series of highly important chemotherapeutics. Among these is the mitotic inhibitor taxol (**5**, paclitaxel), a structurally complex metabolite that is used for the treatment of various types of cancer (i.e., lung, ovarian, breast) (Nicolaou, Dai and Guy 1994). Taxol (**5**) stabilises microtubules of cancer cells during mitosis, thus preventing cell division and inducing apoptosis (Schiff, Fant and Horowitz 1979). Originally, **5** was discovered in the bark of the Pacific yew tree, *Taxus brevifolia* (Wani et al. 1971). This plant had also been the main source of **5** until the mid 1990s, leading to significant destruction of *T. brevifolia* populations. This problem has meanwhile been solved by using plant cell fermentation technology. Most interestingly, Stierle et al. showed in 1993 that taxol and other taxans are indeed produced by an endophytic fungus living inside *T. brevifolia* (Stierle, Strobel and Stierle 1993). This opens the possibility of microbial fermentation to access the compound in an even more efficient biotechnological way. Another important example of a plant terpenoid used in medicine is artemisinin (**6**), a metabolite isolated from the sweet wormwood *Artemisia annua* (Ziffer, Highet and Klayman 1997). Interestingly, the producing plant itself was already used in traditional Chinese medicine for the treatment of malaria (Miller and Su 2011). The active chemical constituent from this plant, artemisinin (**6**), and its (semi-synthetic) derivatives, e.g. artemether or artesunate, meanwhile became the most important biomedical agents for the treatment of this infectious disease, to prevent fast evolvement of resistances preferably used in combination therapy with other antimalarials (Nosten and White 2007).

Natural products from terrestrial microorganisms – the golden age of antibiotics

Another important natural source of biologically active small molecules was discovered by the serendipitous finding of *Penicillium* fungi that efficiently inhibited the growth of bacteria. This milestone work initiated by A. Fleming (1929) and further developed by E. Chain et al. (1940; Chain and Florey 1944) led to the penicillins, e.g. penicillin G (**7**, Figure 17.2), one of the most important families of antibiotics to date. With these compounds being discovered from a microorganism, the microcosm came to the attention of large industrial screening programmes aimed at the development of chemotherapeutics. In particular, bacteria living in terrestrial soil, a habitat with limited resources but huge numbers of diverse competing organisms, were found to be prolific sources of antimicrobials, in particular antibacterial chemicals. These endeavours started a golden age of antibiotic discovery with almost all antibacterial chemical classes from natural sources being identified between the 1940s and 1970s (Table 17.1; cf. Butler and Buss 2006). Most of these natural antibiotics are still used today and are absolutely indispensable

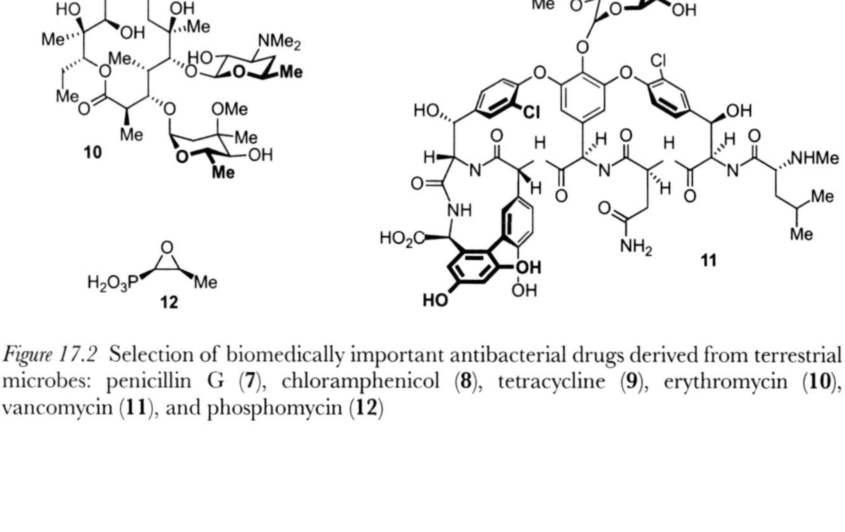

Figure 17.2 Selection of biomedically important antibacterial drugs derived from terrestrial microbes: penicillin G (**7**), chloramphenicol (**8**), tetracycline (**9**), erythromycin (**10**), vancomycin (**11**), and phosphomycin (**12**)

Table 17.1 List of important microbial natural product classes marketed as antibacterial drugs from 1941 to date (for selected chemical structure, see Figure 17.2)

Antibiotic Class	*Approx. Introduction*	*Acting on*	*Example*
Penicillins (β-Lactams)	1941	cell wall synthesis	penicillin G (**7**)
Bacterial polypeptides	1942	cell wall synthesis	bacitracin
Aminoglycosides	1944	protein synthesis	streptomycin
Cephalosporins (β-Lactams)	1945	cell wall synthesis	cephalosporin
Chloramphenicol	1949	protein synthesis	chloramphenicol (**8**)
Tetracyclines	1950	protein synthesis	tetracycline (**9**)
Tuberactinomycins	1951	protein synthesis	viomycin
Macrolides	1952	protein synthesis	erythromycin (**10**)
Lincosamides	1952	protein synthesis	lincomycin
Streptogramins	1952	protein synthesis	pristinamycin
Cycloserine	1955	cell wall synthesis	cycloserine
Glycopeptides	1956	cell wall synthesis	vancomycin (**11**)
Aminocoumarins	1956	DNA synthesis	novobiocin
Ansamycins	1957	RNA synthesis	rifamycin
Fusidanes	1963	protein synthesis	fusidic acid
Phosphonates	1969	cell wall synthesis	phosphomycin (**12**)
Pseudomonic acids	1985	protein synthesis	mupirocin
Lipopeptides	2003	cell membrane function	daptomycin

in combating human bacterial infections. While the global mode of action of these compounds is limited to a small number of bacterial targets, e.g. cell wall, protein, DNA and RNA synthesis, their diversity in terms of chemical structures is extremely remarkable. Some of the most important examples of such antibacterial metabolites include the protein biosynthesis inhibitors chloramphenicol (**8**) from *Streptomyces venezuelae* (Vining and Stuttard 1984), tetracycline (**9**) from *Streptomyces aureofaciens* (Griffin et al. 2010) and erythromycin (**10**) from *Streptomyces erythreusk* (Pal 2006), as well as β-lactams such as penicillin G (**7**) from diverse *Penicllium* species (Chain et al. 1940; Chain and Florey 1944), the glycopeptide vancomycin (**11**) from *Amycolatopsis orientalis* (Hubbard and Walsh 2003) or the very small phosphomycin (**12**) found in *Streptomyces* sp. (Seto and Kuzuyama 1999), all interfering with cell wall biosynthesis in the target organisms.

While Table 17.1 clearly illustrates the success of antibacterial natural products in human medicine, it also reveals a huge innovation gap following the first three decades of microbial metabolite discovery efforts. This strongly correlates with the dwindling interest of big pharmaceutical companies to invest in rather expensive natural product screening programmes, in particular because of increasing difficulties to isolate and characterise truly novel chemical structures from a rather small number of easily cultivable bacteria (e.g. actinomycetes, myxobacteria) using standard natural product discovery tools. Nevertheless, truly new scaffolds continue to be found also in very well-studied organisms, as exemplified by the recently discovered FabF inhibitor platensimycin (**13**, Figure 17.3) isolated from the African strain *Streptomyces platensis*, a compound whose unusual mode of action targets fatty acid biosynthesis (Wang et al. 2006). The pleuromutilins (**14**), protein synthesis inhibitors initially isolated from the basidiomycete *Clitopilus passeckerianus* in 1951 (Novak and Shlaes 2010), by contrast, exemplify that re-investigating long-known secondary metabolites with new methods can also result in the development of innovative antibacterial drugs. Of even greater value, however, is to unlock completely new bacterial producers from unusual habitats to increase the chance of discovering exciting novel chemistry. An impressive example is the polycyclic antibiotic abyssomicin C (**15**), an inhibitor of tetrahydrofolate biosynthesis derived from the new marine deep-sea actinomycete *Verrucosispora marisk* (Bister et al. 2004).

To continuously facilitate the search for unusual bacterial antibiotic producers, new methods to access the full biodiversity of bacteria have to be developed and unusual bacterial habitats have to be protected and preserved. In fact, less than 95% of bacteria known to date can currently be cultured. Gaining deeper insights into the biosynthetic repertoire of this completely untapped source of new chemical entities by traditional methods, i.e. establishing suitable fermentation strategies, and by modern biomolecular approaches (see below), will beyond doubt play a crucial role in fighting the evolving resistance of many serious pathogens against our current arsenal of antibacterial drugs. Given the past and current success of natural antibiotics, the development of new drugs from nature against this extremely serious health threat is an indispensable, highly rewarding task.

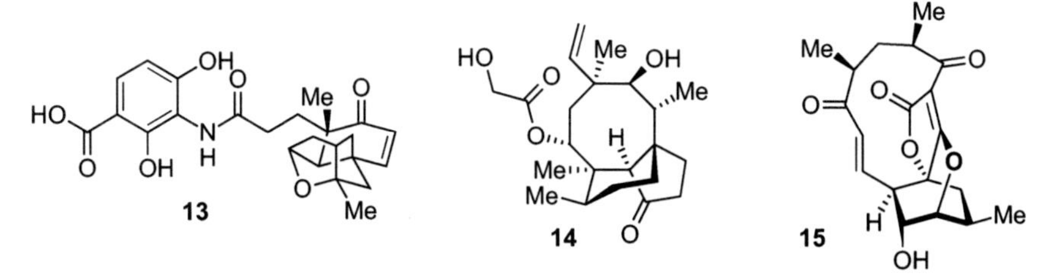

Figure 17.3 Chemical structures of the antibiotically active microbial natural products: platensimycin (**13**), pleuromutilin (**14**) and abyssomicin C (**15**)

Natural products from the sea – diving deeper into hidden pharmaceuticals

As mentioned above, the investigation of largely untapped habitats for the discovery of prolific producers of novel secondary metabolites is a worthwhile task. In particular, biodiversity-rich habitats have proven to be populated by talented natural chemists due to the strong competition for resources found in such areas, as exemplified by soil bacteria in the previous chapter. A very young natural product treasure chest fulfilling these preconditions is the ocean, especially densely populated tropical reefs (Montaser and Luesch 2011). Starting with the development of modern scuba diving equipment, this new natural product world more and more came into the focus of biomedical research. The first important findings in terms of marine natural products were already published in 1951 by W. Bergmann and R. J. Feeney, who discovered the two arabinosyl nucleotides spongouridine (**16a**) and spongothymidine (**16b**) from the marine sponge *Cryptotethia crypta*, which ultimately served as the molecular basis for the development of antiviral and cytostatic drugs such as cytarabine (**17**) and vidarabine (**18**) (Figure 17.4). The significance of marine organisms for human drug research was further established by Weinheimer and Spraggins who

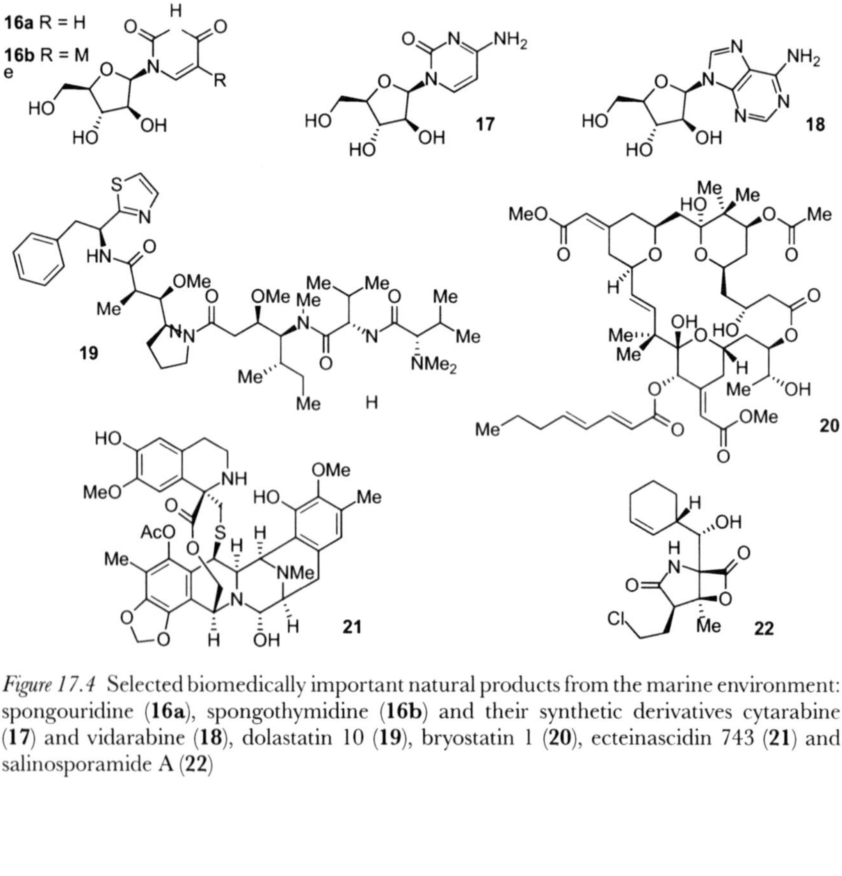

Figure 17.4 Selected biomedically important natural products from the marine environment: spongouridine (**16a**), spongothymidine (**16b**) and their synthetic derivatives cytarabine (**17**) and vidarabine (**18**), dolastatin 10 (**19**), bryostatin 1 (**20**), ecteinascidin 743 (**21**) and salinosporamide A (**22**)

identified two prostaglandin derivatives from the horn coral *Plexaura homomalla* (Weinheimer and Spraggins 1969). This early success spurred intense marine natural product research, in particular aiming at investigations on permanently attached or slow-moving animals, such as sponges, corals, tunicates, bryozoans, sea hares, cone snails, etc. These organisms cannot escape predatory enemies or other threats by simply 'swimming away', but instead in many cases evolved highly efficient small molecules as chemical defence mechanisms that can be utilised in human health applications, in many cases primarily as cytotoxic anticancer drug candidates (Blunt et al. 2013 and earlier review of this series; Petit and Biard 2013).

An interesting example of an unusual marine-derived peptide is dolastatin 10 (**19**) that was first isolated from the Indian Ocean sea hare *Dollobella auricularia* and shows potency against breast and liver cancers, solid tumours and some forms of leukemia (Pettit et al. 1993). The dolastatins interfere with tubulin formation, thereby preventing cell division of cancer cells by mitosis (Kingston 2009). The complex macrolide bryostatin 1 (**20**) was discovered from the bryozoan *Bugula neritina* collected in the Gulfs of California and Mexico (Pettit et al. 1982). This secondary metabolite is a potent modulator of protein kinase C, thus inhibiting cell growth and angiogenesis while accelerating apoptosis (Hale et al. 2002). Interestingly, **20** is not only developed as an anticancer drug candidate, but also as a memory-enhancing agent, e.g. with applications against Alzheimer's disease (Etcheberrigaray et al. 2004). Another structurally extremely challenging marine compound is the tetrahydroisoquinoline alkaloid ecteinascidin 743 (**21**) that was discovered from the Caribbean sea squirt *Ecteinascidia turbinate* (Wright et al. 1990). This compound is meanwhile marketed under the name Yondelis® and is used for the treatment of advanced soft tissue carcinoma and undergoing additional clinical trials for the treatment of various other sarcomas (e.g. breast, prostate, paediatric). The compound likewise efficiently inhibits cell division and consequently induces apoptosis (D'Incalci and Galmarini 2010).

While the above examples nicely illustrate the importance of marine-derived natural products from most diverse macroorganisms as resources in cancer chemotherapy, they likewise clearly underline the problems generally associated with such metabolites, i.e. the supply problem that extremely slows down the development of such complex molecules as drugs. The producing organisms are slow growing and scarce, they are usually only found in small, often remote areas, and the content of the desired bioactive ingredient is often extremely low. This already leads to supply problems during clinical studies and certainly prevents production of sufficient amounts for a drug applied worldwide. In addition, mass collection and harvesting of the producing organism in their natural habitat would most likely lead to fast extinction of the respective species.

One solution to this problem could be the sustainable cultivation of the producers in their natural environment. Such so-called mariculture approaches are technically challenging but have been shown to not only be possible, but also economically valid, e.g. in the case of **20** and **21** (Mendola 2000, 2003). Such endeavours, however, cause massive infrastructural investments and expensive development of new techniques that many pharmaceutical companies dread. In addition, secure compound supply is constantly endangered in the wild, e.g. by attacking predators or infectious pathogens, unpredictable severe weather conditions, etc. The traditional solution to such problems, the total chemical synthesis of the compound of interest in the laboratory, is in almost all cases not economically conductible due to the complex chemical structures of the marine-derived metabolites. Intriguing exceptions are, e.g., the analgesic conotoxins, peptides isolated from cone snails, which can be readily synthesised by automated solid-phase peptide synthesis and are thus in clinical use in the form of fully synthetic drugs (Olivera 2000; Miljanich 2004). The rescue again comes from microbiological biodiversity research. In fact, many of the marine-animal-derived natural products have meanwhile been shown to originate from diverse microorganisms (König et al. 2006). For example, dolastatin 10 (**19**) is produced by the cyanobacterium *Symploca* sp. VP642 that serves as food for *Dollobella auricularia* and thus leads to accumulation of **19** in its tissue (Luesch et al. 2001). In the case of bryostatin (**20**), an endosymbiotic bacterium '*Candidatus* Endobugula sertula' found in *Bugula neritina* has been shown to be capable of production of **20** (Sudek et al. 2007). A secure and sustainable supply of these compounds could thus be granted by fermentation of the respective bacterial producers, which in many cases would still need further improvements in fermentation technologies. For the industrial production of ecteinascidin (**21**) an interesting combined approach was developed (Cuevas et al. 2000). It has been shown that the Japanese soil bacterium *Pseudomonas fluorescens* produces large amounts of a structurally closely related metabolite, cyanosafracin B, which can easily be produced in mass amounts by fermentation (Ikeda et al. 1983). This compound is transformed into the desired drug **21** by efficient semi-synthetic methods in the chemical laboratory.

The potential of marine free-living bacteria and symbiotic microorganisms to produce many of the seemingly animal-derived natural products is meanwhile widely recognised and opens another virtually untapped resource for new chemical entities with potential application in human health. This has in recent years also led to investigations directly using (unusual) bacteria isolated from the marine environment (Gulder and Moore 2009; Fenical and Jensen 2006; Hughes and Fenical 2010). One example was already shown above in the case of the abyssomicins (**15**, cf. Figure 17.3; Bister et al. 2004). Other extremely potent marine bacterial sources of natural products are *Salinispora* species, marine-dwelling bacteria that live in ocean sediments and have within only a few years led to the discovery of numerous structurally novel bioactive compounds (Gulder and Moore 2009; Fenical and Jensen 2006; Hughes and Fenical 2010). The most prominent among these is salinosporamide A (**22**) isolated from *Salinispora tropica*, a natural product scaffold that constitutes the strongest irreversible proteasome inhibitor known to date (Gulder and Moore 2010). **22** is currently in Phase-II clinical trials for the treatment of multiple myeloma.

Concluding remarks

The above examples, which constitute only a small fraction of natural products that play major roles in modern human health applications, clearly showcase the importance of secondary metabolites isolated from organisms as resources for drug development. Such natural products have molecular structures that were generated and optimised over millions of years of evolution and as a result can efficiently and selectively bind their molecular targets, yielding highly effective medicines. Many of the compounds described in this chapter are of such enormous importance to mankind that they are even listed in the WHO Model List of Essential Medicines. These include morphine, quinine, paclitaxel, artemisin derivatives, penicillins, chloramphenicol, tetracycline, vancomycin, erythromycin and cytarabine. To combat quickly evolving resistances against many of our currently used drugs, in particular in the field of antibiotics, but also in other fields, such as cancer chemotherapy, a continued supply of novel chemical entities with new modes of action is absolutely required. This search can now be assisted by sophisticated analytical methods and genomic data to allow the directed identification of truly novel small molecules from minute amounts of biological material. Modern biomolecular techniques can furthermore help to generate sustainable systems for compound production, e.g. by expression in heterologous microbial host systems, and furthermore offer methods to manipulate natural products structures produced by a certain organism to generate even better, tailor-made 'unnatural natural products'. To facilitate this continued development, the protection and preservation of the biodiversity on our planet is of the greatest importance. In particular, biodiversity-rich areas, like tropical rain forest or tropical reefs, are virtually inexhaustible chemical treasure troves. Their importance for mankind, in particular but not exclusively for the development of modern chemotherapeutics, can thus not be overestimated.

References

Bergmann, W. and Feeney, R.J. (1951) 'Contribution to the study of marine products. XXXII. The nucleosides of sponges. I.', *J. Org. Chem.* 16: 981–7.

Bister, B., Bischoff, D., Ströbele, M., Riedlinger, J., Reicke, A., Wolter, F., Bull, A.T., Zähner, H., Fiedler, H.-P. and Süssmuth, R.D. (2004) 'Abyssomicin C – A polycyclic antibiotic from a marine *Verrucosispora* strain as an inhibitor of the p-aminobenzoic acid/tetrahydrofolate biosynthesis pathway', *Angew. Chem. Int. Ed.* 43: 2574–6.

Blunt, J.W., Copp, B.R., Keyzers, R.A., Munro, M.H. and Prinsep, M.R. (2013) 'Marine natural products', *Nat. Prod. Rep.* 30: 237–323.

Butler, M.S. (2004) 'The role of natural product chemistry in drug discovery', *J. Nat. Prod.* 67: 2141–53.

Butler, M.S. and Buss, A.D. (2006) 'Natural products – The future scaffolds for novel antibiotics?', *Biochem. Pharmacol.* 71: 919–29.

Chain, E. and Florey, H.W. (1944) 'Penicillin', *Endeavour* 3: 3–14.

Chain, E., Florey, H.W., Gardner, A.D., Heatley, N.G., Jennings, M.A., Orr-Ewing, J. and Sanders, A.G. (1940) 'Penicillin as a chemotherapeutic agent', *Lancet* 2: 226–8.

Combrinck, J.M., Mabotha, T.E., Ncokazi, K.K., Ambele, M.A., Taylor, D., Smith, P.J., Hoppe, H.C. and Egan, T.J. (2013) 'Insights into the role of heme in the mechanism of action of antimalarials', *ACS Chem. Biol.* 8: 133–7.

Cragg, G.M., Newman, D.J. and Snader, K.M. (1997) 'Natural products in drug discovery and development', *J. Nat. Prod.* 60: 52–60.

Cuevas, C., Pérez, M., Martín, M.J., Chicharro, J.L., Fernández-Rivas, C., Flores, M., Francesch, A., Gallego, P., Zarzuelo, M., de la Calle, F., García, J., Polanco, C., Rodríguez, I. and Manzanares, I. (2000) 'Synthesis of ecteinascidin ET-743 and phthalascidin Pt-650 from cyanosafracin B', *Org. Lett.* 2: 2545–8.

Dewick, P.M. (2009) *Medicinal Natural Products – A Biosynthetic Approach*, Chichester: Wiley.

D'Incalci, M. and Galmarini, C.M. (2010) 'A review of trabectedin (ET-743): A unique mechanism of action', *Mol. Cancer. Ther.* 9: 2157–63.

Egan, T.J., Combrinck, J.M., Egan, J., Hearne, G.R., Marques, H.M., Ntenteni, S., Sewell, B.T., Smith, P.J., Taylor, D., van Schalkwyk, D.A. and Walden, J.C. (2002) 'Fate of haem iron in the malaria parasite *Plasmodium falciparum*', *Biochem. J.* 365: 343–7.

Etcheberrigaray, R., Tan, M., Dewachter, I., Kuipéri, C., van der Auwera, I., Wera, S., Qiao, L., Bank, B., Nelson, T.J., Kozikowski, A.P., van Leuven, F. and Alkon, D.L. (2004) 'Therapeutic effects of PKC activators in Alzheimer's disease transgenic mice', *Proceedings of the National Academy of Sciences (USA)* 101: 11141–6.

Fenical, W. and Jensen, P.R. (2006) 'Developing a new resource for drug discovery: Marine actinomycete bacteria', *Nat. Chem. Biol.* 2: 666–73.

Fleming, A. (1929) 'On the antibacterial action of cultures of a *Penicillium*, with special reference to their use in isolation of *B. influenza*', *Br. J. Exp. Pathol* 10: 226–36.

Gheorghiade, M., Adams Jr., K.F. and Colucci, W.S. (2004) 'Digoxin in the management of cardiovascular disorders', *Circulation* 109: 2959–64.

Griffin, M.O., Fricovsky, E., Ceballos, G. and Villarreal, F. (2010) 'Tetracycline: A pleitropic family of compounds with promising therapeutic properties. Review of the literature', *Am. J. Physiol. Cell Physiol.* 299: C539–48.

Gulder, T.A.M. and Moore, B.S. (2009) 'Chasing the treasures of the sea – Bacterial marine natural products', *Curr. Opin. Microbiol.* 12: 252–60.

Gulder, T.A.M. and Moore, B.S. (2010) 'Salinosporamide natural products: Potent 20S proteasome inhibitors as promising cancer chemotherapeutics', *Angew. Chem. Int. Ed.* 49: 9346–67.

Hale, K.J., Hummersone, M.G., Manaviazar, S. and Frigerio, M. (2002) 'The chemistry and biology of the bryostatin antitumour macrolides', *Nat. Prod. Rep.* 19: 413–53.

Hubbard, B.K. and Walsh, C.T. (2003) 'Vancomycin assembly: Nature's way', *Angew. Chem. Int. Ed.* 17: 730–65.

Hughes, C.C. and Fenical, W. (2010) 'Antibacterials from the sea', *Chemistry* 16: 12512–25.

Ikeda, Y., Idemoto, H., Hirayama, F., Yamamoto, K., Iwao, K., Asao, T. and Munakata, T. (1983) 'Safracins, new antitumor antibiotics. I. Producing organism, fermentation, isolation', *J. Antibiot.* 36: 1279–83.

Iversen, L.L. (1996) 'How does morphine work?', *Nature* 383: 759f.

Kaufman, T.S. and Rúveda, E.A. (2005) 'The quest for quinine: Those who won the battles and those who won the war', *Angew. Chem. Int. Ed.* 44: 854–85.

Kingston, D.G. (2009) 'Tubulin-interactive natural products as anticancer agents', *J. Nat. Prod.* 72: 507–15.

König, G.M., Kehraus, S., Seibert, S.F., Abdel-Lateff, A. and Müller, D. (2006) 'Natural products from marine organisms and their associated microbes', *ChemBioChem* 7: 229–38.

Luesch, H., Moore, R.E., Paul, V.J., Mooberry, S.L. and Corbett, T.H. (2001) 'Isolation of dolastatin 10 from the marine cyanobacterium *Symploca* species VP642 and total stereochemistry and biological evaluation of its analogue symplostatin 1', *J. Nat. Prod.* 64: 907–10.

Mendola, D. (2000) 'Aquacultural production of bryostatin 1 and ecteinascidin 743', in N. Fusetani (ed.) *Drugs from the sea*, Basel: Karger, 120–33.

Mendola, D. (2003) 'Aquaculture of three phyla of marine invertebrates to yield bioactive metabolites: Process developments and economics', *Biomol. Eng.* 20: 441–58.

Miljanich, G.P. (2004) 'Ziconotide: Neuronal calcium channel blocker for treating severe chronic pain', *Curr. Med. Chem.* 11: 3029–40.

Miller, L.H. and Su, X. (2011) 'Artemisinin: Discovery from the Chinese herbal garden', *Cell* 146: 855–8.

Montaser, R. and Luesch, H. (2011) 'Marine natural products: A new wave of drugs?', *Future Med. Chem.* 3: 1475–89.

Newman, D.J. and Cragg, G.M. (2007) 'Natural products as sources of new drugs over the last 25 years', *J. Nat. Prod.* 70: 461–77.

Newman, D.J. and Cragg, G.M. (2012) 'Natural products as sources of new drugs over the 30 years from 1981 to 2010', *J. Nat. Prod.* 75: 311–35.

Newman, D.J., Cragg, G.M. and Snader, K.M. (2003) 'Natural products as sources of new drugs over the period 1981–2002', *J. Nat. Prod.* 66: 1022–37.

Nicolaou, K.C., Dai, W.-M. and Guy, R.K. (1994) 'Chemistry and biology of taxol', *Angew. Chem. Int. Ed.* 33: 15–44.

Nosten, F. and White, N.J. (2007) 'Artemisinin-based combination treatment of *falciparum* malaria', *Am. J. Trop. Med. Hyg.* 77: 181–92.

Novak, R. and Shlaes, D.M. (2010) 'The pleuromutilin antibiotics: A new class for human use', *Curr. Opin. Investig. Drugs* 11: 182–91.

Olivera, B.M. (2000) 'ω-conotoxin MVIIA: From marine snail venom to analgesic drug', in N. Fusetani (ed.) *Drugs from the sea*, Basel: Karger, 75–85.

Pal, S. (2006) 'A journey across the sequential development of macrolides and ketolides related to erythromycin', *Tetrahedron* 62: 3171–200.

Petit, K. and Biard, J.F. (2013) 'Marine natural products and related compounds as anticancer agents: An overview of their clinical status', *Anticancer Agents Med. Chem.* 13: 603–31.

Pettit, G.R., Herald, C.L., Doubek, D.L., Herald, D.L., Arnold, E. and Clardy, J. (1982) 'Isolation and structure of bryostatin 1', *J. Am. Chem. Soc.* 104: 6846–8.

Pettit, G.R., Kamano, Y., Herald, C.L., Fujii, Y., Kizu, H., Boyd, M.R., Boettner, F.E., Doubek, D.L., Schmidt, J.M., Chapuis, J.-C. and Michel, C. (1993) 'Isolation of dolostatins 10–15 from the marine mollusk *Dolabella auricularia*', *Tetrahedron* 49: 9151–70.

Ruetsch, Y.A., Böni, T. and Borgeat, A. (2001) 'From cocaine to ropivacaine: The history of local anesthetic drugs', *Curr. Top. Med. Chem.* 1: 175–82.

Schiff, P.B., Fant, J. and Horowitz, S.B. (1979) 'Promotion of microtubule assembly *in vitro* by taxol', *Nature* 277: 665–7.

Seto, H. and Kuzuyama, T. (1999) 'Bioactive natural products with carbon-phosphorus bonds and their biosynthesis', *Nat. Prod. Rep.* 16: 589–96.

Stierle, A., Strobel, G. and Stierle, D. (1993) 'Taxol and taxane production by *Taxomyces andreanae*, an endophytic fungus of Pacific yew', *Science* 260: 214–16.

Sudek, S., Lopanik, N.B., Waggoner, L.E., Hildebrand, M., Anderson, C., Liu, H., Patel, A., Sherman, D.H. and Haygood, M.G. (2007) 'Identification of the putative bryostatin polyketide synthase gene cluster from "*Candidatus* Endobugula sertula", the uncultivated microbial symbiont of the marine bryozoan *Bugula neritina*', *J. Nat. Prod.* 70: 67–74.

Tilley, L., Dixon, M.W.A. and Kirk, K. (2011) 'The *Plasmodium falciparum*-infected red blood cell', *Int. J. Biochem. Cell. Biol.* 43: 839–42.

Vining, L.C. and Stuttard, C. (1984) 'Chloramphenicol: Properties, biosynthesis and fermentation', in E.J. Vandamme (ed.) *Biotechnology of industrial antibiotics*, New York: Marcel Dekker, 387–411.

Wang, J., Soisson, S.M., Young, K., Shoop, W., Kodali, S. et al. (2006) 'Platensimycin is a selective FabF inhibitor with potent antibiotic properties', *Nature* 441: 358–61.

Wani, M.C., Taylor, H.L., Wall, M.E., Coggon, P. and McPhail, A.T. (1971) 'Plant antitumor agents. VI. The isolation and structure of taxol, a novel antileukemic and antitumor agent from *Taxus brevifolia*', *J. Am. Chem. Soc.* 93: 2325–7.

Warren, J.V. (1986) 'William Withering revisited: 200 years of the foxglove', *Am. J. Cardiol.* 58: 189f.

Weinheimer, A.J. and Spraggins, R.L. (1969) 'The occurrence of two new prostaglandin derivatives (15–epi-PGA2 and its acetate, methyl ester) in the gorgonian *Plexaura homonella*', *Tetrahedron Lett.* 10: 5185–8.

Wright, A.E., Forleo, D.A., Gunawardana, G.P., Gunasekera, S.P., Koehn, F.E. and McConnell, O.J. (1990) 'Antitumor tetrahydroisoquinoline alkaloids from the colonial ascidian *Ecteinascidia turbinata*', *J. Org. Chem.* 55: 4508–12.

Ziffer, H., Highet, R.J. and Klayman, (1997) 'D.L. Artemisinin: An endoperoxidic antimalarial from *Artemisia annua* L', *Prog. Chem. Org. Nat. Prod.* 72: 114–221.

18 Protection of biodiversity for the sake of science?

Gesine Schepers

Introduction

Authors in the ethics of nature debate argue for the protection of biodiversity in different ways. For example, Spaemann (1986: 196–8) and Norton (1988: 205) reason that biodiversity is indispensable for human life, Norton (1986: 128f.) and Alho (2008: 1116f.) point out that biodiversity is aesthetically valuable, and Ehrenfeld says that biodiversity just *is* valuable ('[v]alue is an intrinsic part of [bio]diversity') (1988: 214). Another argument says that biodiversity should be protected because it is valuable for science. I shall try to show that this particular argument fails. I do not maintain that there are no reasons to protect biodiversity. I only maintain that it should not be protected for the sake of science. We will, in the following, first look at the 'science argument', as I call it. Secondly, I touch upon some difficulties with this argument. Thirdly, I put forward two main objections against it: first, that it contradicts our understanding of the natural sciences; second, that science does not rely on biodiversity. It could be asked in addition whether science is valuable at all, but I shall not take up this issue here.

The science argument

Before we turn to the science argument, we will take one step back and place it in a larger context. In a more general sense one may ask about the interrelation between science and the protection of biodiversity as follows: Are there 'purely scientific reasons' (Norton 1987: 14) for the protection of biodiversity?[1] According to Norton, there are two kinds of answers to this question.[2] Answers of the first kind justify the protection of biodiversity with ecological insights; for example, on how the extermination of species influences ecosystems. Answers of the second kind, in contrast, regard biodiversity 'as an important object and source of knowledge' (Norton 1987: 14–18).

This article is not about answers of the first kind. Thus, we do not deal with the question of whether ecological insights can be reasons for the protection of biodiversity. Rather, it is about answers of the second kind; that is, about whether biodiversity should be conserved 'as an important object and source of knowledge'. So this is the science argument. It says that biodiversity should be protected because it is valuable for science as an object of science.[3] For example, Takacs ascribes to biodiversity 'scientific value', and he writes: 'As the raw material for biological study, biodiversity is essential for the scientific endeavor to continue unhindered. If we recognize science and its goals as unquestionable, overarching goods, then this value transcends mere subjective preference' (Takacs 1996: 197). Elliot argues:

> The preservation of species affords researchers the opportunity of studying unique forms of life in their natural states thereby contributing something to the growth of human knowledge. … [I]t is the knowledge which is … valuable; the continued existence of species is just a way of increasing it.
>
> (Elliot 1980: 12f.)

Lanzerath points to the dependence of biodiversity research on the occurrence of biodiversity, to the gaining of knowledge that this research yields, and to 'the purely scientific value' of this research. Against this background he thinks that '[t]he extinction of species leads … to an irreversible loss which narrows living mankind's and future generations' urge for knowledge' (2008: 207; my translation).[4]

To start with, biology explores biodiversity. However, other disciplines deal with it as well: 'Biodiversity research is extremely varied, both in its scope and approach' (Alexiades and Laird 2002: 4). Among these disciplines are geography, pharmacognosy, and forestry (Alexiades and Laird 2002: 4), for example, as well as economics, the social and the cultural sciences (Eser 2002: 113). Within biology, many partial disciplines investigate biodiversity; Alexiades and Laird (2002: 4) name botany, zoology, and ecology, for example. If I restrict myself to the natural sciences, to biology or even to its partial disciplines taxonomy and systematics in the following, this does not preclude that the respective considerations also apply to other (partial) disciplines.

As already seen, along with the science argument there is sometimes talk of the 'scientific value' of biodiversity.[5] However, such talk is not strictly pertinent, because things have 'scientific value' which are valuable *within the framework* of science – a successful scientific exploration or an additional argument, for example. However, this is not what the science argument means. It says that biodiversity should be protected *for* science, and not everything that is valuable *for* science is *scientifically* valuable. Financial support, for example, is valuable for science, but not scientifically valuable. Understood as the existing, natural biological diversity in the world, biodiversity cannot be scientifically valuable at all, because it is not a product of a scientific process, with all its standards and criteria. One could only call the concept, or a concept, of 'biodiversity' scientifically valuable. Therefore, instead of the 'scientific value' of biodiversity I prefer to speak of its 'value for science'.[6]

Which biodiversity is the argument about?

It is difficult to determine what biodiversity is.[7] In this regard we take the diversity of species, genes, and ecosystems, mentioned by Loreau (2010: 13f.), as representative for other conceptions and components of biodiversity. However, the science argument as an argument for the protection of biodiversity in general is not aimed at protecting the idea of biodiversity for science, with the help of books, for example. (Like the idea of the antique polity, which we can scientifically deal with still today.) Rather, the science argument is aimed at protecting the *real* biodiversity; that is, the biodiversity that really exists.

If one assumes that ordinary, immediately noticeable objects are real,[8] the real biodiversity first of all consists of the ordinary components of biodiversity; that is to say, of the different animals and plants as well as of the non-living, concrete elements of ecosystems (stones, for example). However, whether theoretical things like species, genes, and ecosystems are real is controversial.[9] Nevertheless we will particularly use the concept 'species' sometimes in the following. We do this in accordance with ordinary scientific procedure, with the classical understanding of biodiversity (diversity of species),[10] and with the fact that some creatures are more closely related, and more similar, to one another than others. Similar considerations apply to other kinds of diversity.

Furthermore, one can assume that the science argument is first of all aimed at protecting the (more or less) *natural* biodiversity (in rainforests, for example), not at protecting biodiversity in man-made surroundings (zoos, for example)[11] and in cultured form. Moreover, one can assume that as regards animals and plants, the science argument is targeted primarily at the protection of *animate* biodiversity, not at the protection of dead creatures as in scientific collections or in the form of fossils. (We will later see that it is important to be aware of this.)

'Object of science' versus laboratory beings

The science argument says that biological diversity should be protected for science, as an object of science. Presumably, 'object of science' does not mean here entities used in the laboratory, like peas, mice, and apes. For if 'object of science' were understood in that way, the science argument would not be particularly effective, at least as regards species diversity, because compared to the abundance of all species only a few of them are used in laboratories.[12] Furthermore, some laboratory entities are bred only for scientific purposes (Kohler 1994: 6); they are 'technological artifacts' (Kohler 1994: 5). Thus, this is not *natural* biodiversity. Above all, if law does not prevent it, creatures used in the laboratory are precisely not protected, but worn down like chemicals or technical equipment. They are merely means of research, or with Kohler's expression 'living instruments' (1994: 6). Not only does this contradict the aim of protection, it would also be cynical to use the science argument this way.

Precritical considerations

The destructive role of science in history

According to White (1967), science played a crucial role in the destruction of nature and the environment in history. Wigner as well writes:

> Science is less than 300 years old. This number has to be compared with the age of Man, which is certainly greater than 100,000 years. … Man's increased mastery of the Earth can be directly traced to his increased knowledge of the laws of nature. The surface of the Earth, as a whole, has not been affected by Man for 99,700 years but vast areas were deforested or the surface's store of some minerals depleted since the birth of Science. For 99,700 years, a man equipped with a good telescope on the moon might not have discovered Man's existence on the Earth. He could not have overlooked it during the last three-hundred years.
>
> (Wigner 1950: 422)

If this is correct, how can one call for the protection of biodiversity for the sake of science? On the other hand, science can contribute to the protection of biodiversity.[13] However, is this enough to reduce the tension?

Scientific fieldwork destructive of nature

The concrete scientific proceedings in nature often reduce diversity. 'Destructive sampling, … removal or transplantation of vegetation, and the collection and sacrifice of specimens are common methods in field research' (Farnsworth and Rosovsky 1993: 464). For example, Wilson describes a case in which entomologists act as follows.

> Walking into the rain forest in the evening, they select a tree for sampling and lay out a grid of 1-meter-wide funnels beneath it. The funnels feed into bottles partly filled with 70 percent alcohol, the specimen preservative of choice. In the predawn hours next morning, when the wind in the treetops dies down to a minimum, the crew forces the insecticide upward into the canopy from a motor-driven "cannon." They continue the treatment for several minutes. Then they stand by for five hours while the dead and dying arthropods rain down in the thousands, many falling into the funnels. Finally, the collected specimens are then sorted, roughly classified to major taxonomic group (such as ants, leaf beetles, or jumping spiders), and sent to specialists for further study.
>
> (Wilson 1992: 137)

If scientific fieldwork is accomplished in such a manner, why should one protect biodiversity for science? However, Wheeler holds that scientific gathering does not affect a great deal of bugs: 'The simple fact is that a single automobile on many stretches of highway kills more insects in a given year than an insect taxonomist is likely to collect for the purpose of scientific study' (Wheeler 2009: 363). In addition, Reichholf states that no one can prove 'that scientific collecting' reduces 'the occurrences of protected species' (2003: 13; my translation). However, it certainly pays to look into the science argument from this angle as well.

Devaluing knowledge

Finally, the protection of biodiversity (of species or ecosystems, for example) is sometimes supported by the fact that we know nothing or too little about biodiversity. It is possible that later, more comprehensive knowledge will show that biodiversity is useful for us. Hence, to be on the safe side, biodiversity should be protected, at least as a start.[14] However, one will eventually take the wind out of the sails of this argument, if one protects biodiversity as an object of science as in the science argument, because scientists may find out that a species, for example, is of no use for us. Indeed, this danger does not refute the science argument, because we face a strategic problem here, not an argumentative one. Also, it makes no sense to insist upon not scrutinising species because of such a risk (Schepers 2010: 192), and the same applies to biodiversity in general. Furthermore, the possibility of future, yet more comprehensive, maybe even corrective knowledge in each case might be a reason for a permanent protection. Moreover, the argument for the protection of biodiversity that is based on our lack of knowledge is contested.[15] Nevertheless, a further investigation could also be worthwhile here.

Why is the science argument not convincing?

My major critique of the science argument is, first, that it contradicts our understanding of the natural sciences, and second, that science does not depend on biodiversity.

The science argument contradicts our understanding of the natural sciences

'The task and the aim of the natural sciences' is to explore 'the entities in nature and the changes involving them' as well as the corresponding connections and principles (Hartmann 1959: 114; my translation). We understand the natural sciences as passive. Nature is as it is, and the natural sciences explore it – afterwards, thus passively. With 'passively' I mean two things: 1) A natural science cannot dictate norms, it is *non-normative*. This view goes back to Max Weber. He writes: 'An empirical science cannot tell people what they *ought* to do or to be' (Weber 1968: 151; my translation). 2) The natural sciences *react* to the world.[16] As the fire brigade was formed in answer to fires occurring again and again, and did not guard or even spark them first, so the natural sciences were formed in answer to existing nature – and did not protect or create it first.

To call for protection of biodiversity for the sake of science is thus to contradict this passivity: 1) The natural sciences in this way become the reason for biodiversity protection and so their role becomes *normative.* 2) The natural sciences do *not react* any longer to the world. If they did, they would just study – among other things – the decrease of biodiversity, the extinction of species, for example.[17] In other words: The science argument is not persuasive because 1) the natural sciences cannot be a reason to interrupt the dwindling of biodiversity and 2) according to their basic understanding they would just have to explore this dwindling.

Against this critique of the science argument you might raise four objections. Firstly, you could object that humans constitute the active element in the science argument and science does not: Humans finally are the reason and give the norm to protect biodiversity, and they make science stop reacting. Hence, the critique does not succeed.

However, this objection is not convincing. To begin with, the science argument only states that *science in general* has value for humans, not that *biodiversity research* does. Thus, the science argument cannot justify the protection of *biodiversity*. If the argument stated that biodiversity research has value for humans, other difficulties would arise; see the section 'The borders of biodiversity research ...' below and the following two sections for this. Furthermore, that science is valuable for humans does not preclude and perhaps especially requires that science takes place according to its rules, thus passively. Which value would science have for humans if it did not observe its rules, thus would be bad science or no science at all? Hence, that science is valuable for humans does not imply that humans influence science in the above-mentioned way.

Secondly, you could object that today's decrease of biodiversity is artificial, as man is responsible for it. Hence, protecting biodiversity is precisely not about manipulating nature for science, but about keeping it as it really is. Thus, science does not lose its passivity at all.

This objection does not succeed because even if you were to keep nature as it is you would not avoid doing that for the sake of science, and then you have the active understanding of science again. Thus, it is not decisive whether a certain state of the world is natural or not, but whether it is as it is because it is kept that way for the sake of science or not. As soon as it is, we have an active understanding of science.

It could be argued, thirdly, that the exploration of the decrease in biodiversity is no longer a *natural* science, because today's decrease of biodiversity is artificial.

This objection is not convincing either, because there are many ways to 'preserve' biodiversity, and not all of them yield nature in terms of the non-manmade. For example, it is possible to save a species from extinction by breeding offspring. Thus, preserved biodiversity is artificial in a manner in some cases as well.

Above all it will be hardly relevant whether the exploration of the biodiversity decrease is or is not a natural science. The natural sciences concern themselves with several natures, so to speak. If biodiversity gets lost, the nature that chemistry and physics deal with, like salts or the universe, does not get lost.

Hence, with the loss of biodiversity not all of the natural sciences are at risk, but only some of them.

Apart from that, the natural sciences are not necessary for science, as the section 'The borders of biodiversity research ...' shows. It explains why *biodiversity research* is not necessary for science, and the reason applies here as well.

Finally, you could object that the natural sciences do not only *react to* the world, but – without being unscientific – also *act in* it, namely when they experiment.[18] Janich, for example, thinks that experimenting resembles rather '*the constructing of a machine than the discovering of something*' uncreated (1997: 101; my translation). The experiment reveals '*effects*, whose *causes* lie in construction, preparation and start of the experiment' (Janich 1997: 99; my translation). Hence, it does not count against the science argument that it rests on the view that science acts in the world.

However, this objection does not solve the problem of normativity. Furthermore, the purpose of experimenting is not to play god and create a piece of *new* world, so to speak – which one has to explore with scientific means again. Rather, the aim is to build the experiment in a way that allows optimally to perceive the *existing* world. Thus, science *at large* remains reactive.

However, Janich holds that for naturalists the validity of theories does not rely on the fact that the theories apply to nature or picture it. Rather, it is crucial for these researchers whether the experiment works or not (1997: 60, 70).

That an experiment works, though, does not preclude that it makes perceivable a part of the *existing* world. It is even questionable whether it can do anything else at all. Even failing experiments make perceivable a part of the existing world. They demonstrate that under these conditions – the wrong conditions in the experimental sense – the matter just goes *that* way. Thus, experimenting as well is reactive (at large).

Nevertheless, the following thought remains to be considered. We said that science at large *re*acts, but also that it *acts* with its experiments. Thus, one could avoid a flawed understanding of science, flawed with regard to acting and reacting, if one imagined the existing biodiversity as a constituent of an outdoor experiment.[19] For example, Weingart, Carrier and Krohns' '*Natural Experiment*' (2007: 148) could be such an outdoor experiment. What they have in mind with theirs is an experiment to investigate natural or social developments which works without the arrangement of circumstances, but solely through the installation of watching equipment (Weingart, Carrier and Krohn 2007: 148). For example, according to them, Darwin believed that the Galapagos Islands would be a natural experimental setting as regards research of the local species of finches (Weingart, Carrier and Krohn 2007: 148).

However, it is improbable that the existing biodiversity as a whole meets the conditions of a constituent of an experiment, or that most parts of the existing biodiversity do. One just avails oneself of experiments to respond to a certain question the answer to which nature does not give away without further ado. To answer it, the conditions need to be sharpened, and biodiversity as it is cannot be expected to meet these sharpened conditions. If only *some* parts of the existing biodiversity meet the conditions of a constituent of experiment, the right to protect may be more or less restricted to these parts. Furthermore, the question arises what happens to the elements of biodiversity after the experiment. Their protection is merely reasonable as long as they play a role in the experiment. Apart from these basic difficulties it is questionable whether cases such as the above-mentioned '*Natural Experiment*' (Weingart, Carrier and Krohn 2007: 148) are experiments at all: Pure watching – even if it is done by means of appliances – does not yet make an experiment. Rather, a certain degree of design is necessary.[20]

Science does not depend on biodiversity

One might also think as follows of the connection between biodiversity protection and science. 1) Biodiversity as viand: Some authors[21] say that biodiversity is indispensable to human life. One could further argue that without living humans there is no science and that science therefore depends on biodiversity. 2) Biodiversity as precondition of our mental development: According to Shepard (1969: 4), a manifold nature was and is needed for the formation of our mental abilities. These abilities are indispensable for science. Thus, science depends on biodiversity. 3) Biodiversity as source of inspiration: Biodiversity inspires.[22] Inspiration is indispensable for science. Thus, biodiversity can be conducive to science.

However, the science argument is about protecting biodiversity as an *object* of science. Thus, it does not concern biodiversity as a viand, as a precondition of our mental development, and as a source of inspiration. Biodiversity as a source of inspiration is excluded here because not all objects of science inspire and because there are inspiring objects that are not objects of science. In other words: Being an object of science is neither a necessary nor a sufficient condition for being inspiring. The same is true of biodiversity: Not all of it inspires, and other things inspire as well. Newton could have conceived gravity if a book had fallen instead of an apple. Thus, the above-mentioned argument would be a very weak one. In the end, the question arises whether things inspire *to research*. Otherwise, the connection between biodiversity protection and science would barely hold.

The infinity of science

The science argument says that biodiversity should be protected because it is valuable for science as an object of science. The argument seems to be based on the assumption that along with the loss of biodiversity, objects of science get lost, and that this threatens science. However, in which way is science threatened? Wilson points to the information content of biodiversity, and as an example he writes about the house mouse's DNA: 'The full information contained therein, if translated into ordinary-sized printed letters, would just about fill all 15 editions of the *Encyclopaedia Britannica* published since 1768' (Wilson 1985: 701). Thus when biodiversity gets lost, a lot of information gets lost. Yet this large quantity of information is not directly accessible knowledge of the sort one finds in a lexicon, but raw data that research has to make accessible first, thus *potential* knowledge. However, in which way does it threaten science if potential knowledge gets lost? Baumgärtner and Becker ascribe to biodiversity 'an important role as a source of new insights' (2008: 79; my translation). And Wilson thinks that species are '[n]ew sources of scientific information' (1992: 347). On this view, the danger to science seems to consist in a sort of drying-out.

Thus, behind the science argument there appears to be the assumption that science ends when everything is explored. Otherwise, the complete or partial disappearance of biodiversity as a source of knowledge would not threaten science. This assumption also becomes apparent in the belief that biodiversity is more likely to deliver new knowledge the less it has been investigated. Thus, Sarkar thinks that 'species that we do not know, living in habitats that have not been fully explored ... are most likely to contribute novel insights into the biological world' (2005: 105). Regan shares this kind of belief when he states that 'where we must choose between two similar species ... we have reason to save the species about which we know *less* in order that, after study, we may eventually end up with the greatest aggregate knowledge of the two species' (1986: 207). Accordingly, Gunn holds that 'the extinction of an unknown species represents a considerable loss of knowledge' (1984: 319).

Beliefs like this can be seen as cases of what Rescher calls the assumption of 'Nature Exhaustion'. Rescher explains this assumption with the help of a metaphor: 'The ... potential scientific discoveries' in nature are 'like the apples on a tree' – when all of them are plucked, harvesting ends (Rescher 1978: 8f.). This attitude becomes apparent in Glass, for instance:

> The uniformity of nature and the general applicability of natural laws set limits to knowledge. ... There is a finite number of species of plants and of animals – even of insects – upon the earth. We are as yet far from knowing all about the genetics, structure and physiology, or behavior of even a single one of them. Nevertheless, a total knowledge of all life forms is only about 2×10^6 times the potential knowledge about any one of them. Moreover, the universality of the genetic code, the common character of proteins in different species, the generality of cellular structure and cellular reproduction, the basic similarity of energy metabolism in all species and of photosynthesis in green plants and bacteria, and the universal evolution of living forms through mutation and natural selection all lead inescapably to a conclusion that, although diversity may be great, the laws of life, based on similarities, are finite in number and comprehensible to us in the main even now. We are like the explorers of a great continent who have penetrated to its margins in most points of the compass and have mapped the major mountain chains and rivers. There are still innumerable details to fill in, but the endless horizons no longer exist.
>
> (Glass 1971: 24)

However, the protection of biodiversity would not solve this problem, but merely postpone it, because sooner or later also the protected biodiversity would be explored completely. Furthermore, according to Sarkar (2005: 85), the extinction of biodiversity can yield an advance in knowledge as well. Above all, the assumption that science ceases when all objects are explored is wrong in principle. This is what Rescher shows. The following three aspects are important for our concern.

One could think that science ends when it has depicted the world entirely. However, according to Rescher this assumption is erroneous: 'Science simply cannot describe the world completely.' Firstly, after Rescher, the numerousness of objects renders a survey impracticable. Secondly, 'the number of true descriptive remarks that can be made about a thing … is theoretically inexhaustible'. For example, one can 'describe … a stone' not only in view of 'its physical features – its shape, its surface texture, its chemistry, and so on', but also in view of 'its causal background: its historical genesis and evolution' and in view of 'its functional aspects as relevant to its uses by the stonemason, the architect, the landscape decorator, et cetera' (Rescher 1984: 14f.). Thus, biodiversity does not need to be protected so that science has something to depict as long as possible.

One could think that with a finite number of research objects science runs out of questions someday, and therefore ends. Yet, Rescher shows, appealing to Kant, that this is not right either: Every novel level of insights brings with it several novel questions. The replies to them form the basis for still more questions. This results in an incessant sequence (Rescher 1984: 28–30). Thus, biodiversity does not need to be protected, either, for science to find questions for as long as possible.

In the end, one could think that science comes to lack of objects when it has explored everything up to the last corner, and therefore ceases. However, this again is not correct, as Rescher demonstrates: For natural science to continue permanently it is not necessary that nature is of 'infinite *structural complexity*'. For example, it is not necessary that always novel worlds loom beyond outer space and minute elements into the boundless big and the tiny, as Rescher says. Rather, an 'endless *functional or operational complexity*' of nature is enough.

However, even if there is a limit to the '*functional or operational complexity*' of nature, natural science is not in danger. Rather, the '*cognitive* … complexity' on the part of us is enough for it to continue. The infinity results from '*conceptual depth*' that allows us to explore nature 'from various *perspectives of consideration*'. To say it in other words, 'there will always be facts … that we do not *know*' as we do not at all hold the representational inventory to comprehend them. Scientific advancement, however, entails terminological renewal, thereby rendering these facts graspable, as Rescher says. The better our terms capture the items, the better we conceive the items. 'Every time we contrive a different mode of description for actual processes … we find … that new laws of nature become accessible to our intellect.' For example, 'uranium-containing substances were radioactive before Becquerel', but not until the term 'radioactivity' existed – so one has to understand Rescher – radioactivity could become part of our ken. To illustrate the infinity of our perception of nature, Rescher compares nature with 'a typewriter': 'Even a finite nature can, like a typewriter with a limited keyboard, yield an endlessly varied text' (Rescher 1984: 47–56, 135f.). Thus, biodiversity does not have to be protected for science to keep having objects of research as long as possible.

Rescher relates the three mentioned aspects as follows.

> Even if we view the world as inherently finitistic, and espouse a Principle of Limited Variety which has it that nature can be portrayed descriptively with the materials of a finite taxonomic scheme, there can be no *a priori* guarantee that the progress of science will not engender an unending sequence of changes of mind regarding this finite register of descriptive materials.
>
> (Rescher 1984: 136)

Even if there is no longer nature at all, science will still be possible, because non-natural objects of research will still exist. Even if they no longer exist, science will still be possible, because scientists are a precondition of science, and as soon as one has scientists, one has objects of research. If the scientists do not consider it a good idea to explore one another, they still can do mathematics and similar object-independent research. Thus, the loss of biodiversity does not endanger science as such, at least as far as a lack of research objects is concerned.

Nevertheless one could adhere to the idea that it is better for science to have more options than fewer, that means to have mathematical contents *and* biodiversity as items of research, for example, instead of mathematical contents only. Thus, biodiversity should be protected for the sake of science.

However, if science is an endless matter, as we have just seen, the dwindling of items of research is no dwindling of options at all – at least as long as, radically spoken, one item of research exists, a kind of 'initial' or 'spark' for endlessness. As scientists are necessary for science and potential objects of research themselves, they guarantee that science is endless. As regards theoretical, object-independent issues, an 'initial' item is not even necessary. Thus, the dwindling of biodiversity does not reduce the number of options for science. Therefore, the objection does not convince.

One could object, though, that not only the quantity of options is of importance for science, but also their quality. For example, it might be better for science to have different kinds of items, mathematical ones *and* biodiversity, for instance, than to have only one kind of item, only mathematical ones, for instance.

However, if science is endless, if all items of research afford endless scientific progress, why should some items be better for science than others? The investigation of some items might be boring and/or useless for people, indeed, but boredom and uselessness for people are no demerits in science. Furthermore, one could suppose that the investigation of some items yields insights that advance the undertaking of science and that the investigation of other items does not or at least not to that extent. Yet if science is endless and has therefore an infinite number of chances of advancement, or at least an equal probability of having such chances, no matter which direction it takes, this kind of quality fades.

The borders of biodiversity research and why they do not rescue the science argument

One could object, though, that if not science at large, *biodiversity research* depends on the occurrence of biodiversity, because without objects of investigation there is no biodiversity research. Lanzerath, for instance, thinks that '[b]iodiversity research ... is dependent on the existence of biodiversity' (2008: 207; my translation).[23] With regard to species, Kaule, too, states that 'intact research objects' are necessary for 'basic research and applied research' (1986: 16; my translation).

However, this objection is not convincing for two reasons. Firstly, as we saw in the beginning, the aim is to find reasons for the protection of the *natural* and the *animate* biodiversity. Thus, it would have to be shown that biodiversity research depends on the occurrence of *natural* and *animate* biodiversity – but it does not. Rather, it is possible on non-natural and/or non-animate sources as well. For example, according to Gudo and Steininger, fossils are objects of biodiversity research (2001: 32). Cotterill mentions an example of a non-animate source which additionally is to be found in a non-natural setting: 'Natural history collections ... are' one of 'the key resources of systematics' (Cotterill 1995: 184). He states: 'Preserved specimens are stable sources of reliable information ... from which biologists can assemble knowledge' (Cotterill 1995: 188).[24] Without natural and animate biodiversity, indeed, biodiversity research misses a whole field: 'The dynamics of population change, interspecific competition, and the like can often be observed and understood only in large areas' (Gunn 1984: 327), and 'studying preserved specimens or fossils is a poor substitute for studying living and behaving creatures', as Norton (1987: 18) describes a sceptical position. Nevertheless, biodiversity research does not *disappear* along with natural and animate biodiversity.

Secondly, biodiversity research is not necessary for science. Of course, biodiversity research can make a contribution to science. Norton, for example, states that results 'from the study of a particular species ... might plug a gap in scientific knowledge and allow an important theoretical breakthrough' (1987: 19).[25] Moreover, results from biodiversity research can be useful for realms of research that do not immediately concern themselves with biodiversity. For example, according to Lanzerath, molecular biology needs these results for comprehending evolution (2008: 207).

Nevertheless, if biodiversity research ceased to exist, science would not be restricted. The considerations above regarding the borders of science also apply here: If the results from biodiversity research do not exist, science will just advance on other paths. It has enough possibilities of depiction, enough questions, and enough – and good enough – objects of research.

The scientist argument

If biodiversity research were to come to an end, this would solely limit scientists in their personal predilection for exploring biodiversity. These personal predilections could be a good reason to protect biodiversity. The argument occurs in this form as well.[26] However, then the argument has nothing important to do with science any more. Indeed, scientists are those who practise science, but not everything scientists do is scientific. When a scientist dedicates himself to biodiversity research, this is not a scientific act. His motives are not scientific, but personal ones. His proceeding is comparable to the choice of a course of studies. Indeed, it is possible that the occupation with biodiversity arises out of the scientific activity. For example, a pharmaceutist who studies something else in fact may notice that in order to answer a question in his original field of research, he would have to investigate biodiversity. However, as soon as biodiversity research arises in this way, the previous considerations take effect again: Biodiversity research can make a contribution to science, but if biodiversity research does not exist – because biodiversity does not exist – scientific progress will not be stalled or of lower quality.

Now one may object that science never proceeds neutrally but that values and valuations always influence it. Science is never entirely objective. Why is the personal decision to explore biodiversity not part of science then and why is it not possible to hold on to the science argument accordingly?

It is true that an entirely value-free science does not exist. However, one has to distinguish between scientific and non-scientific values. The scientific criterion 'verifiability', for example, contains a valuation, but one with respect to our knowledge.[27] In contrast, the personal decision to research biodiversity contains a valuation, but none which is conducive to knowledge. The following contrasting examples help to see this. Think about a scientist who does not like to carry heavy things. For this reason, he reads only lightweight books. Or about a scientist who is sensitive to noise and breaks off laboratory experiments because he feels they are too loud. Finally, imagine a scientist who determines the species of an area and includes only the pretty ones. One could justifiably call the proceeding of these scientists as barely conducive to knowledge and therefore as unscientific.

Secondly, regarding non-scientific values especially: To recognise that we cannot keep values out of science is different from exploiting this insight to let values flow into science. That we cannot exclude values from influencing science does not mean that this impact is good. Rather, we should try to prevent it. To say it with Goethe, 'these difficulties, one may arguably say this hypothetical impossibility, must not prevent us from doing the best possible' (1977: 6; my translation). (Goethe says this on the difficult undertaking to keep personal valuations out of the perception of nature.) The difference between science and non-science has to be preserved, otherwise the complete enterprise lapses.

However, as mentioned already, that scientists exist who commit themselves to the exploration of biodiversity might be a good reason to protect biodiversity. One would have to take a closer look, then, at *how much* this interest concerns these scientists, how important it is in comparison to the concerns of other people, and how difficult the protection of biodiversity is.

How many biodiversity researchers exist could also be of importance. For example, according to di Castri and Younès (1996: 7) the number of taxonomists is declining. Additionally, Wuketis points out that numerous biologists have no interest in the variety of species (1999: 140). How the small number of interested scientists can be explained, though, should also be considered. Do they really have no interest, or do appointments and other endorsement just lack? For instance,

Wilson believes: 'Taxonomy … is a declining profession through lack of support' (Wilson 1984: 136). Alexiades and Laird add that 'in general biodiversity research budgets are small relative to other areas of science' (2002: 4). Beyond that, the question arises whether the present state of interests is decisive at all. According to Willmann (2003), humans have again and again tried to gain knowledge about biological diversity since antiquity. One could take this as an indication that time and again there will also be scientists who are interested in biodiversity in the future.

As the present considerations are concerned with the personal interest in biodiversity, one could widen the argument at this point and also consider interested laymen. Biological diversity could be protected not only for the sake of scientists, but also for the sake of interested laymen then. This broad version of the argument can also be found sometimes.[28] Sometimes it is also argued that *all* people are interested in biodiversity.[29] This line would certainly need to be examined.[30]

Talk about the 'scientific value' of biodiversity sometimes occurs in the scien*tist* argument as well.[31] As mentioned at the beginning, the expression is not apt. Those things have 'scientific value' which are valuable *within* science. The scientist argument, though, is about protecting biodiversity *for* scientists. However, not everything that is valuable for scientists is scientifically valuable. Concentration and even science-independent things like beautiful music, for example, are valuable for scientists, but not scientifically valuable. Furthermore, biodiversity cannot be scientifically valuable at all. At most the concept 'biodiversity' can. Beyond these two initial objections we have now seen that the decision for biodiversity research is a personal one, not a scientific one. In this respect the term 'scientific value' does not fit either. For the same three reasons it neither fits in the case of interested laymen and humans collectively.

Useful knowledge

However, what seems above all to speak in favour of biodiversity research is the fact that it can be useful for us in a different way. Biodiversity research can yield numerous insights that are of particular importance for us, for example in the pharmaceutical, the bionic or the food domain. Science is a means to the social end here. The reason for the protection of biodiversity would no longer be science as in the science argument, but the diverse social uses of biodiversity, which are merely elucidated by science.[32] Thus, once again the argument has nothing to do with science, and once again the term 'scientific value' is inappropriate here.

It needs to be considered here as well how much effort the protection of biodiversity takes. It is also important how probable the discovery of something socially relevant is, and *how* relevant it is for society. Is it an important medical agent or a further culinary delicacy? If other social concerns – the abolition of poverty and disease, for example – are regarded as more pressing than the exploration of biodiversity, one also must not forget that results from biodiversity research may contribute to the solution of social problems – medical agents, for instance.

To sum up: In both cases, in the scientist argument (including the variants 'laymen' and 'all people') and in the general social variant, the argument has nothing important to do with science. It is about the protection of biodiversity for interested people or for society in general. Science is merely a means to a personal or broadly social end.

Thus, I neither want to say that science always has to serve a purpose, nor that the interests of people, and society in general, cannot be good reasons for the protection and exploration of biodiversity. I only want to say that the science argument is not convincing – and 'scientific value' is not the right term for the issue under consideration.

Conclusion

In sum, this means: It is possible that there are good reasons for protecting biodiversity, and maybe one of these reasons is that people are interested in biodiversity. Beyond that it could be right to preserve biodiversity because it is useful for us. Biodiversity research can point out in what way biodiversity can be useful for us. However, to protect biodiversity for science as an object of science contradicts our understanding of the natural sciences and is unnecessary; and to ascribe 'scientific value' to biodiversity is inappropriate.

Acknowledgements

Thanks to Rüdiger Bittner and to the anonymous reviewers for helpful comments on previous drafts, and to Christian Byrnes for keeping an eye on my English.

Notes

1 We will soon see that the use of the adjective 'scientific' can be problematic in the science argument. To be consistent with the literature, we will ignore this difficulty until then.
2 Norton refers to the protection of species. The application to biodiversity here and in the description of the two groups is mine.
3 Peters, McCloskey and Rolston describe the argument in alternative ways, namely as saying that one must protect nature 'because many things in nature are still unexplored, the destruction of nature therefore hinders science' (Peters 1980: 14f.; my translation) and as saying that 'the extinction of a species is something that cannot be ignored if we are concerned with the corpus of natural science' (McCloskey 1980: 66f.), respectively. Rolston describes the science argument as follows. 'Nature' has '[*s*]*cientific value*' because '[n]ature is a laboratory for the pursuit of science, good not just because individuals like it …, but because society gains pure knowledge, which enlarges our understanding of the world and our roles in it' (Rolston 1985: 27).
4 Other variants of the science argument are put forward by Godfrey-Smith (1980: 31), Ott (1999: 54), and Alho (2008: 1118).
5 In Norton (1987: 19f.), for example, and in Takacs (1996: 197).
6 Takacs (1996: 197) uses the term 'value for science' as well; similarly Alho (2008: 1118) ('value for Science').

7 According to Norton, 'there seems to be a near-consensus' that 'there is no generally accepted definition of … "biodiversity"' (2008: 369). See also Hertler (1999: 40) and 'the diversity of … definitions' (Norton 2008: 371) biologists give in Takacs (1996: 46–50).
8 For this assumption and its usage by scientific realists as well as scientific anti-realists, see Hüttemann (1997: 39).
9 As regards ecosystems, see Trepl (1988). As regards species, see Richards (2008: 185f.) and Grasshoff and Weingarten (1999: 85) as illustrations, for example. For an explanation of scientific realism and anti-realism in general, see Hacking (1983: 21).
10 After Loreau, diversity of species is 'by far the most widely used definition of biodiversity in both the natural and the social sciences' (2010: 13).
11 As regards arguments for the protection of species, a similar appraisement is to be found in Elliot (1980: 12).
12 Similarly Marcel Weber (2008: 473).
13 Conway (1988) and Dresser (1988), for example, deal with the relevance of science *for* biodiversity protection. However, most of the methods they describe yield no natural biodiversity.
14 Such an argument is to be found in Godfrey-Smith (1980: 31) and Lovejoy (1988: 724). A description of the argument concerning species diversity is also to be found in Polasky and Solow (1995: 298).
15 For a critique see Elliot (1980: 12) and Sober (1986: 175). For critical discussions see Meyer (2003: 56–64) and Maclaurin and Sterelny (2008: 154–7, 171).
16 At least on the whole they do; see the third objection below.
17 Wilson (1988: 13) as well thinks that the extermination of species is a research topic, and he even bewails its disregard in ecological research.
18 Of course, it is not the sciences, but scientists who do it. Yet, we ignore this here.
19 The biologist Dobzhansky proceeded the other way round, so to speak. According to Kohler, 'Dobzhansky's lab was … an extension of the field' as 'Dobzhansky altered domestic laboratory practices to accommodate hoards of wild flies' (Kohler 1994: 292). This model is out of the question for us, among other things because we concern ourselves with the protection of biodiversity in *natural* environments, and a laboratory is no natural environment.
20 For Marcel Weber (2008: 472), too, intervening is an important characteristic of experimentation. Even more than that may be necessary to speak of an experiment. What about cases in which people intervened, but not for scientific purposes? For example, Wucherer and Breckle report on incomparable occurrences on the desiccating ground of the Aral Sea: Creatures move into the novel site – 'the new desert called Aralkum' – for the first time. Wucherer and Breckle call this 'a huge experiment, an experimental set' (2001: 52f.). However, the Aralkum did not originate in the framework of some crazy scientist's experiment, but – according to Breckle et al. – because of giant watering measures in Soviet days (2001: 27). Is this really an experiment then?
21 Spaemann (1986: 197f.), Norton (1988: 205) and Wilson (1992: 133) (Wilson does not refer to biodiversity as such but merely to terrestrial arthropods).
22 For example, the National Research Council (1999: 61) maintains that the variety of living beings inspires us to figure novelties.
23 Similarly Ott (1999: 54).
24 Similarly Wheeler (2009: 361f.).
25 Similarly Wilson (1985: 702) and the National Research Council (1999: 62).
26 For example, Passmore (1974: 102), Ehrlich and Ehrlich (1982: 45f.) and Rolston (2001: 404) give variants of this argument.
27 Rescher (1984: 207) as well thinks that knowledge-oriented values exist in science.
28 For example, in Elliot (1980: 10f.), Ehrlich and Ehrlich (1982: 46f.) and Sarkar (2005: 86).

29 Regan at least thinks that all humans ('we') can and should be interested ('take pleasure in knowledge') in biodiversity (1986: 201f.), and Wilson (1984: 1) finally thinks that they are interested in biodiversity, namely in the variety of creatures.
30 For a critique of Regan's statement see Galert (1998: 74).
31 For instance, in Passmore (1974: 102).
32 As regards this difference, see also Norton (1987: 19).

References

Alexiades, M.N. and Laird, S.A. (2002) 'Laying the Foundation: Equitable Biodiversity Research Relationships', in S.A. Laird (ed.) *Biodiversity and Traditional Knowledge. Equitable Partnerships in Practice*, London: Earthscan Publications, 3–15.

Alho, C.J.R. (2008) 'The Value of Biodiversity. Valor da biodiversidade', *Brazilian Journal of Biology* 68: 1115–8.

Baumgärtner, S. and Becker, C. (2008) 'Ökonomische Aspekte der Biodiversität', in Deutsches Referenzzentrum für Ethik in den Biowissenschaften (DRZE) (ed.) *Biodiversität*, Freiburg: Alber, 75–115.

Breckle, S.-W., Wucherer, W., Agachanjanz, O. and Geldyev, B. (2001) 'The Aral Sea Crisis Region', in S.-W. Breckle, M. Veste and W. Wucherer (eds) *Sustainable Land Use in Deserts*, Berlin: Springer, 27–37.

Conway, W. (1988) 'Can Technology Aid Species Preservation?', in E.O. Wilson (ed.) *Biodiversity*, Washington, D.C.: National Academy Press, 263–8.

Cotterill, F.P.D. (1995) 'Systematics, Biological Knowledge and Environmental conservation', *Biodiversity and Conservation* 4: 183–205.

Di Castri, F. and Younès, T. (1996) 'Introduction: Biodiversity, the Emergence of a New Scientific Field – Its Perspectives and Constraints', in F. di Castri and T. Younès (eds) *Biodiversity, Science and Development. Towards a New Partnership*, Wallingford: CAB, 1–11.

Dresser, B.L. (1988) 'Cryobiology, Embryo Transfer, and Artificial Insemination in Ex Situ Animal Conservation Programs', in E.O. Wilson (ed.) *Biodiversity*, Washington, D.C.: National Academy Press, 296–308.

Ehrenfeld, D. (1988) 'Why Put a Value on Biodiversity?', in E.O. Wilson (ed.) *Biodiversity*, Washington, D.C.: National Academy Press, 212–6.

Ehrlich, P. and Ehrlich, A. (1982) *Extinction. The Causes and Consequences of the Disappearance of Species*, London: Gollancz.

Elliot, R. (1980) 'Why Preserve Species?', in D.S. Mannison, M.A. McRobbie and R. Routley (eds) *Environmental Philosophy*, Canberra: Australian National University, 8–29.

Eser, U. (2002) 'Zwischen Wissenschaft und Gesellschaft: Ökologische Gegenstände als Grenzobjekte', in A. Lotz and J. Gnädinger (eds) *Wie kommt die Ökologie zu ihren Gegenständen? Gegenstandskonstitution und Modellierung in den ökologischen Wissenschaften. Beiträge zur Jahrestagung des Arbeitskreises Theorie in der Ökologie in der Gesellschaft für Ökologie vom 21.–23. Februar 2001 im Kardinal-Döpfner-Haus Freising (Bayern)*, Frankfurt am Main: Lang, 107–16.

Farnsworth, E.J. and Rosovsky, J. (1993) 'The Ethics of Ecological Field Experimentation', *Conservation Biology* 7: 463–72.

Galert, T. (1998) *Biodiversität als Problem der Naturethik. Literaturreview und Bibliographie*, ed. Europäische Akademie zur Erforschung von Folgen wissenschaftlich-technischer Entwicklungen, Bad Neuenahr-Ahrweiler.

Glass, B. (1971) 'Science: Endless Horizons or Golden Age?', *Science* 171: 23–9.

Godfrey-Smith, W. (1980) 'The Rights of Non-humans and Intrinsic Values', in D.S. Mannison, M.A. McRobbie and R. Routley (eds) *Environmental Philosophy*, Canberra: Australian National University, 30–47.

Goethe, J.W. (1977) 'Der Versuch als Vermittler von Objekt und Subjekt', in M. Böhler (ed.) *Johann Wolfgang Goethe. Schriften zur Naturwissenschaft. Auswahl*, Stuttgart: Reclam, 5–15 (orig. 1823).

Grasshoff, M. and Weingarten, M. (1999) 'Für eine pragmatische Taxonomie', in C. Görg, C. Hertler, E. Schramm and M. Weingarten (eds) *Zugänge zur Biodiversität. Disziplinäre Thematisierungen und Möglichkeiten integrierender Ansätze*, Marburg: Metropolis, 71–90.

Gudo, M. and Steininger, F.F. (2001) 'Der Beitrag der Paläontologie zur Biodiversitätsdebatte', in P. Janich, M. Gutmann and K. Prieß (eds) *Biodiversität. Wissenschaftliche Grundlagen und gesetzliche Relevanz*, Berlin: Springer, 31–114.

Gunn, A.S. (1984) 'Preserving Rare Species', in T. Regan (ed.) *Earthbound. New Introductory Essays in Environmental Ethics*, Philadelphia, PA: Temple University Press, 289–335.

Hacking, I. (1983) *Representing and Intervening. Introductory Topics in the Philosophy of Natural Science*, Cambridge: Cambridge University Press.

Hartmann, M. (1959) *Die philosophischen Grundlagen der Naturwissenschaften. Erkenntnistheorie und Methodologie*, 2nd rev. edn, Stuttgart: Gustav Fischer.

Hertler, C. (1999) 'Aspekte der historischen Entstehung von Biodiversitätskonzepten in den Biowissenschaften', in C. Görg, C. Hertler, E. Schramm and M. Weingarten (eds) *Zugänge zur Biodiversität. Disziplinäre Thematisierungen und Möglichkeiten integrierender Ansätze*, Marburg: Metropolis, 39–52.

Hüttemann, A. (1997) *Idealisierungen und das Ziel der Physik. Eine Untersuchung zum Realismus, Empirismus und Konstruktivismus in der Wissenschaftstheorie*, Berlin: de Gruyter.

Janich, P. (1997) *Kleine Philosophie der Naturwissenschaften*, München: Beck.

Kaule, G. (1986) *Arten- und Biotopschutz*, Stuttgart: Ulmer.

Kohler, R.E. (1994) *Lords of the Fly. Drosophila Genetics and the Experimental Life*, Chicago, IL: University of Chicago Press.

Lanzerath, D. (2008) 'Der Wert der Biodiversität: Ethische Aspekte', in Deutsches Referenzzentrum für Ethik in den Biowissenschaften (DRZE) (ed.) *Biodiversität*, Freiburg: Alber, 147–213.

Loreau, M. (2010) *The Challenges of Biodiversity Science*, Oldendorf (Luhe): International Ecology Institute.

Lovejoy, T.E. (1988) 'Will Unexpectedly the Top Blow Off?', *BioScience* 38: 722–6.

Maclaurin, J. and Sterelny, K. (2008) *What Is Biodiversity?*, Chicago, IL: University of Chicago Press.

McCloskey, H.J. (1980) 'Ecological Ethics and its Justification: A Critical Appraisal', in D.S. Mannison, M.A. McRobbie and R. Routley (eds) *Environmental Philosophy*, Canberra: Australian National University, 65–95.

Meyer, K. (2003) *Der Wert der Natur. Begründungsvielfalt im Naturschutz*, Paderborn: Mentis.

National Research Council (1999) *Perspectives on Biodiversity. Valuing Its Role in an Everchanging World*, Washington, D.C.: National Academy Press.

Norton, B.G. (1986) 'On the Inherent Danger of Undervaluing Species', in B.G. Norton (ed.) *The Preservation of Species. The Value of Biological Diversity*, Princeton, NJ: Princeton University Press, 110–37.

Norton, B.G. (1987) *Why Preserve Natural Variety?*, Princeton, NJ: Princeton University Press.

Norton, B.G. (1988) 'Commodity, Amenity, and Morality: The Limits of Quantification in Valuing Biodiversity', in E.O. Wilson (ed.) *Biodiversity*, Washington, D.C.: National Academy Press, 200–5.

Norton, B.G. (2008) 'Biodiversity: Its Meaning and Value', in S. Sarkar and A. Plutynski (eds) *A Companion to the Philosophy of Biology*, Malden, MA: Blackwell, 368–89.

Ott, K. (1999) 'Zur ethischen Bewertung von Biodiversität', in M.E. Hummel, H.-R. Simon and J. Scheffran (eds) *Konfliktfeld Biodiversität: Erhalt der biologischen Vielfalt – Interdisziplinäre Problemstellungen. Arbeitsbericht/Working Paper Ianus 7/1999*, Darmstadt, 45–64.

Passmore, J. (1974) *Man's Responsibility for Nature. Ecological Problems and Western Traditions*, New York: Scribner's Sons.

Peters, D.S. (1980) 'Warum und wozu Naturschutz?', in Senckenbergische Naturforschende Gesellschaft (ed.) *Landschaft als Lebensraum. Ziele und Möglichkeiten der Naturschutzarbeit*, Frankfurt am Main, 11–21.

Polasky, S. and Solow, A.R. (1995) 'On the Value of a Collection of Species', *Journal of Environmental Economics and Management* 29: 298–303.

Regan, D.H. (1986) 'Duties of Preservation', in B.G. Norton (ed.) *The Preservation of Species. The Value of Biological Diversity*, Princeton, NJ: Princeton University Press, 195–220.

Reichholf, J.H. (2003) 'Biologische Vielfalt: Sammeln, Sammlungen und Systematik. Einführung in das Rundgespräch', in Bayerische Akademie der Wissenschaften (ed.) *Biologische Vielfalt: Sammeln, Sammlungen und Systematik. Rundgespräch am 14. Oktober 2002 in München*, München: Pfeil, 13f.

Rescher, N. (1978) *Scientific Progress. A Philosophical Essay on the Economics of Research in Natural Science*, Pittsburgh, PA: University of Pittsburgh Press.

Rescher, N. (1984) *The Limits of Science*, Berkeley, CA: University of California Press.

Richards, R.A. (2008) 'Species and Taxonomy', in M. Ruse (ed.) *The Oxford Handbook of Philosophy of Biology*, Oxford: Oxford University Press, 161–88.

Rolston, H. (1985) 'Valuing Wildlands', *Environmental Ethics* 7: 23–48.

Rolston, H. (2001) 'Biodiversity', in D. Jamieson (ed.) *A Companion to Environmental Philosophy*, Malden, MA: Blackwell, 402–15.

Sarkar, S. (2005) *Biodiversity and Environmental Philosophy. An Introduction*, Cambridge: Cambridge University Press.

Schepers, G. (2010) 'Der Wert biologischer Vielfalt', in Bundesamt für Naturschutz (ed.) *Treffpunkt Biologische Vielfalt IX. Aktuelle Forschung im Rahmen des Übereinkommens über die biologische Vielfalt vorgestellt auf einer wissenschaftlichen Expertentagung an der Internationalen Naturschutzakademie Insel Vilm vom 24.–28. August 2009*, Bonn, 189–97.

Shepard, P. (1969) 'Introduction: Ecology and Man – a Viewpoint', in P. Shepard and D. McKinley (eds) *The Subversive Science. Essays Toward an Ecology of Man*, Boston, MA: Houghton Mifflin, 1–10.

Sober, E. (1986) 'Philosophical Problems for Environmentalism', in B.G. Norton (ed.) *The Preservation of Species. The Value of Biological Diversity*, Princeton, NJ: Princeton University Press, 173–94.

Spaemann, R. (1986) 'Technische Eingriffe in die Natur als Problem der politischen Ethik', in D. Birnbacher (ed.) *Ökologie und Ethik*, Stuttgart: Reclam, 180–206 (orig. 1979).

Takacs, D. (1996) *The Idea of Biodiversity. Philosophies of Paradise*, Baltimore, MD: Johns Hopkins University Press.

Trepl, L. (1988) 'Gibt es Ökosysteme? Do Ecosystems Exist?', *Landschaft + Stadt* 20: 176–85.

Weber, Marcel (2008) 'Experimentation', in S. Sarkar and A. Plutynski (eds) *A Companion to the Philosophy of Biology*, Malden, MA: Blackwell, 472–88.

Weber, Max (1968) 'Die "Objektivität" sozialwissenschaftlicher und sozialpolitischer Erkenntnis', in J. Winckelmann (ed.) *Gesammelte Aufsätze zur Wissenschaftslehre von Max Weber*, 3rd, rev. and ext. edn, Tübingen: Mohr (Siebeck), 146–214 (orig. 1904).

Weingart, P., Carrier, M. and Krohn, W. (2007) *Nachrichten aus der Wissensgesellschaft. Analysen zur Veränderung der Wissenschaft*, Weilerswist: Velbrück.

Wheeler, Q.D. (2009) 'The Science of Insect Taxonomy: Prospects and Needs', in R.G. Foottit and P.H. Adler (eds) *Insect Biodiversity. Science and Society*, Chichester: Wiley-Blackwell, 359–80.

White, L. (1967) 'The Historical Roots of Our Ecologic Crisis', *Science* 155: 1203–7.

Wigner, E.P. (1950) 'The Limits of Science', *Proceedings of the American Philosophical Society held at Philadelphia for promoting useful knowledge* 94: 422–7.

Willmann, R. (2003) 'Die Erfassung der Artenvielfalt vor Linné', in S.R. Gradstein, R. Willmann and G. Zizka (eds) *Biodiversitätsforschung – Die Entschlüsselung der Artenvielfalt in Raum und Zeit*, Stuttgart: Schweizerbart'sche Verlagsbuchhandlung (Nägele u. Obermiller), 13–26.

Wilson, E.O. (1984) *Biophilia*, Cambridge, MA: Harvard University Press.

Wilson, E.O. (1985) 'The Biological Diversity Crisis', *BioScience* 35: 700–6.

Wilson, E.O. (1988) 'The Current State of Biological Diversity', in E.O. Wilson (ed.) *Biodiversity*, Washington, D.C.: National Academy Press, 3–18.

Wilson, E.O. (1992) *The Diversity of Life*, Cambridge, MA: Belknap Press.

Wucherer, W. and Breckle, S.-W. (2001) 'Vegetation Dynamics on the Dry Sea Floor of the Aral Sea', in S.-W. Breckle, M. Veste and W. Wucherer (eds) *Sustainable Land Use in Deserts*, Berlin: Springer, 52–68.

Wuketis, F.M. (1999) *Die Selbstzerstörung der Natur. Evolution und die Abgründe des Lebens*, Düsseldorf: Patmos.

Index

Page references in *italic* indicate figures and tables. These can also be found listed in full after the Contents.